GINA RIPPON

Gina Rippon is an international researcher in the field of cognitive neuroscience based at the Aston Brain Centre at Aston University in Birmingham. Her research involves the use of state-of-the-art brain imaging techniques to investigate developmental disorders such as autism. She is a regular contributor to events such as the British Science Festival, New Scientist Live and the Sceptics in the Pub series. In 2015 she was made an Honorary Fellow of the British Science Association for her contributions to the public communication of science.

She is also an advocate for initiatives to help overcome the under-representation of women in STEM subjects. As part of a European Union Gender Equality Network, she has addressed conferences all over the world. She belongs to WISE and ScienceGrrl, and is a member of Robert Peston's Speakers4Schools programme and the Inspiring the Future initiative.

The Gendered Brain is her first book for the general reader.

'Rippon takes a scalpel to the research surrounding sex differences in the brain with precision and humour, exposing everything... Enjoyable and enlightening'
Financial Times

'A smart and witty addition to the literature on sex differences. Gina Rippon is one of the most outspoken scientists in this area, and she debunks a whole host of sexist stereotypes in her new book'
Angela Saini, author of *Inferior*

'Evidence of brain plasticity is key to Gina Rippon's new book... Where the book really shines – not surprisingly – is in the details about the science of the brain: what we know and what we do not. Rippon's explanation of how we've studied the brain in the past, and how recent technological advances are giving us increasingly precise tools to do so, is endlessly interesting'
Emily Oster, *New York Times Book Review*

'A brilliant and thorough debunking of the popular myths around sex differences in brains and behaviour'
Dr Emily Grossman, broadcaster

'The history of sex-difference research is rife with innumeracy and misinterpretation... Rippon, a leading voice against the bad neuroscience of sex difference, uncovers so many examples in this ambitious book that she uses a whack-a-mole metaphor to evoke the eternal cycle... A juicy history... The book accomplishes its goal of debunking the concept of a gendered brain'
Nature

'A fresh and much-needed perspective on the gender debate'
...es

GINA RIPPON

The Gendered Brain

The new neuroscience that shatters the
myth of the female brain

VINTAGE

1 3 5 7 9 10 8 6 4 2

Vintage
20 Vauxhall Bridge Road,
London SW1V 2SA

Vintage is part of the Penguin Random House group of companies
whose addresses can be found at global.penguinrandomhouse.com

First published in Vintage in 2020
First published in hardback by The Bodley Head in 2019

penguin.co.uk/vintage

A CIP catalogue record for this book is available from the
British Library

ISBN 9781784706814

Diagrams on p. 2, p. 123 and p. 285 by Liane Payne.
Diagram on p. 75 by Jeremy Kemp and used under Creative
Commons licence.

Printed and bound in Great Britain by Clays Ltd, Elcograf S.p.A.

Penguin Random House is committed to a sustainable future for
our business, our readers and our planet. This book is made from
Forest Stewardship Council® certified paper.

For Jana and Hilda – two indomitable grandmothers who certainly over-rode their Inner Limiters.

For my parents, Peter and Olga – whose love and support gave me so many of the opportunities I have had in my life's journey – and for my twin brother, Peter, who has been with me along the way.

For Dennis – partner, sounding board, sommelier and horticulturalist extraordinaire, with thanks for his tireless patience and support (and lashings of gin).

For Anna and Eleanor, for your future, whatever it might hold.

Few tragedies can be more extensive than the stunting of life, few injustices deeper than the denial of an opportunity to strive or even to hope, by a limit imposed from without, but falsely identified as lying within.

Stephen Jay Gould,
The Mismeasure of Man

Contents

Introduction:
Whac-a-Mole Myths

This book is about an idea that has its roots in the eighteenth century and still persists today. That is the notion that you can 'sex' a brain, that you can describe a brain as 'male' or 'female' and that you can attribute any differences between individuals in behaviour, abilities, achievements, personality, even hopes and expectations to the possession of one or the other type of brain. It is a notion that has inaccurately driven brain science for several centuries, underpins many damaging stereotypes and, I believe, stands in the way of social progress and equality of opportunity.

The question of sex differences in the brain is one that has been debated, researched, encouraged, criticised, praised and belittled for over 200 years, and can certainly be found in different guises for long before that. It is an area of entrenched opinion and has been the ongoing focus of just about every research discipline from genetics to anthropology, mixed with history, sociology, politics and statistics. It is characterised by bizarre claims (women's inferiority comes from their brains being five ounces lighter) which can readily be dismissed, only to pop up again in another form (women's inability to read maps comes from wiring differences in the brain). Sometimes a single claim lodges itself firmly in the public consciousness as a fact and, despite the best efforts of concerned scientists, remains a deeply entrenched belief. It will be frequently referred to as a well-established fact and triumphantly re-emerge to trump arguments about sex differences or, more worryingly, to drive policy decisions.

I think of these seemingly endlessly recurring misconceptions as 'Whac-a-Mole' myths. Whac-a-Mole is an arcade game which involves repeatedly hitting the heads of mechanical moles with a mallet as they pop up through holes in a board – just when you think you've dispatched them all, another pesky mole pops up elsewhere. The term 'Whac-a-Mole' is used nowadays to describe a process where a problem keeps recurring after it is supposedly fixed, or any discussion where some type of mistaken assumption keeps popping up despite it supposedly having been dispatched by new and more accurate information. In the context of sex differences, this might be the belief that newborn baby boys prefer to look at tractor mobiles rather than human faces (the 'men are born to be scientists' mole), or that there are more male geniuses and more male idiots (the 'greater male variability' mole). 'Truths' such as these have, as we shall see in this book, been variously whacked over the years but can still be found in self-help manuals, how-to guides and even in twenty-first-century arguments about the utility or futility of diversity agendas. And one of the oldest and apparently hardiest of moles is the myth of female and male brains.

The so-called 'female' brain has suffered centuries of being described as undersized, underdeveloped, evolutionarily inferior, poorly organised and generally defective. Further indignities have been heaped upon it as being the cause of women's inferiority, vulnerability, emotional instability, scientific ineptitude – making them unfit for any kind of responsibility, power or greatness.

Theories about women's inferior brains emerged long before we were actually able to study the human brain, other than when it was damaged or dead. Nevertheless, 'blame the brain' was a consistent and persistent mantra when it came to finding explanations for how and why women were different from men. In the eighteenth and nineteenth centuries it was generally accepted that women were socially, intellectually and emotionally inferior; in the nineteenth and twentieth centuries the focus shifted to women's supposedly 'natural' roles as carers, mothers, womanly companions of men. The message has been consistent: there are 'essential' differences between men's and women's brains, and these will determine their different capacities and characters and their different places in society. Even though we could not test these assumptions, they remained the bedrock on which stereotypes were firmly and immutably grounded.

But at the end of the twentieth century the advent of new forms of brain imaging technology offered the possibility that we could, at last, find out if there really were any differences between the brains of women and those of men, where they might come from, and what they might mean for the brains' owners. You might think that the possibilities offered by these new techniques would be seized on as 'game changers' in the arena of research into sex differences and the brain. The development of powerful and sensitive ways for studying the brain, together with a chance to reframe a centuries-old quest for differences, *should* be revolutionising the research agenda and galvanising discussion in media outlets. If only that were the case …

Several things went wrong in the early days of sex differences and brain imaging research. With respect to sex differences, there was a frustrating backward focus on historical beliefs in stereotypes (termed 'neurosexism' by psychologist Cordelia Fine). Studies were designed based on the go-to list of the 'robust' differences between females and males, generated over the centuries, or the data were interpreted in terms of stereotypical female/male characteristics which may not have even been measured in the scanner. If a difference was found, it was much more likely to be published than a finding of no difference, and it would also breathlessly be hailed as an 'at last the truth' moment by an enthusiastic media. Finally the evidence that women are hard-wired to be rubbish at map reading and that men can't multi-task!

The second difficulty with early brain imaging research was the images themselves. The new technology produced wonderfully colour-coded brain maps that gave the illusion of a window into the brain – the impression that this was an image of the real-time workings of this mysterious organ, now available for inspection by all. These seductive images have fed a problem which I have called 'neurotrash' – the sometimes bizarre representations (or misrepresentations) of brain imaging findings that appear in the popular press and in piles of brain-based self-help books. These books and articles are frequently illustrated with beautiful brain maps, which are considerably less frequently accompanied by any kind of explanation of what such maps are really showing. Understanding the differences between women and men has been a particular target for such manuals or headlines, bringing us apparently enlightening links to crowbars, polka dots and clams, and, of course, compounding the idea that 'Men are from Mars, Women are from Venus'.

So the advent of brain imaging at the end of the twentieth century did not do much to advance our understanding of alleged links between sex and the brain. Here in the twenty-first century, are we doing any better?

*

New ways of looking at the brain focus on connections between structures rather than just the size of the structures themselves. Neuroscientists today have started decoding the brain's 'chatter', the way in which different frequencies of brain activity seem to pass on messages and bring back answers. We are getting better models of how the brain does what it does, and we are beginning to have access to huge data sets, so comparisons can be made and models can be tested using hundreds if not thousands of brains, rather than the handfuls that were available previously. Could these advances shed any light on the vexed question of the myth or the reality of the 'female' and the 'male' brain?

One major breakthrough in recent years has been the realisation that the brain is much more 'proactive' or forward-thinking with respect to information gathering than we had first realised. It doesn't just respond to the information when it arrives, it generates predictions about what might be coming next, based on the kind of patterns it has identified on previous occasions. If it turns out that things didn't quite work out as planned, then this 'prediction error' will be noted and the guidelines adjusted accordingly.

Your brain is continuously making guesses as to what might be coming next, building templates or 'guide images' to help us take shortcuts to get on with navigating our lives. We could think of the brain as some kind of 'predictive texter' or high-end satnav, helpfully completing our words or sentences, or finishing off a visual pattern to let us get on with life quickly, or guiding us down the safest paths for 'people like us'. Of course, in order to make predictions you need to learn some kind of rules about what usually happens, about the normal course of events. So what our brain does with our world very much depends on what it finds in that world.

But what if the rules our brains are picking up are actually just stereotypes, those pervasive shortcuts that lump together past truths or half-truths or even untruths? And what might this mean for understanding sex differences?

This brings us into the world of self-fulfilling prophecies. The brain doesn't like making mistakes or prediction errors – if we are confronted with a situation where 'people like us' aren't commonly found or where we are clearly unwelcome, then our brain-based guidance system may drive us to withdraw ('Make a U-turn when possible'). If we are expected to make mistakes, then the additional stress makes it highly likely that mistakes will be made and we will lose our way.

Until the twenty-first century it was generally held that with regard to the brain, biology was destiny. The bottom line had always been that, apart from the known flexibility in very young, developing brains, the brains we ended up with were pretty much the ones we were born with (only bigger and a bit more connected). Once you were an adult, your brain had reached its developmental endpoint, reflecting the genetic and hormonal information with which it had been programmed – no upgrades or new operating systems were available. This message has changed in the last thirty years or so – our brains are plastic and malleable and this has significant implications for our under-standing of how entangled our brain is with its environment.

We now know that, even in adulthood, our brains are con-tinually being changed, not just by the education we receive, but also by the jobs we do, the hobbies we have, the sports we play. The brain of a working London taxi driver will be different from that of a trainee and from that of a retired taxi driver; we can track differences among people who play videogames or are learning origami or to play the violin. Supposing these brain-changing experiences are different for different people, or groups of people? If, for example, being male means that you have much greater experience of constructing things or manipulating complex 3D representations (such as playing with Lego), it is very likely that this will be shown in your brain. Brains reflect the lives they have lived, not just the sex of their owners.

Seeing the life-long impressions made on our plastic brains by the experiences and attitudes they encounter makes us realise that we need to take a really close look at what is going on outside our heads as well as inside. We can no longer cast the sex differences debate as nature versus nurture – we need to acknowledge that the relationship between a brain and its world is not a one-way street, but a constant two-way flow of traffic.

Perhaps an inevitable consequence of looking at how the outside world is entangled with the brain and its processes is a greater focus on social behaviour and on the brains behind it. There is an emerging theory that humans have been successful because we evolved to be a co-operative species. We can decode invisible social rules, 'mind-read' our fellow humans to know what they might do, what they might be thinking or feeling, or what they might want us to do (or not to do). Mapping the structures and networks of this social brain has revealed how it is involved with forging our self-identity, with spotting members of our in-group (are they male or female?), and with guiding our behaviour to be appropriate to the social and cultural networks to which we belong ('girls don't do that'), or to which we wish to belong. This is a key process to monitor in any attempt to understand gender gaps, and it appears to be a process that starts from birth, or even before.

Even the very youngest members of our world, highly dependent newborn babies, are in fact much more sophisticated socialites than we ever realised. Despite their fuzzy vision, rather rudimentary hearing and absence of pretty much all basic survival skills, babies are quickly picking up on useful social information: as well as key facts such as whose face and voice might signal the arrival of food and comfort, they start to register who is part of their in-crowd, to recognise different emotions in others. They appear to be tiny social sponges, quickly soaking up the cultural information from the world around them.

A story that neatly illustrates this comes from a remote village in Ethiopia, where computers had never been seen. Some researchers dropped off a pile of boxes, taped shut. The boxes contained brand-new laptops, preloaded with some games, apps and songs. And no instructions. The scientists videoed what happened next.

Within four minutes, one child had opened a box, found the on–off switch and powered the computer up. Within five days, every child in the village was using forty or more of the apps they found and singing the songs the researchers had preloaded. Within five months, they had hacked the operating system in order to reboot the camera that had been disabled.

Our brains are like these children. Unguided, they will work out the rules of the world, learn the applications, go beyond what was initially thought possible. They work by a combination of astute detection and self-organisation. And they will start very young!

And one of the first things they will turn their attention to is the rules of the gender game. With the relentless gender bombardment coming from social and mainstream media, it is an aspect of these little humans' world that we should be watching very carefully. Once we acknowledge that our brains are not only rule-hungry scavengers, with a particular appetite for social rules, but that they are also plastic and mouldable, then the power of gender stereotypes becomes evident. If we could follow the brain journey of a baby girl or a baby boy, we could see that right from the moment of birth, or even before, these brains may be set on different roads. Toys, clothes, books, parents, families, teachers, schools, universities, employers, social and cultural norms – and, of course, gender stereotypes – all can signpost different directions for different brains.

*

Resolving arguments about differences in the brain really matters. Understanding where such differences come from is important for everyone who has a brain and everyone who has a sex or a gender (more on this later) of some kind. The outcomes of these debates and research programmes, or even just anecdotes, are embedded in how we think about ourselves and others, and are used as yardsticks against which to measure self-identity, self-respect and self-esteem. Beliefs about sex differences (even if ill-founded) inform stereotypes, which commonly provide just two labels – girl or boy, female or male – which, in turn, historically carry with them huge amounts of 'contents assured' information and save us having to judge each individual on their own merits or idiosyncrasies. As well as providing a list of the contents themselves, these labels may carry an additional nature or nurture stamp. Is this a 'natural' product, based on pure biology, with its characteristics fixed and unchangeable, or is it a socially determined creation, manured by the world around it, with its characteristics quickly adjustable by the flick of a policy switch or an added sprinkling of environmental input?

With input from exciting breakthroughs in neuroscience, the neat, binary distinctiveness of these labels is being challenged – we are coming to realise that nature is inextricably entangled with nurture. What used to be thought fixed and inevitable is being shown to be plastic and flexible; the powerful biology-changing effects of our physical and our social worlds are being revealed. Even something that is 'written in our genes' may come to express itself differently in different contexts.

It has always been assumed that the two distinct biological templates that produce different female and male bodies will also produce differences in the brain, which will underpin sex differences in cognitive skills, personalities and temperament. But the twenty-first century is not just challenging the old answers – it is challenging the question itself. One by one, we will see that past certainties are being dismantled. We will see what is happening to those well-known differences in masculinity and femininity, in fear of success, in nurturance and caring – even the very notion of female and male brains. Revisiting the evidence that supported these conclusions suggests that these characteristics do *not* neatly match the male/female labels they have been given.

So, yes, this *is* another book about sex differences in the brain, in the wake of many influential and hugely well-informed predecessors. It is a book that I believe is needed, as the old misconceptions still keep popping up in new guises, Whac-a-Mole style. There are still problems to solve – we will see how big the gender gaps are in key areas of achievement – and there are still gender paradoxes to explain, such as why do the most gender equal countries have the lowest proportion of female scientists?

The message at the heart of this book is that a gendered world will produce a gendered brain. I believe that understanding how this happens and what it means for brains and their owners is important, not just for women and girls, but for men and boys, parents and teachers, businesses and universities, and for society as a whole.

Sex, Gender, Sex/ Gender or Gender/ Sex: A note on gender and sex

We need to address the issue of whether we should talk about 'sex' or 'gender' or neither or both or some sort of combination. This book will be about sex differences in the brain but it will also be about gender differences in the brain. So are these the same thing – does your biologically determined sex bring with it all the characteristics that define your socially constructed gender? Will being the possessor of two X chromosomes, or an XY pair, determine your place in society, the roles you will play, the choices you will make?

For centuries, the answer to this was an unequivocal 'yes'. As well as bestowing on you the appropriate reproductive gear, your biological sex allegedly gave you an appropriately distinct brain, and thus determined your temperament, your skills, your fitness to lead or be led. The term 'sex' was commonly used to refer to both biological and social characteristics of women and men.

Towards the end of the twentieth century, in the light of feminist concerns, there was a move to challenge this deterministic approach. There was an emerging insistence that the term 'gender' be used when referring to matters that were solely to do with social matters, distinct from 'sex', which should be reserved for any reference to biology. Fast-forward a few years and, as we shall see, it became clear that it was getting harder and harder to sustain this neat distinction between sex and gender. Our emerging understanding of how much the brain can be influenced by social pressures meant that we needed a term to reflect this entanglement; in academic circles, the use of 'sex/gender' or 'gender/sex' has been offered as a solution. But this is not widespread in everyday usage and is rarely to be found in the popular media or in more populist articles about females and males.

The solution there seems to be to use 'sex' or 'gender' pretty interchangeably, with perhaps a greater tendency to use 'gender' to avoid the impression that you believe whatever you are talking about is actually all down to biology. You never see articles on 'sex pay gaps' or 'sex imbalances', for example, in business leadership. But when it comes down to it, it is clear that the term 'gender' now bundles together all aspects of females and males in just the same way that 'sex' used to. Recently browsing through the BBC's popular online revision guides for sixteen-year-olds (not, I hasten to add, for tips for this book) I noted that there was a section on the determination of gender. It was actually about the production of XX and XY chromosome pairs, headed by the statement 'So a human baby's *gender* [my emphasis] is determined by the sperm that fertilises the egg cell'. So even august institutions such as the BBC are cheerily contributing to this linguistic confusion.

What does this mean for how I will label the brain differences (or lack of them) that are at the heart of this book? Are they 'sex differences' or 'gender differences' or both? Given that many of the arguments are about the core role of biology, I shall use the term 'sex' or 'sex differences' as the default option when talking about the brain or about individuals clearly being divided according to whether they are biologically female or male. 'Gender differences' will mainly be reserved for when we are looking at socialisation issues such as, for example, the pink and blue tsunami which washes over newly arrived humans. The title *The Gendered Brain* aims to acknowledge that we are looking at the brain-changing effects of social processes.

xxiv   The Gendered Brain

Gendered pronouns can also be a fraught topic. If you don't know the sex (or gender) of the person you are writing about, the default option has, historically, been the male version, 'he'. In a book where part of the story is to challenge default options, doing so would clearly be unacceptable. Although 'he or she' or 's/he' can be alternatives, this can become awkward and distracting in a lengthy tome like this. My solution has been to try and redress the balance by, where appropriate, deliberately using 'she' rather than 'he'.

PART ONE

Cross-section of the Brain

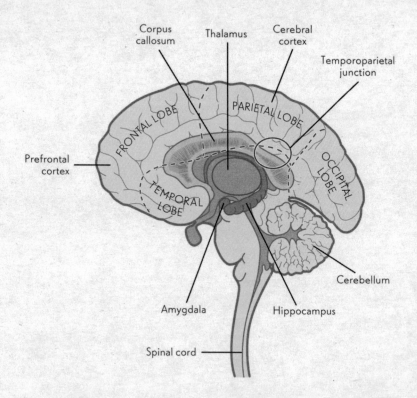

Chapter 1:
Inside Her Pretty Little Head – The hunt begins

> Women ... represent the most inferior forms of human evolution and ... are closer to children and savages than to an adult, civilized man.
>
> Gustave Le Bon, 1895

For centuries, women's brains have been weighed and measured and found wanting. Part of women's allegedly inferior, deficient or fragile biology, their brains were at the heart of any explanation as to why they were lower down any scale, from the evolutionary to the social and the intellectual. The inferior nature of women's brains was used as the rationale for frequently proffered advice that the fairer sex should focus on their reproductive gifts and leave education, power, politics, science and any other business of the world to men.

While views about women's capabilities and their role in society varied somewhat over the centuries, a consistent theme throughout was 'essentialism', the idea that differences between female and male brains were part of their 'essence', and that these brains' structures and functions were fixed and innate. Gender roles were determined by these essences. It would be going against nature to overturn this natural order of things.

An early version of this story starts, but unfortunately does not end, with a seventeenth-century philosopher, François Poullain de la Barre, bravely questioning the alleged inequality of the sexes.[1] Poullain was determined to have a clear-eyed look at the evidence behind the assertion that women were inferior to men, and was careful not to accept anything as true just because it was how things had always been (or because some appropriate explanation could be found in the Bible).

His two publications, On the Equality of the Two Sexes: a physical and moral discourse in which is seen the importance of undoing prejudice in oneself (1673) and On the Education of Women, to guide the mind in sciences and manners (1674), show a startlingly modern approach to issues of differences between the sexes.[2] Poullain even tries to show how women's skills can be equated with those of men; there's a charming section in his treatise on sexual equality where he muses that the skills required of embroidery and needlework are as demanding as those required to learn physics.[3]

Based on his studies of findings from the then new science of anatomy, he made a startlingly prescient observation: 'Our most accurate anatomical investigations do not uncover any difference between men and women in this part of the body [the head]. The brain of women is exactly like ours.'[4] His close examination of the different skills and dispositions of men and women, boys and girls, drew him to the conclusion that, given the opportunity, women would be just as capable of benefiting from the privileges which were then only offered to men, such as education and training. For Poullain, there was no evidence that women's inferior position in the world was due to some biological deficit. 'L'esprit n'a point de sexe,' he declared: the mind has no sex.[5]

Poullain's conclusions were strongly against the prevailing ethos; at the time of his writing, the patriarchal system was firmly entrenched. The 'separate spheres' ideology, with men fit for public roles and women for private, domestic ones, determined a woman's inferiority, necessarily subordinate to her father and then to her husband, and physically and mentally weaker than any man.[6]

It was downhill all the way after that. Poullain's views were largely, to his disappointment, ignored when they were first published (at least in France), and had little impact on the established view that women were essentially inferior to men, and would be unable to benefit from educational or political opportunities (which was, of course, a self-fulfilling prophecy as they were not, with notable exceptions, given access to education or political opportunities).* This remained the prevailing view throughout the eighteenth century, with little attention to it as a matter worthy of debate.

The woman question

In the nineteenth century, with the emergence of interest in science and scientific principles, there was a focus on linking society's structures and functions to biological processes, as characterised by early forms of social Darwinism. Among the intellectuals of the day, there were continuing concerns about the 'woman question', the increasing demands from women for rights to education, property and political power.[7] This feminist wave served as a rallying call for scientists to provide evidence in favour of the status quo, and to demonstrate how harmful it would be to give power to women – not only for the women themselves but for the whole framework of society. Even Darwin himself weighed in, expressing his concern that such changes would derail mankind's evolutionary journey.[8] Biology was destiny and the different 'essences' of men and women determined their rightful (and different) places in society.

The views expressed by other scientists indicated that they were likely to be less than objective in their approach to this

---

* Poullain's 'feminist' ideas were allegedly widely plagiarised (without acknowledgement) in England (e.g. *Female Rights Vindicated: or the equality of the sexes morally and physically proved*, by 'a Lady' (G. Burnet, 1758). His works started to attract attention in France at the beginning of the twentieth century, in the context of debates about women's equality. Simone de Beauvoir quoted him in her book *The Second Sex*.

issue. A favourite quote of mine comes from one Gustave Le Bon, a Parisian interested in anthropology and psychology. His main focus was on demonstrating the inferiority of non-European races, but he clearly had a special place in his heart for women:

> Without a doubt there exist some distinguished women, very superior to the average man but they are as exceptional as the birth of any monstrosity, as, for example, of a gorilla with two heads; consequently, we may neglect them entirely.[9]

Brain size was an early focus in this campaign to prove the inferiority of women and their biology. The fact that the only brains that researchers had access to were dead ones did not stand in the way of trenchant brain-based observations on women's lesser mental capacities (and, while they were at it, on those referred to at the time as 'coloured people, criminals and the lower classes'). In the absence of direct access to brains inside the skull, head size was initially adopted as a stand-in for brain size. Le Bon again was an eager exponent of this 'research', developing a portable cephalometer which he took around with him to measure the heads of those whose 'mental constitutions' would be more or less likely to stand up to the rigours of independence and education. Here we have another example of his penchant for ape comparisons: 'There are a large number of women whose brains are closer in size to those of gorillas than to the most developed male brains ... This inferiority is so obvious that no-one can contest it for a moment.'[10]

Skull capacity was another eagerly adopted index in the hunt for ways of proving the link between brain size and intellect. Bird seed or buckshot was poured into empty skulls and the amount required to fill it was weighed.[11] An early finding that, on average, women's brains were five ounces lighter than men's by this measure was enthusiastically seized upon as all the proof that was needed. Clearly, Nature had awarded men five extra ounces of brain matter, and this was the secret of their superior abilities and their right to positions of power and influence.

However, there was a flaw in this argument, as the philosopher John Stuart Mill pointed out: 'a tall and large-boned man must on this showing be wonderfully superior in intelligence to a small man and an elephant and a whale must prodigiously excel mankind'.[12] Various contortions followed, including a brain size–body size calculation, but that didn't come up with the 'right' answer either.[13] This is known in the business as the Chihuahua paradox: if you claim that a brain/body weight ratio as a measure of intelligence, then Chihuahuas should be the most intelligent dogs of all.

Perhaps more details about the brain's container, the skull itself, might help to produce the 'right' answer? This is where the science of craniology, or skull measurement, stepped in. Based on detailed measurements of every possible angle, height, ratio, forehead perpendicularity and jaw juttedness, craniology seemed to offer a suitable answer.[14] The twists and turns of craniology and its measurements were complex and varied. Facial angles were particularly popular, calculated by looking at the angle in profile between a line drawn horizontally from the nostril to the ear and one from the chin to the forehead. A nice big angle, with the forehead pretty much in line with the chin, was a measure of what was termed 'orthognathism'; a small acute angle, with a jutting chin way in advance of a receding forehead, was a measure of 'prognathism'. By devising a scale from orangutans through central Africans to European males, craniologists produced the satisfying finding that orthognathism was characteristic of the evolutionarily superior, higher races. However, with respect to fitting women on this scale, a problem emerged: women, on average, turned out to be more orthognathic than men. Fortunately, help was at hand.

The German anatomist Alexander Ecker, whose paper reported this disturbing observation, noted that *advanced* orthognathism was also characteristic of children, so on this basis women could be characterised as infantile (and, thus, inferior).[15] These suggestions were backed up by the findings of one John Cleland who, writing in 1870, reported on his painstaking catalogue of thirty-nine different measurements of ninety-six different skulls, which were all either 'civilised' or 'uncivilised', some male, some female, one a 'Hottentot chief', some 'cretins and idiots', another a 'savage Spanish pirate', and one the skull of a Fife man named Edmunds executed for the murder of his wife.[16] (We are told that Edmunds was from Fife and that he carried out the murder 'under circumstances of provocation'. We are not told whether either of these two facts earned him a 'civilised' or an 'uncivilised' classification). One particular measure in Cleland's catalogue, the ratio of the arch of the skull to its baseline, neatly ensured that adult females were distinct from adult males, and (mainly) distinguishable from members of 'uncivilised' nations.

There was to be no stone unturned (or skull unexamined) in the hunt for the proof of women's inferiority. One paper used over 5,000 measurements on a single skull.[17] There were seemingly infinite ways of measuring the skull, with the focus on those that not only best differentiated men from women, but also ensured that women were reliably characterised as inferior, either childlike or similar to reviled 'lower' races.

A group of mathematicians at University College London soon got involved in the great measuring game, and their findings would end up leaving craniology in disrepute.[18] This group of researchers, headed by Karl Pearson, the father of statistics, also included Alice Lee, one of the first women to graduate from London University. Lee created a mathematically based volumetric formula to work out skull capacity, which she intended to correlate with intelligence. She used this measurement on a group of thirty women students from Bedford College, twenty-five male staff at UCL and (a good move, this) a group of thirty-five leading anatomists who attended a meeting of the Anatomical Society in Dublin in 1898.

The results of her study were the nail in the coffin for craniology; she found that one of the most eminent of these anatomists had one of the smallest heads and, indeed, that one of her future examiners, a Sir William Turner, was eighth from the bottom. The discovery that these eminent men's heads were on the smaller side magically created a large number of instant converts to the conclusion that linking skull capacity to intelligence was obviously ludicrous (especially as some of the Bedford students had greater cranial capacities than the anatomists). A series of other such studies followed and in a 1906 paper Pearson declared that measure of head size was *not* an effective indication of intelligence.[19]

So craniology had had its day, but there were plenty of other sex difference explainers waiting in the wings. Another technique soon evolved out of craniology, which focussed on the mapping of different 'skill areas' onto the brain (though, again, without access to the means of directly measuring these). Moving from buckshot to bumps, scientists now focussed on the surfaces of skulls, scrutinising them for evidence of different-sized protuberances, which were taken to reflect the different landscapes of the underlying brains. This led to the infamous 'science' of phrenology, developed by Franz Joseph Gall, a German physiologist, who claimed that personality characteristics such as 'benevolence', 'cautiousness' or even the capacity to produce children could be assessed by measuring the relevant bit of a person's skull.[20] This technique was popularised by Johann Spurzheim, a German physician who was initially a student of Gall's but, after a disagreement with him, established his own career as an exponent of phrenology.[21] The claim of this system was that the different-sized bumps on the skull reflected the different sizes of the many different 'organs' of the brain, and that these organs controlled different individual characteristics such as combativeness, philoprogenitiveness or cautiousness. Again, there was, perhaps unsurprisingly, a neat matching of the bigger bumps on male skulls with more superior faculties.

Phrenology became particularly popular in the United States and, in some circles, was enthusiastically adopted by women. In an odd sort of early self-help movement, women were encouraged to 'know thyself' by getting their phrenological profile read.[22] One strange outcome was the simpering claim that this 'science' provided proof that 'we women' were indeed lower down a social hierarchy than our differently bumped male counterparts and that we should, with relief, acknowledge our place in the pecking order.

Phrenology eventually fell into disrepute by the middle of the nineteenth century, partly because of the unreliability of the measurements and the lack of any systematic testing of its theories.[23] But the notion that specific psychological processes could be localised to discrete areas of the brain lived on, partly supported by the emergence of neuropsychology, matching parts of the brain to specific aspects of behaviour. Scientists began to study patients who had suffered significant injuries to specific parts of the brain in the hope that their 'before and after' behaviour would reveal the exact function of those parts.

In the mid-nineteenth century, the French physician Paul Broca established a link between localised damage in the left frontal lobe and speech production.[24] His first clue came from the post-mortem examination of the brain of a patient called 'Tan', thus named because that was all he could say, although it was clear he could understand speech. The area of damage that was discovered, on the left-hand side of Tan's frontal lobe, is still called Broca's area.

More powerful evidence of the links between brain and behaviour was shown by the reported changes in behaviour of one Phineas Gage, an American railway worker who, in 1848, while preparing to blast rocks by tamping down some dynamite with an iron rod, set off an explosion which blew the rod through his left cheek and out of the top of his head, taking a substantial chunk of his frontal lobes with it. He was treated and subsequently studied by the physician John Harlow, who wrote up his observations in two papers with the informative titles of 'Passage of an Iron Rod through the Head' (1848) and 'Recovery from the Passage of an Iron Bar through the Head' (1868).[25] The reported changes in Gage's behaviour – sober and industrious before the accident; surly, impulsive, uninhibited and unpredictable after – were interpreted as showing that the frontal lobes were the seat of 'higher intellect' and civilised conduct. Forming as they do some thirty per cent of the human brain, as compared to about seventeen per cent in chimpanzees, the suggestion that within these lobes lay the higher powers that make us human made intuitive sense.

Enthusiastic bouts of cortical map making followed, with a focus on pinpointing *where* in the brain things were happening, more than when or how. Early models of the brain thought of it as a collection of specialised units or modules, each almost solely responsible for some particular skill. So if you wanted to find out where a skill was localised in the brain, you usually studied someone who had lost that skill following a brain injury. Broca's and Harlow's patients are probably the best-known examples of this. The loss of a particular part of language by Tan and the change in personality shown by Gage 'localised' these aspects of human behaviour to the frontal lobes.

In looking for sex differences, neurologists cheerily matched their assumptions about which bits of the brain were the most important to their findings about which bits of the brain were largest in males, even if it meant reversing earlier conclusions. For example, a paper in 1854 reported that women often had more extensive parietal lobes than men, whose brains were characterised by larger frontal lobes, thus earning the former the generic title of *Homo parietalis* and the latter *Homo frontalis*.[26] However, during a brief fashion for identifying the parietal lobes as the seat of human intellect, neurologists had to quickly back-pedal and report that female parietal lobes had in fact been mismeasured and women actually had larger frontal areas than had previously been thought.[27] It was not scientific research's finest hour.

As the turn of the century approached, declarations of inferiority gave way to references to the 'complementary' nature of women's alternative attributes (as defined, of course, by men). This was a concept that had its roots in eighteenth-century philosophy and ideas that justified the unequal distribution of citizens' rights. As Londa Schiebinger summarises:

Henceforth, women were not to be viewed merely as *inferior to* men but as fundamentally *different from*, and thus *incomparable to*, men. The private, caring woman emerged as a foil to the public, rational man. As such, women were thought to have their own part to play in the new democracies – as mothers and nurturers.[28]

The 'complementary roles' set aside for women ensured their inferior position in (if not, indeed, their absence from) most spheres of influence. A classic example of this approach is Jean-Jacques Rousseau's enthusiasm for the 'domestication' of woman, her weaker constitution and unique mothering skills rendering her unfit for any kind of education or political activism.[29] This was reflected in the opinions of other leading intellectuals such as anthropologist J. McGrigor Allan, who claimed when talking to the Royal Anthropological Institute in 1869:

In reflective power, woman is utterly unable to compete with man; but she possesses a compensating gift in her marvellous faculty of intuition. A woman will (by a power similar to that sort of semi-reason by which animals avoid what is hurtful, and seek what is necessary to their existence) arrive instantaneously at a correct opinion on a subject to which a man cannot attain, save by a long and complicated process of reasoning.[30]

As well as only being blessed with animal-like semi-reason, women's inferior biology was also identified as further justification for exclusion from the corridors of power. The vulnerability caused by the demands of their reproductive system was a constant thread in the assertions. McGrigor Allan again, also apparently an expert on the effects of menstruation, declared:

At such times, women are unfit for any great mental or physical labour. They suffer under a languor and depression which disqualify them for thought or action, and render it extremely doubtful how far they can be considered responsible beings while the crisis lasts ... Much of the inconsequent conduct of women, their petulance, caprice, and irritability, may be traced directly to this cause ... Imagine a woman, at such a time, having it in her power to sign the death-warrant of a rival or a faithless lover![31]

The contention of a direct link between biology and brain meant that overtaxing one could damage the other. In 1886, William Withers Moore, then president of the British Medical Association, warned of the dangers of overeducating women, asserting that their reproductive systems would be affected and they would succumb to the disorder 'anorexia scholastica', becoming more or less sexless and certainly unmarriageable.[32] Although the importance of 'mate choice', a keystone of Darwin's theory of sexual selectivity, was not much in vogue at this time, a woman's status was certainly closely determined by who she was married to, so diminishing your chances on the marriage market was a significant social threat.

The century came to a close with brain differences still a given, with the added acknowledgement of the fragility and vulnerability of the female. This was helpfully illustrated by the many 'mad, bad or sad' heroines in the literature of the time; women like Charlotte Brontë's Lucy Snowe, the heroine of *Villette*, George Eliot's Maggie Tulliver, from *The Mill on the Floss*, or Catherine Earnshaw, the heroine of Emily Brontë's *Wuthering Heights*, were all doomed by their wilful attempts to overturn the natural order of things.[33]

The birth of imaging

With respect to studying the brain itself, the twentieth century saw a continued focus on the consequences of brain injury, with the ravages of the First World War sadly providing many more case studies. But the models being built were based on the assumption that there is a direct mapping from a particular structure to a particular function, and that you can 'reverse-map' what a specific structure does by seeing which function is disrupted when that structure has been damaged. Now we know much more about how different parts of the brain interact with each other and how different networks are being formed and dismantled all the time, it means that we are rarely able to assume a direct connection between a specific brain structure (the hardware) and a specific brain function. Just because a particular skill or piece of behaviour is lost when a particular part of the brain is damaged does not mean that that part of the brain is solely responsible for controlling that particular skill. Unfortunately for us neuroscientists (but fortunately for us brain owners), there is not a neat one-to-one relationship between any one skill and any one part of the brain.

To get a better handle on how the brain supports different behaviours, we need to be able to access an intact healthy brain, to measure what is happening in real time, while the brain's owner is carrying out the task in which we are interested. Activity in the brain is a mixture of electrical and chemical activity within and between our nerve cells. In non-human animals, or during specific types of brain surgery on human brains, we can see this at the level of single cells, but generally, in the type of cognitive neuroscience research discussed in this book, the activity has to be measured from outside the head, as changes in the electrical status of the cells that make up the brain's different pathways, in the tiny magnetic fields associated with these electric currents, or in the characteristics of the blood flowing to and from busy areas of the brain. It was the development of technologies that could pick up these tiny biological signals which is the foundation of today's brain imaging systems.

The first breakthrough in measuring brain activity came in 1924 when the German psychiatrist Hans Berger, by taping small metal discs to the skull, was able to demonstrate patterns of electrical activity that changed depending on whether the person was relaxed, paying attention or carrying out specific tasks.[34] Berger showed that the signal he was picking up had varying frequencies and amplitudes, depending on where they were coming from and what the person was doing – the 'alpha wave' is most evident when people are alert and paying attention, while the very slow and relatively large 'delta wave' is most evident when they are asleep. He called his device the 'electroencephalogram'.

Electroencephalography or EEG is the oldest human brain imaging technique of all and the basis of much of the early knowledge of brain imaging research. In 1932, a multi-channel ink writing machine was developed which meant that the output of electrodes pasted over different parts of the skull could be transferred onto a moving paper roll and inspected for changes associated with, for example, flashing lights or intermittent sounds.[35] These changes could be plotted on millisecond timescales, so were a very good measure of the speed at which things were happening in the brain. But because the electrical signals were distorted by their passage through brain tissue, brain membranes and the skull itself, scientists were not always able to get a reliable picture of *where* these changes were taking place.

EEG remained the primary source of information about activity in the intact human brain up until the 1970s, when the first positron-emission tomography (PET) system was developed. PET made use of the fact that when activity in a particular part of the brain increases, there is an increase in the amount of blood flowing there. In PET systems, a small amount of a radioactive tracer is injected into the bloodstream; this can flag up the amount of glucose uptake from blood flowing to different parts of the brain, a measure of the amount of activity going on there.[36] PET was a much better indicator than EEG of the location of activity in the brain, but the use of radioactive isotopes raised ethical issues, and also limited who could be tested – children and females of childbearing age were generally excluded from research-only projects.

These problems were overcome by functional magnetic resonance imaging (fMRI), which emerged in the 1990s and works in a similar way to PET. Increased brain activity, as well as causing increased glucose uptake, also creates an increased demand for oxygen. As with glucose, this is supplied by more blood flowing to the relevant part of the brain and oxygen being absorbed to supply its needs; as the activity increases then oxygen levels in the brain will change. The changes in oxygen levels in the blood will result in changes in the blood's magnetic properties. If you put a brain (or, actually, the head of the brain's owner) in a strong magnetic field, you can measure these blood-oxygen-level-dependent (BOLD) responses. Following a very lengthy and complex chain of statistical analyses, the output of the scanner can be converted to colour-coded patches which are superimposed on a structural scan, usually in the form of a characteristic grey and white horizontal or vertical cross-section of a brain in its skull, producing what appears to be an image of what's happening inside our heads.[37]

The first fMRI studies of the human brain promised to provide stunning insights into processes we had previously only been able to guess at.

Size still matters

You might think that the superior technology on offer would have lifted the age-old debate to a higher plane. No more 'missing five ounces' or '*Homo parietalis*' taunts, no more agonising over tiny jaw angles?

Well, I'm afraid you would be disappointed. The 'blame the brain' mantra continued unabated and the 'size matters' emphasis remained just as evident in the scrutiny of brain imaging data as it was in the days of 'bumps and buckshot', with women's brains still being found wanting. As biologist and gender studies expert Anne Fausto-Sterling has pointed out, this issue is perfectly encapsulated in the 'corpus callosum wars'.[38] I use the term 'wars' advisedly here – one recent commentary from a researcher in the area was titled 'In the Trenches with the Corpus Callosum'.[39]

The corpus callosum is the bridge of nerve fibres, about ten centimetres long, that connects the right and left halves of the brain; it is the largest white matter structure in the brain, containing the projections from over 200 million nerve cells. It can clearly be seen in cross-section pictures of the brain as something rather like an elongated cashew nut, its even, pale grey shape in easily viewed contrast with the whorls of darker grey matter surrounding it.[40]

In 1982, American anthropologist Ralph Holloway and his student, Christine DeLacoste-Utansing, a cell biologist, reported finding sex differences in the size of the corpus callosum based on a very small cohort of subjects (fourteen males, five females).[41] The difference wasn't found across the whole corpus callosum, but only in the most posterior part of the brain, which was demonstrated to be wider or 'more bulbous' in females. It also wasn't actually a statistically significant difference, although some follow-up studies did add some additional cases which supported the initial finding. The size of the cohort and the really low level of statistical difference would mean that Holloway and DeLacoste-Utansing's paper would never have seen the light of day if it had been produced today, yet it has left a lasting legacy in the study of sex differences of the brain.

This tiny thread of a finding has resulted in a veritable tug-of-war between different researchers over the years, and provides a great case study of how finding the answer to what you are looking for in the brain might just be a function of how you ask the question. Multiple studies, of varying cohorts and using different measurement techniques, have been carried out – and a consensus has not yet been reached. Why so, you might ask?

First of all, it might be worth noting that measuring an awkwardly shaped three-dimensional structure buried inside two halves of an even more awkwardly shaped blob of organic matter is not straightforward. Early studies were based on autopsied brains which were neatly dissected into two halves, revealing the cross-section of the corpus callosum. Photos were taken and the resultant images back-projected onto a glass table. These images were then drawn around (yes, by hand) and various measures taken, of the length, the area and the width of the different substructures. Measures of length could be calculated by drawing a straight line from tip to tip, or a curved line, following the shape of the corpus callosum.[42] These manual methods have partly been overtaken by automated procedures nowadays, but the basic 'tracing' principle remains much the same.

The number of ways these different measures have been put to use in order to make a point about the corpus callosum in relation to sex differences is extraordinary, and alarmingly similar to the way craniology was discussed back in the nineteenth century. For example, an 1870 paper explains a craniology measurement as follows:

> The skull is suspended in a horizontal frame by means of two pointed screws, one on each side, which work in fixed supports; and by other screws moving on slides it may be set with any two points on a level. A vertical bar, which can be slipped up and down, slides along the side of the frame, and bears a sliding horizontal bar directed inwards, to which a needle may be attached at right angles if necessary, in either a vertical or longitudinal direction. The frame, the bars, and the needle are all marked off in inches and tenths, and by this means the vertical and horizontal distance of any point on the skull from the place of suspension is easily determined and marked on paper, so that by a series of such points a diagram may be constructed. With the assistance of a sheet of ruled paper such a diagram may be constructed in a few minutes from a series of figures not occupying more than a couple of lines.[43]

Now let's compare this with a 2014 explanation of a corpus callosum measurement:

> The contours of both corpus callosums were outlined by one rater (M.W.), and the top and bottom edges were defined relative to anterior and posterior end points. The middle line of N's corpus callosum (i.e. that courses rostro-caudally through the centre of the corpus callosum approximately parallel to its superior and inferior edges) was defined by the Symmetry-Curvature Duality Theorem (Leyton, 1987) and then sectioned into 400 equidistant points, with 400 corresponding points on the top edge and bottom edge. The distance between corresponding points

at the top and bottom edges was defined as the thickness of the corpus callosum at that level. The value of the 400 thicknesses were coded in colour and mapped onto N's left callosal space. The 400 values were averaged and defined as the mean thickness of the corpus callosum, whereas the summed distances between the 400 adjacent points was defined as the length of the middle line of the corpus callosum.[44]

It doesn't look as if things have moved on a lot in nearly 150 years, does it? It makes you wonder if we're just looking at extraordinary attention to detail, or at a rather desperate search for a way of spotting a difference, any difference.

The second lesson to be learned from the corpus callosum wars is that, when you are comparing brains, describing something as 'bigger' is not as straightforward as you might think. The key issue is that, on average, men's brains are bigger than women's brains, which has implications for all the structures within those brains. A bigger brain has a bigger corpus callosum, as is true of all structures in the brain, including key ones such as the amygdala and the hippocampus, over which similar such wars have been fought (and where the significance of such size differences has similarly been drafted in to support arguments about the 'natural' dispositions and abilities of women and men).

To settle such arguments there needs to be an *agreed* way of 'correcting' for differences in brain size. And the devil is in the word 'agreed'. Early studies took brain weight as a good indication of size and statistically corrected for that; others thought that brain area was more appropriate; later studies thought brain volume was a better variable to control for. But others felt that it was more of a scaling issue, so you needed to report corpus callosum size as a proportion of some aspect of the brain.[45] But proportional to what?

Everyone seemed to have a favourite bit of the brain with which they wanted to compare the corpus callosum. And woe betide you if you disagreed with their choice. These types of arguments elicited a rather exasperated rhetorical question from two researchers in the field:

> On what basis does a researcher select an organ against which to assess proportionality of the corpus callosum? Brain size seems obvious, but what about the volume of the occipital lobe or the ventricles, the length of the spinal cord, pupil size when dilated, or the volume of the left big toe raised to the 0.667 power?[46]

In my more irreverent moments, I am reminded of *Monty Python's Life of Brian*, where the crowd are exhorted to 'follow the gourd' only for a different holy sign to emerge, with the exhortation to 'follow the shoe'.

But even if some kind of correction consensus could be reached, what might any difference really mean? What does it mean if you have a larger or wider corpus callosum? If the female corpus callosum *was* different to the male version, how might you link it to the sex differences in behaviour, explanations for which were the point of the exercise in the first place? Very few of these studies actually measured any kind of behavioural differences alongside their motley array of size measurements.

A bigger bridge between the two hemispheres, in theory, must mean greater inter-communication between them. Early neuropsychology studies had proposed that the right side of the brain supported emotional and global processing skills, as these were more likely to be deficient in patients with right hemisphere damage.[47] And, as we know from Broca and his followers, the left side of the brain was in charge of language and logic. So, of course, if women generally have a larger corpus callosum, that must be why they are good at spotting the emotional undertones of a conversation, or why they are often able to tell what is going on without someone spelling it out for them (in other words, intuition). Less easy communication between the hemispheres would mean that each could be left to get on with its USP skills; the coolly logical left hemisphere of a male could face the world without distraction from noisy emotional inter-lopers, while the stunningly efficient spatial abilities of his right hemisphere could be focussed, laser-like, on the task in hand. Hence men's more efficient callosal filtering mechanism explained their mathematical and scientific genius (with chess brilliance thrown in for good measure), their right to be captains of industry, win Nobel Prizes and so on and on. In this instance, in the 'size matters' wars, with respect to the corpus callosum, small is beautiful.

However, as I've said before, the fundamental problem with this is that we are still somewhat uncertain about the relation-ship between the size of *any* brain structure and the expression of *any* behaviour with which it might be involved. At a very basic level, we know that the more sensitive the part of our body (for example our lips as opposed to our back), the bigger the area of the sensory cortex dedicated to processing informa-tion from that particular body part.[48] We know from training studies that areas of the brain associated with particular skills can be shown to increase in size with the acquisition of the skill.[49] Quantitatively a correlation, qualitatively an association, but we're a very long way off modelling any kind of causal relationship. As we will see later in the book, quite often the link between a particular structure and some particular aspect of behaviour is assumed as a 'given', possibly without the behav-iour itself having been part of any investigation of said structure. Women have wider callosal highways? Well, that is why they are ace multi-taskers! Female right hemispheres crowded with linguistic gossips? No wonder women can't read maps!

And there is a twenty-first-century issue which will be brought to bear here: what about the brain's plasticity in all of these arguments about who has the biggest corpus callosum? Bearing in mind that brain pathways can continue developing until the age of about thirty, and that increases in the corpus callosum have been shown until well into adoles-cence, there is an awful lot of scope for the world to impinge during this time. For example, one study has shown that the transfer rates in corpus callosum nerve fibres are faster in string-playing musicians (where the involvement of the two hands is asymmetrical) than in piano players (symmetrical use of the hands) or in non-musicians.[50] So, even if the various factions in the corpus callosum wars agree on what measure they might use, any conclusions about sex differences that might then result would need to take social or experiential factors into account.

The corpus callosum story encapsulates many of the issues surrounding attempts to measure sex differences in the brain. Not only are there intricate arguments about how measurements should be taken, but there are then ensuing disagreements about the origins of any differences that might be found and even more vehement arguments about what those differences might mean. Yet, in the populist 'sex differences' literature there are still bald statements that the corpus callosum is bigger in females than males, earnestly cited in continued support for right/left brain myths.[51]

Another measure that is enthusiastically debated is the ratio of grey matter (GM) to white matter (WM) in the brain, namely the balance between the overall volume of nerve cells in the brain (GM) and the pathways connecting them (WM). A 1999 report of this particular sex difference in the brain, using early structural MRI technology, came from the lab of Ruben and Raquel Gur, whence many such reports have since emanated.[52] The results were that females had a higher percentage of GM volume, whereas males had a higher percentage of WM volume. Four subsequent studies corrected for brain volume, as both grey and white matter can be affected by scaling issues, with GM distributed more widely in bigger brains, which additionally would require longer communicating pathways.[53] Two studies reported higher grey/white matter ratios in females; two reported no difference between males and females. A later review of this research looked at over 150 studies and concluded that actually males had a higher percentage of overall GM volume (the reverse of the original finding).[54] It is also evident that there are marked regional variations across the brain where these sex differences can be found. So this GM/WM measure wouldn't appear to be a useful way of distinguishing women's and men's brains.

That didn't stand in the way of its continued use as evidence in the ongoing debate. The issue of sex differences in grey and white matter has become another factoid which has morphed into a brain myth in populist literature. A study in 2004 looked at correlations between IQ scores and measures of grey and white matter in brains from twenty-one men and twenty-seven women.[55] The researchers reported that men had more significant brain–IQ correlations in their GM (6.5 times more than women, in fact) whereas women had nine times more significant brain–IQ correlations in their WM. There was no real discussion of what these correlations might actually mean, just that these two measures happened to go together. It's not hard to detect here shades of jaw-juttedness and forehead slopes.

The research was reported in the science press as demonstrating that women's IQ performance was related to integrating and assimilating information (using more pathways in the brain), whereas men were more locally focussed. Headlines such as 'Intelligence in men and women is a grey and white matter' and (of course) 'Men and women really do think differently' ensured that this early, small-scale study using a crude and rather mysterious measure of structure–function relationships has been quoted nearly 400 times to date, often in the context of discussions about single-sex schooling or the underrepresentation of women in science.

We have tracked the 'blame the brain' campaign down the ages, and seen how diligent was the scientists' pursuit of those brain differences that would keep women in their place. If a unit of measurement didn't exist to characterise those inferior female brains, then one must be invented! This measurement frenzy continued in the twentieth century, with imaging techniques clearly more sophisticated than craniometry's calipers or phrenology's bumps, but certainly with some of the same kind of debates about what types of measures to use. The whole campaign began with the assertion of differences and the hunt to find them, and this drive continued to motivate research programmes throughout the ensuing decades.

With the dawn of the twentieth century, scientists turned their attention to yet another potential source of evidence of women's vulnerable biology, their so-called 'raging hormones'. A whole new search was to begin.

Chapter 2:
Her Raging Hormones

In any discussion about sex differences in human brains and any link to behaviour, a frequently asked question is 'What about hormones?' The belief that sex differences in behaviour are as much linked to the action of these chemical messengers as to that of the brain is firmly entrenched in popular biological explanations of our skills, aptitudes, interests and abilities. Financial success (or failure), leadership skills, aggression and even promiscuity have been attributed to men's high levels of testosterone, whereas women's nurturing skills, great memory for birthdays and talent for needlework are, apparently, down to their oestrogen levels.[1] Indeed, it is claimed that hormones are directly responsible for sex differences in the brain, with the presence or absence of prenatal exposure to testosterone setting brain development on divergent male or female pathways.[2]

With the discovery of the first hormone at the beginning of the twentieth century, attention became focussed on the chemical control of behaviour, with gonads and glands being measured and manipulated to see how this affected their owners' behaviour.

It was a Mauritian-French physiologist, Charles-Edouard Brown-Séquard, who was the first to speculate that there were some kinds of chemicals, secreted into the bloodstream, that could control organs at a distance.[3] He tested this by putting together a cocktail of ground guinea pig and dog testicles, which he bravely drank himself, subsequently reporting an increased feeling of vitality and mental clarity. Secretin, the first such chemical to be identified, was discovered in 1902 by an English doctor, Ernest Starling, while working with a physiologist, William Bayliss.[4] They demonstrated that this chemical, which they now termed a hormone (from the Greek for 'stir into action'), was made by glands in the small intestine and could stimulate the pancreas. Discovery of the many sites of production and action of these chemical control agents, or bioregulators, speedily followed. As you might expect, investigating the control of sex-related behaviour and sex differences was high up on the list of early research projects.

Androgens, oestrogens and progestogens, the hormones which determine the development of sex organs and control reproductive behaviour, were identified in the late 1920s and early 1930s, although the effects of transplanting testes into various animals had been studied since the eighteenth century.[5] Similarly, at the end of the nineteenth century, ovarian extract had been found to be effective in treating hot flushes, indicating the existence of some specifically female secretion linked to menstruation.[6]

A key androgen, testosterone, was named in 1935 when it was isolated from bull testes. The chemistry professor who discovered testosterone, Fred Koch (no, really), showed that castrated roosters or rats could be remasculinised if this hormone was injected into them. For example, the shrivelled cockscomb of a castrated rooster was shown to spring back to its former glory.[7] This was the basis for some rather bizarre treatments claiming to improve virility (in a spare moment you might like to find out what being 'Steinached' entailed).[8]

With respect to so-called female hormones, in 1906, secretions from the ovaries had been shown to produce cyclic sexual activity in non-human females.[9] These were named oestrogens, from the Greek terms *oistrus* (mad desire) and *gennan* (to produce). (You can probably guess the gender of the scientists who named them thus.) The different oestrogens (oestrone, oestriol and oestradiol) were isolated as hormones and synthesised at the beginning of the 1930s. They were shown, for example, to induce the onset of puberty in non-human female animals and could induce female-like sexual behaviour in male rats.[10]

One thing we should note is that, although androgens are described as male hormones, and oestrogens and progestogens as female hormones, they are found in all of us, both male and female alike (although there was an early suggestion that the oestrogen found in men actually came from their consumption of rice and sweet potatoes – thus, presumably, freeing the field to attribute the negative aspects of oestrogen just to the natural and unchangeable version found in women).[11] It is the levels of each that vary between men and women; the range of testosterone is naturally generally higher in men than women, and oestrogen higher in women than men, but it is worth bearing this dual possession in mind when considering explanations of hormone-related sex differences in behaviour.

As with early brain studies, there was an enthusiasm to explore the link between this newly discovered chemical means of controlling behaviour and sex differences, especially as the 'sex' hormones were clearly linked to well-differentiated facets of behaviour in non-human animals, namely their different roles in reproduction. But how to investigate them in humans? Heroic ingestion of testicular or ovarian secretions was quickly (and fortunately) established as somewhat limited in its usefulness in the quest for evidence. Similarly, it would be tricky to find a human parallel for the effects of early castration in male rats followed by injections of oestrogen.

Additionally, what aspects of behaviour were to be examined? If you were interested in explaining the status quo social phenomenon of superior, high-achieving human males versus inferior, emotionally labile human females, then comparing the reproductive practices of both sexes would probably not prove as politically revealing as you might hope. Attention focussed on the 'well-known' monthly cycle of increases and decreases in women's fundamental irrationality and emotional lability which, as we saw in the last chapter, had been so fervently detailed by nineteenth-century male experts on the topic. Perhaps Brown-Séquard didn't try out a matching cocktail based on female organs in case he experienced a devastating loss of mental clarity? The 'raging hormones' problem, already hinted at by McGrigor Allan's concerns about menstruation in the nineteenth century, became the explanation *du jour* of the inadvisability of giving women any positions of power.

The menstrual cycle: mean, moody or mythical?

Tracking behavioural changes in women during the menstrual cycle has been a popular source of such data – and historically of course has been held up as a reason why women should be kept away from positions of power and influence. In 1931, a gynaecologist called Robert Frank handed scientific credibility to this notion by suggesting a link between the newly discovered hormones and incidences of 'premenstrual tension' (now known commonly as PMT) in his women patients who showed 'foolish and ill-considered actions' just prior to menstruation. This was the birth of the now notorious 'premenstrual syndrome' (PMS).[12]

It was Katherina Dalton, a UK endocrinologist working in the 1960s and 1970s, who really gave PMS the identity of a medical syndrome by packaging together many associated physical and behavioural symptoms, linking them firmly to the premenstrual phase and identifying a clear biological cause, a hormonal imbalance.[13] PMS has become a widely accepted phenomenon in Western cultures, where the days prior to the onset of menstruation are allegedly associated with dramatic outbursts of negative mood, poor performance in school or at work, overall decline in cognitive competence, and increases in accident rates. It has been estimated that eighty per cent of the women in the United States experience premenstrual emotional or physical symptoms.[14] PMS has a well-established place in popular culture, where we find a general consensus about premenstrual frenzy and hormonal roller-coasters, with out-of-control women suffering weeks of hell.[15]

Interestingly, World Health Organization surveys suggest that there are cultural variations in the kinds of complaint that are associated with the premenstrual phase. The emotional changes reported above are almost exclusively found in western Europe, Australia and North America, whereas women in Eastern cultures such as China are more likely to note physical symptoms such as water retention but rarely mention emotional problems.[16]

In 1970, Dr Edgar Berman, then a member of the US Democratic Party's Committee on National Priorities, declared that women were unfit for positions of leadership because of their 'raging hormonal imbalances'. By his reasoning, only the pre-menarcheal and post-menopausal could be depended upon not to be irrational several days a month. Imagine, he said, a female bank president 'making loans at that particular period. Or, worse, a menopausal woman in the White House faced with the Bay of Pigs, the Button and – hot flashes.'[17] Women were initially barred from the space programme as it was felt that it would be inadvisable to have such 'temperamental psycho-physiologic humans' on board a spacecraft.[18]

In the West, the concept of PMS is so well established that it can become a kind of self-fulfilling prophecy, used to explain or be blamed for events which could just as well be attributed to other factors. One study showed that women were more likely to blame their own menstrually related biological problems for negative moods, even when situational factors could equally well be the source of difficulties.[19] Another study showed that if women were 'tricked' into thinking they were premenstrual by being given artificial feedback from a realistic-looking physiological measure, they reported significantly more occurrences of negative symptoms than the women who had been tricked into believing they were intermenstrual.[20]

But what exactly is premenstrual syndrome? How do you know if you have it? And what causes it? The answers to these questions are not straightforward. With respect to its definition, it has been noted that this is 'vague and various'.[21] There appears to be no agreed definition of what behavioural changes might be investigated. One hundred or more 'symptoms' (*sic*) have been identified: some physical, such as 'pain' or 'water retention'; some emotional, such as 'anxiety' or 'irritability'; some cognitive, such as 'lowered work performance'; some even more ill-defined, such as 'lowered judgement'. There is a heavy emphasis on negative events. Indeed, the most frequently used questionnaire to gather data on such events is unsubtly entitled the 'Moos Menstrual Distress Questionnaire' (MDQ, with the name Moos referring to its author rather than to those using it).[22] The questionnaire asks women to rate forty-six different symptoms on a scale from 'no experience' to 'acute or partially disabling'. Almost all are behavioural, for example, 'forgetfulness', 'distractibility' or 'confusion', and only five are positive, such as 'bursts of energy', 'orderliness' and 'feelings of wellbeing'. Interestingly, studies have found that individuals who have never experienced menstruation come up with profiles that were indistinguishable from women who had menstruated when asked to fill out the MDQ.*[23]

More recent work has shown that there may in fact be a link between female hormones and *positive* behavioural changes (which, of course, would not be the focus of attention of those of the school of Gustave Le Bon, J. McGrigor Allan and Edgar Berman). An emerging consensus is that the most reliable findings are of *improved* cognitive and affective processing associated with the ovulatory and post-ovulatory phases, rather than of the alleged deficits that have been claimed to emerge premenstrually. In a recent systematic review of cognitive functioning and emotion processing throughout the menstrual cycle, which included fMRI measures as well as hormone assays, improved performance in verbal and spatial working memory was found

---

* It is also the case that retrospective measures do not anyway provide reliable data, particularly if the context of the questions is known to those filling out the answers. A safer approach is to use daily prospective measures of behavioural changes, throughout at least one complete cycle, thereby avoiding an obvious focus on the premenstrual phase and its 'reputation', and ideally obscuring the purpose of the enquiry or keeping it as low-key as possible. A recent survey looked at the methodology of research studies linking mood and the menstrual cycle to gauge the extent to which these kinds of pitfalls had been avoided. Of the 646 studies identified, only 47 met the criteria of using prospective measures for at least one cycle. Of these, only seven reported the classic pattern of negative mood in the premenstrual phase; eighteen of them didn't show any relationship between mood and the menstrual cycle when measured in this way.

to be associated with high oestradiol levels.[24] Emotion-related changes, such as better emotion recognition accuracy and enhanced emotional memory, were found when both oestrogen and progesterone levels were high. These were associated with increased reactivity in the amygdala, part of the brain's emotion-processing network. I haven't, as yet, come across an Ovulation Euphoria Questionnaire!

The PMS story provides a nice case study of the role of self-fulfilling prophecies in linking biology to behaviour. A vague phenomenon, defined by highly biased self-report measures, has become a useful hook on which to hang behavioural events, tellingly labelled as 'symptoms', with, additionally, an emphasis on the problems this biological phenomenon can cause women (and those around them). What looked like the ideal way to establish cause and effect by tracking how behavioural changes linked to menstrual-cycle-related hormonal changes became more of an example of how stereotypical beliefs can become so firmly established that even those to whom they refer can come to believe in them.*

Other ways of establishing cause and effect take us back to animal studies. The early work had shown that hormones could determine key physical differences in female and male organisms and that, in non-human animals at least, they also controlled reproduction-relevant behaviour, with females in oestrus presenting themselves to males (who obligingly mounted them) and the mothers of newborns showing the appropriate pup-caring skills.[25] It was suggested that these different aspects of masculine and feminine behaviour were linked to the action of different hormones on brain pathways. An even more radical suggestion was that hormones had a more fundamental role and that they actually organised the brain differently, with male hormones causing brains to develop along masculine lines, producing a 'male

---

* This is not to deny that some women may have negative physical and emotional problems linked to hormone fluctuations, but simply shows that the stereotype of PMS being a near-universal phenomenon is a good example of the 'blame game' aspect of biological determinism.

brain', and female hormones producing a 'female brain'. This is known as the brain organisation theory.[26]

We now know that hormone activity in mammalian foetuses is crucial to determining their sex. In humans, up to about five weeks after conception, male and female foetuses are indistinguishable, gonadally speaking. At this point, the female (XX) foetus will develop ovaries whereas the male (XY) will develop testes. Shortly afterwards, there is a surge of testosterone production from the testes, which continues until about the sixteenth week of pregnancy. From then until birth, testosterone levels are pretty similar in boys and girls. At birth, the effects of this difference in prenatal hormones are normally immediately evident by looking at the external genitalia of the newborn – penis for boys, clitoris for girls. The brain organisation theory proposes that this prenatal hormonal activity in male foetuses is not just limited to individuals' gonads, but will also 'masculinise' their brains, determining particular kinds of neural real estate in males and distinguishing them from females, who haven't experienced this testosterone marinade. These brain differences will then determine differences in their cognitive skills and their emotional characteristics, as well as, quite possibly, their sexual preferences and occupational choices.

The basis for the brain organisation theory was an early study on guinea pigs. In 1959, Charles Phoenix, a graduate student of endocrinology at the University of Kansas, working with his supervisor, William Young, and his team, published a paper demonstrating that administering testosterone prenatally to female guinea pigs caused them to show characteristic male, rather than female, mating behaviour when they reached puberty, enthusiastically trying to mount other female guinea pigs.[27] This suggested that hormones could exert a very long-lasting effect if administered early enough.

One implication of the brain organisation theory was that, just as the structure and function of the feminine or masculine genitalia were fixed and permanent, so too were the feminine or masculine characteristics of the brain. A further refinement of this theory referred to an activational or 'switching-on' process; prenatal organisation having guided relevant structures in the brain to fixed, sexually distinct endpoints, these would then form the substrate for any future effects of variations in hormones, most commonly associated with the onset of puberty. So the brain's masculinised or feminised structures would respond differently to male or female hormones, resulting in 'sex-appropriate' behaviour.

The brain organisation theory appeared to be the 'missing link' in the chain of argument that the biological differences between males and females determined their behavioural differences. Males and females were different because the chemicals that determined their reproductive apparatus also determined key structures and functions in their brains. The theory was to be extended further into the realms of types of sex differences other than those associated with reproduction, such as 'rough and tumble play', or spatial or mathematical skills, allegedly associated with testosterone exposure, and nurturance or doll play linked with oestrogen levels.[28]

Testing such assertions would not only require monitoring hormones, brains and behaviour in the different sexes but would also involve trying out various kinds of within- and between-sex hormone manipulations, both prenatally and postnatally. The foundational evidence for the theory so far had been based on the manipulation of hormone levels in animals by severe physical interventions such as ovariectomy or gonadectomy, and subsequently watching the effect on behaviours such as frequency of copulation, mounting or lordosis (the posture assumed by some animals indicating sexual receptivity). As indicated above, this isn't something that could be tried out on humans in quite the same way. Either it would have to be accepted that what was being carried out with non-human animals was an appropriate proxy for the study of humans, or researchers would have to make use of typical or atypical fluctuations in hormone levels.

Of mice and men?

For biologists in the first half of the twentieth century, the use of so-called 'animal models' was not seen as incongruous. There was an assumption of some kind of physiological equivalence between all mammals which could justify extrapolating conclusions about biological measurements from one group (rats, monkeys) to another (humans).

You might think that behavioural equivalence could be a bit more of a problem. Could you equate, for example, the maze-learning behaviour of a rat with the spatial cognition skills of a human male? The prevailing psychological thinking at the time was that of behaviourism, a school of thought based on the idea that it was appropriate to draw parallels between human and non-human behaviour. Behaviourism stated that the only acceptable subject matter for psychology was activities and events that could be clearly observed, objectively measured and recorded and then interpreted according to agreed rules.[29] There was no appeal to internal thoughts or feelings; the rules of behaviour could be extracted by setting up carefully controlled tasks and observing the consequences of manipulating hypothesised variables. How did learning come about? Set up a learning situation, manipulate key variables and see which worked. Could you increase response rates? Manipulate some rewards (or 'positive reinforcements'). Could you reduce response rates? Try some punishments (or 'negative reinforcements'). It was not considered important what kind of species was producing the responses you were conditioning – no messy introspection was to be allowed to interfere with the generation of scientific theories of behaviour. So what was true for pigeons or white rats could be taken as true for humans, and it was perfectly acceptable to extrapolate from animal to human behaviour.

Animal models were used to test for many different aspects of behaviour, not just simple learning processes but also high-level cognitive skills such as spatial cognition (maze learning) or social skills such as nurturance (care of pups). Parallels between non-human and human types of behaviour were sought so that you could measure the effects of direct intervention on the former, given that for ethical reasons it might be tricky to carry out the necessary experiments on the latter. Is there a biological reason for boys being more active than girls (setting aside for the moment whether or not these levels of activity really are different)? You could measure the effect on 'rough and tumble' play of exposing female embryos to high levels of testosterone. Is it hormones that give females a 'maternal instinct'? Try manipulating oestrogen in female rats and see what happens to their 'pup retrieval' or 'anogenital licking'.[30]

This is why much of our early understanding about the link between hormones and behaviour (and, indeed, between the brain and behaviour) came from the study of non-human animals. 'Well-established' findings of links between sex differences in brain and behaviour may, in fact, be referring to research on the size of song-control nuclei in zebra finches and canaries (the males are the singers and have bigger nuclei).[31] Sometimes this gets lost in translation and there can be an imperceptible sleight-of-hand where you have to look hard to realise that the studies on sexually dimorphic behaviours, which are allegedly of relevance to understanding Alzheimer's disease and autism, have actually been carried out on mice.[32] You'd be surprised how often some of the more careless of populist science writers somehow forget to mention that the research they are quoting in support of their particular sex difference meme was carried out in songbirds or prairie voles, and not people.[33]

But suppose you wanted to test for sex/gender differences in personality characteristics, mathematical abilities or career choice? Or interests rather than abilities? Or gender identity? Here, no parallel animal models can be offered. We are unable to carefully titrate changes in the behavioural measure against changes in hormone levels, which *should*, of course, make us increasingly cautious about making the kind of causal assertions we find in laboratory-based animal studies. We need to make use of unusual or atypical hormonal levels in humans which may naturally or accidentally occur.

The normal patterns of prenatal exposure to different hormones can be quite profoundly disrupted; if a male foetus doesn't get the expected amount of testosterone at the right time, or is insensitive to its effects, the baby will be born with feminised genitalia.[34] Similarly, if a developing female foetus is exposed to high levels of androgens prenatally, then she will arrive with masculinised genitalia. 'Intersex' is the generic term for such conditions; they are rare and those evident at birth generally require immediate and ongoing medical treatments. They are also the kind of 'natural experiment' which allows researchers to study the effects in females and males of exposure to 'cross-sex' hormones.

A wild romping girl

Congenital adrenal hyperplasia (CAH) is an inherited enzyme deficiency which causes overproduction of androgens in a developing baby.[35] In girls it is often immediately identifiable at birth because of their ambiguous genitalia. A lifetime of treatments ensues, including surgical corrections of the genitalia and hormone treatment. Girls with CAH are usually reared as girls and, as well as the medical interventions, they and their families will often be asked to participate in research studies, with the principal line of enquiry being the effects of early exposure to masculinising hormones.*[36] Researchers look out for early sex differences in behaviour, such as toy preference or levels of activity, cognitive skills such as spatial ability, and specific gender-related issues such as gender identity and sexual orientation. These CAH children are seen as the ideal cohort for testing the potency and primacy of biology.

One of the most frequently reported outcomes of such studies is about gender-typed play, with CAH girls reportedly more likely to play with male-type toys, more likely to want to play with boys and more likely to be described as 'tomboyish' by

---

* Given that testosterone determines the development of male genitalia, this aspect of CAH will obviously only affect girls and identify them as test cases for any other effects of testosterone. Testosterone levels in CAH boys will be high, but commonly within the normal range. Both groups are affected by other side-effects of the condition, requiring life-long medical treatment, so boys with CAH can provide a 'control' group for the additional effects of these kind of factors.

their families and teachers.[37] Definitions of the term 'tomboy' tend to include descriptors such as 'wild', 'romping', 'boisterous', or 'a girl who acts like a spirited boy'. Just to ensure scientific credibility, there is a Tomboy Index, which includes questions about preferring 'climbing trees and playing army to ballet or dressing up', preferring 'shorts or jeans to dresses', and taking part in 'traditionally male sports such as football, baseball, basketball'.[38] You might have spotted that behind such questions appears to be a pretty fixed assumption about what constitutes appropriate behaviour for girls. That could be related to the fact that the index was partly developed by studying the activities of females who thought themselves to be tomboys, and partly by asking people what they thought was typical tomboy behaviour. So it's likely this is not a totally objective, context-free measure of this particular label.

Similarly, there is strong evidence of working backwards from stereotypes when you read the kind of ways in which researchers characterised the tomboyishness of the girls they were studying. The features they identified as indicative of tomboyishness were a lack of interest in self-adornment, a lack of interest in 'rehearsal of maternalism' (meaning negligible doll play) and a lack of interest in marriage.[39] Although these early studies were conducted back in the 1950s and 1960s, and we might hope that things have moved on a bit since then, the Tomboy Index continues to be used in studies today, suggesting there is still a firmly entrenched yardstick against which girls' behaviour is measured.

As well as this reported tomboy behaviour, much has been made of the 'masculinised' cognitive skills and behavioural profiles revealed by research with girls with CAH. However, there are also clear flaws in the methodology and interpretation, and a lack of consistency in some of the research findings. For example, if males have superior visuospatial skills, which are supposedly the result of prenatal brain organisation driven by testosterone, then shouldn't CAH females show similar abilities? Or at least be better than unaffected females? In 2004, in her book *Brain Gender*, neuroscientist Melissa Hines looked at seven studies directly addressing this issue and found that only three supported this notion, two found no differences, and one showed that women with CAH were actually worse.[40] Only two of the studies used mental rotation, a task that, it has been claimed, most reliably demonstrates sex differences in performance. In the standard version of a mental rotation task you are shown a two-dimensional image of an abstract three-dimensional object and asked to imagine rotating it in space, then pick out which two of four alternatives would match the rotated original. One study showed that CAH girls did better at mental rotation; the other showed no difference. A later meta-analysis of studies of mental rotation skills in CAH girls showed clearer evidence that, on this particular measure, CAH girls did outperform unaffected girls.[41] But how strong is such evidence in debates about the link between brains and behaviour?

Rebecca Jordan-Young, an American sociomedical scientist based in Barnard College, Columbia University, has carried out a hugely detailed systematic review of research into brain organisation theory, with a focus on research into intersex individuals, such as CAH girls.[42] Her work demonstrates how the existing research has been used to offer a unidirectional biological explanation for allegedly sex-specific behaviours. She argues that overliteral applications of brain organisation theory have led to an oversimplistic view of the connection between human hormones and human brains. In particular, the core notion that prenatal hormones have a permanent, lasting effect entirely ignores our more up-to-date understanding of the plasticity and mouldability of the human brain: 'The problem is that the data have never fit the model so well in the case of brains as the case of genitals ... Brains, unlike genitals, are plastic.'[43] She also points out that many of the hypotheses about and interpretations of hormone effects seem to be based on the assumption that development is context-free, that the outcomes will be inevitable, regardless of social expectations or cultural influences.

Terrible accidents can offer up evidence in favour of the organisational hypothesis. Just as the injuries suffered by Broca's

Tan and Harlow's Phineas Gage gave early clues about the role of the brain in language, executive functioning and memory, a similar kind of unhappy event was studied in the quest to determine whether masculinity or femininity was fixed before birth, with no amount of subsequent socialisation apparently being able to divert this predetermined route.

This is the now-notorious case of a seven-month-old baby boy whose penis was damaged beyond repair after a bungled circumcision in 1966.[44] Some twelve months later, on advice from John Money, a psychologist and 'sexologist', the parents agreed that the child should be raised as a girl. This included removing the boy's testicles and administering female hormones from the age of eighteen months. Sex reassignment surgery was also offered for the child, involving the construction of a vagina, but the parents refused it.

Money believed that gender could be imposed, or learned independently of biology; he was convinced that socialisation experiences, if they began early enough, could ensure the emergence of an appropriate 'gender' identity. Despite the steer given to the brain by prenatal testosterone, Money believed he could prove that behaviour could be reset by determined environmental input. This unfortunate boy offered the perfect way to test his theory, especially as the baby also had an identical twin brother, which offered the ideal control comparison.

At the time, the so-called 'John/Joan' case, as these were the pseudonyms that Money gave to the child (although we now know the boy was originally called Bruce and his name was changed to Brenda), was hailed as living proof for the success of the reassignment process and of the independence of gender from its biological origins. However, in 1997, now aged thirty-one, Brenda went public and revealed a different version of her story.[45] It emerged that she had had what she described as an extremely unhappy childhood, very much tied up with confusions about her gender identity and unhappiness with 'being a girl'. There was also disturbing evidence about interactions with John Money and his attempts to ensure that Brenda retained her female identity, including repeated insistence that she should have full sex reassignment surgery. She described how, once her original reassignment had been revealed to her when she was fourteen years old, she insisted on reverting to her biological sex and renaming herself. Now as David Reimer, he had testosterone injections, a double mastectomy and penile construction surgery. But he remained deeply troubled and was outraged at the discovery that Money was still publishing papers claiming the success of the John/Joan experiment. David committed suicide in 2004, aged thirty-eight.

Much has been made of this tragic case as evidence that gender identity has fixed biological origins that cannot be overridden. However, it's crucial to note here that Bruce was actually over eighteen months old before any kind of sex or gender reassignment took place, which is time enough for a developing child to have absorbed all kinds of social information, especially as he had an identical twin brother. But the individual difficulties associated with this story mean it can really only remain just that, a story. We need to seek out evidence about the potency, or otherwise, of hormone effects on the brain elsewhere.

The measurement of prenatal hormone levels is not currently a standard measure taken from babies before birth, but there is research based on assessments of testosterone in amniotic fluid acquired during amniocentesis. This is linked to work by Simon Baron-Cohen, Director of the Autism Research Centre at the University of Cambridge. One ongoing research programme is a longitudinal study into the effects of foetal testosterone (fT) and how it might be associated with later brain and behaviour characteristics.[46] Baron-Cohen suggests that the masculinisation of the brain that comes about due to prenatal exposure to testosterone will vary as a function of the level of exposure.[47] The kind of masculine behaviour he identifies as affected is a tendency to systemise, to prefer rule-based ways of dealing with the world, rather than the more emotional, empathic approach allegedly characteristic of female behaviour.

So here we do have a possibility of looking at a brain–behaviour relationship, even if it is only correlational, between prenatal levels of masculinising hormones and what are claimed to be characteristically masculine aspects of behaviour.

The results could be described as 'promising but mixed' and certainly suggest the hormone–behaviour relationship is not as straightforward in humans as it is in guinea pigs. For example, there does appear to be a link between a measure of restricted interests (perhaps obsessions with wheeled toys) and fT, but this was only in boys (then aged four). There was a link between fT and social relationships, but this time more strongly in girls than boys. With respect to empathy in slightly older children, when this was measured by a questionnaire, there was a negative correlation between this and fT, again in boys but not girls, whereas when it was measured by an emotion recognition task there was a negative correlation between fT in both boys and girls. At the very best, we would have to conclude that if fT studies can report anything about a relationship between brain and behaviour as moderated by hormones, it is a rather variable and complex one, and may actually be a function of what behavioural measure you use. As the authors of one of the studies observed: 'It is worth keeping in mind that testosterone is not the only factor that varies between males and females.'[48]

These are intriguing findings and were certainly hailed by Baron-Cohen's lab as clear evidence of the organisational effects of prenatal hormones. However, we should remember that the world starts steering children's brains in different directions from a very early age, so the boys and girls who were being tested here may well have had different experiences which could contribute to their different scores as much as their fT.

Another attempt at finding a measure of prenatal testosterone in humans involves our fingers. If your index finger (known as 2D, for second digit) is longer than your ring finger (4D) you have a high 2D:4D ratio. If the reverse is true, you have a low 2D:4D ratio. A range of endocrinology studies indicated that higher levels of testosterone exposure were correlated with lower 2D:4D ratios.[49] So, taking this finger measurement as a biomarker for prenatal androgen exposure, researchers then explored the correlation with behaviour, specifically the kinds of behaviour that were supposed to differentiate the sexes, ranging from spatial skills to aggression in adults, and sex-typed play and toy preferences in children, as well as sexual orientation and leadership skills.[50]

In 2011, psychologists Jeffrey Valla and Stephen Ceci from Cornell University carried out a major review of the use of the 2D:4D measure by exploring sex differences in particular behaviours, specifically those linked to abilities and preferences associated with science subjects such as maths, computer science and engineering.[51] Their overall summary pointed to 'myriad inconsistencies, alternative explanations, and outright contradictions'. One key issue was the validity of this finger measure as an accurate proxy for prenatal testosterone, as the endocrinology evidence was not consistent. Another aspect was the nature of the relationship between this measure and the various abilities being explored. In some cases it was linear (with low ratios associated with higher spatial/mathematical ability) but in some cases there was an inverted-U shape (with both high and low ratios associated with higher levels of cognitive skill); in other cases there was no relationship with cognitive measures that normally reliably differentiated males and females, such as mental rotation; and in many cases, there were relationships that were true in males but not females or vice versa. The conclusion was that this nice simple measure of prenatal hormone levels was not really fit for purpose and that its continued use, particularly when linked to cognitive measures which in themselves were not always consistent, was unlikely to resolve the issue of hormone effects on human behaviour.

The hormone factor: cause and effect

The twentieth-century focus on hormones as the biological driving force that would determine both brain and behaviour differences between men and women did not provide the neat solution that the early animal studies promised. Hormones will, of course, exert strong influences on other biological processes, and hormones linked to sex differences are no exception to this. It is evident that different hormones determine differences in the physical apparatus associated with mating and reproduction, so a nice clear-cut male–female divide in explanations for this aspect of the human condition is generally justified.

But the assertion that this extends to brain characteristics and thence to behaviour as well is proving harder to defend. The ethical issues associated with replicating the original animal-based hormone manipulation studies in humans are obviously insuperable. The various attempts to test the clear unidirectional hypotheses arising from the brain organisation model by studying individuals with anomalous hormonal profiles have not provided clear-cut answers. Nor has the use of indirect clues as to the extent of prenatal hormonal influence proved any more useful. Sometimes this could be ascribed to methodological issues, such as the inevitably small numbers involved, the variability in the different groups, the somewhat subjective ways of measuring behaviour. Crucially, the work to this point did not fully take social and cultural influences into account, if it did so at all, and, as we shall see, these influences can exert effects not only on patterns of behaviour but also on brains and on hormones themselves.

Recent work by Sari van Anders, a neuroscientist at the University of Michigan, and others shows that in the twenty-first century the link between hormones and behaviour, particularly with respect to the supposed potency of testosterone in determining male aggression and competitiveness, is undergoing a radical rethink.[52] Just as we are seeing the power of society and its expectations as brain-changing variables, it is clear that the same effect is evident with respect to hormones. And hormones are, of course, themselves entangled with the relationship between the brain and its environment.

Hormones appear to have become yet another biological process co-opted in the hunt for evidence that women's biology is not only different and generally inferior but periodically extremely deficient. The chemicals that were linked to women's capacity for motherhood became linked to their maladaptive emotionality, irrationality and, via the effects these chemicals had on the developing brain, lack of certain key cognitive skills. On the other hand, extra doses of testosterone were not only linked to men's capacity for fatherhood, but also to the necessary forceful personality characteristics and leadership skills allegedly essential for success in social, political and military circles and, again via particular effects on the developing brain, to the required cognitive capacity to be great thinkers and creative scientists.

All of this is predicated, of course, on the accuracy of claims that men's and women's behavioural profiles *are* actually different. Good science needs to go beyond anecdote and personal opinion and provide strong evidence based on sound methodology. Let us now have a look at how well the study of human behaviour has lived up to these expectations.

Chapter 3:
The Rise of Psychobabble

The emergence of psychology in the twentieth century provided another avenue to explore in the quest for sex differences. How did this new science inform our understanding of men's and women's brains and behaviour?

Helen Thompson Woolley, a psychologist herself and a pioneer in studies of gender differences, summarised in 1910:

> There is perhaps no field aspiring to be scientific where flagrant personal bias, logic martyred in the cause of supporting a prejudice, unfounded assertions, and even sentimental rot and drivel, have run riot to such an extent as here.[1]

This is mirrored in the words of Cordelia Fine, speaking in 2010:

> But when we follow the trail of contemporary science we discover a surprising number of gaps, assumptions, inconsistencies, poor methodologies, and leaps of faith – as well as more than one echo of the insalubrious past.[2]

These two trenchant statements about psychology's studies of sex and gender differences, exactly one hundred years apart, suggest that the one discipline that should be able to throw some objective light on the fraught issue of differences in aptitude, ability and temperament, backed up by some decent empirical data, objectively interpreted, hasn't quite lived up to these expectations.

Psychology's involvement with the story of the sex differences hunt comprises two key contributions. The first is linked to the emergence of the theory of evolution, emphasising our adaptability as the basis for our past and continued success. At heart a theory about individual differences in biological characteristics, evolution quickly extended its range to explanations not only of different individual skills but also of the functions of different social roles, as determined by differences in biology. The line was that sex differences were there for a purpose, and it was the role of evolutionary theorists to explain that purpose.

The second is the role of the emerging discipline of experimental psychology with its emphasis on numerical data. Rightly uneasy about the rather anecdotal nature of early case studies and clinical observations, a psychometric 'industry' emerged, developing elaborate tests and questionnaires to generate numerical scores to attach not only to measures of ability but also to rather more amorphous concepts such as 'masculinity and femininity'. The numbers game offered a sheen of objectivity to the go-to list of sex differences that were being generated.

The evolution of evolution

The publication of Charles Darwin's *On the Origin of Species* in 1859 and *The Descent of Man*[3] in 1871 offered a whole new framework for explaining human characteristics. These ground-breaking works provided insights into the biological origins of individual differences, both physical and mental, and naturally were the ideal source of explanations of differences between men and women. And, of course, Darwin had specifically addressed such issues via his theory of sexual selection, effectively about the dance of sexual attraction and mate choice. Members of one sex flaunt their assets to attract a mate, with members of the other sex taking their pick according to a set of species-specific criteria – most eyes in your tail if you are a peacock, deepest croak if you are a frog – which supposedly signal your 'reproductive fitness'. Assets in humans could include top-of-the-range physical equipment but also the associated behaviours and character types – competitive and combative for men, submissive and conciliatory for their women. Similarly, there were key differences in roles and their associated skill set; the dominant male required the greater strength and intellectual superiority needed to tackle the outside world, whereas the home-based females just needed 'calm mother-love and unruffled housewifeliness'.[4]

Darwin was quite clear that one key difference between men and women was that women, by virtue of being less highly evolved than men, were inferior members of the human race. It is rather chilling to think that the author of one of the most important scientific theories had these views about half of the population he was studying:

> The chief distinction in the intellectual powers of the two sexes is shewn by man attaining to a higher eminence, in whatever he takes up, than woman can attain – whether requiring deep thought, reason, or imagination, or merely the use of the senses and hands.[5]

With respect to the different functions that men and women might have in society, Darwin's views were that women's reproductive capacity was the key determining factor for their place in the pecking order. As a fundamental, but basic, physiological process, it required none of the higher mental attributes that evolution had granted to males; indeed, his concern was that any attempt to expose the female of the species to the demands of any kinds of education or independence could well damage this process.

Darwin wasn't even troubled with the niceties of complementarity, a view (met in Chapter 1) that was based on the idea that men's and women's roles in society were determined by certain inherited traits, with women's gentle, nurturing, softly practical nature the perfect foil to the powerful, public-facing, fiercely rational persona characteristic of men. Although somewhat politer than the Darwinian perspective, we should be under no illusion that this was the dawning of some kind of progress towards gender equality:

> The idea of complementarity – that is, the belief that the traits, strengths and weaknesses of one group are compensated for or enhanced by the traits, strengths and weaknesses of another – is an exceptionally powerful way to maintain power inequities between groups, as it implies that any perception of inequity is illusory and that the actual basis for discriminating between groups is based on each group's relative strengths and weaknesses.[6]

While reviewing psychology's contribution to the construction of gender differences in the late nineteenth century, psychologist Stephanie Shields wrote of this complementarity trap, showing how it became linked to evolutionary theory and was then used as a justification for existing social hierarchies. A major focus was on women's role as mothers and homemakers, meaning they needed to be nurturing, practical and able to focus on everyday detail, which apparently rendered them incapable of the kinds of abstract thought, creativity, objectivity and impartiality that were needed for great thinking and scientific achievement. Emotionally, women were more likely to be sensitive and unstable as compared to men's 'passionate force evident in the drive to achieve, to create, and to dominate'.[7]

This particular aspect of psychology's input to the study of sex differences wasn't based on any kind of measurement but on opinions voiced by the likes of Herbert Spencer and Havelock Ellis. As Shields acerbically points out: 'It goes without saying that the lists of traits assigned to each sex were not derived from systematic empirical research but drew heavily on what was already believed to be true about women and men.'[8]

The notion of complementarity has persisted and found a home in the field of evolutionary psychology, a discipline that emerged in the twentieth century and that merges the biological bases of society and the study of human psychological characteristics.[9] Human behaviour is assumed to be made up of many sets of functions or 'modules', each of which has evolved to solve the kind of problems that we might encounter at any stage of our lives. This has been called the 'Swiss Army knife' model of the mind, with thousands of specialised components, each underwritten by associate brain structures, which have emerged over evolutionary time as required.[10] And there appear to be two types of knife: one (presumably pink) kitted out with the tools for the brow-soothing, household management, child-rearing-type tasks for the female of the species, whereas the other (a martial navy blue), apart from being bigger and more resilient, has the essentials for the life of spear throwing, political power and scientific genius that is the lot of the male of the species.

Evolutionary psychologists come firmly under the heading of scientists-as-explainers-of-the-status-quo. Effectively, they work backwards from what appears to be a well-established fact today; they find an explanation in evolutionary history that could fit this fact and offer it up as the reason for the status quo. An example that we will meet later is of women's alleged preference for pink, reported by visual neuroscientists Anya Hurlbert and Yazhu Ling in 2007.[11] The evolutionary-psychology explanation they offered was that, as the gatherer half of a hunter-gatherer team, women have evolved a differential preference for pink in order to be better equipped for berry finding, as opposed to their mammoth-hunting other halves, who are more attuned to the blue end of the spectrum to enable them to scan the horizon effectively. In addition, men are better at running (to follow said mammoths) and at visuospatial tasks such as targeted spear throwing (to kill the same).

A key take-home message from evolutionary psychology is that our abilities and behavioural characteristics are innate, biologically determined and (now) fixed (though it is less clear why skills that were clearly flexible and adaptive enough in the past have now become immutable). Although the need for these skills and abilities is in our evolutionary past, they can still have consequences for our twenty-first-century lives.

Empathisers and systemisers

One contemporary psychological theory that has a foot (actually, probably both feet) in the evolutionary psychology camp is British psychologist Simon Baron-Cohen's empathising-systemising theory, mentioned briefly in the last chapter.[12] Baron-Cohen nominates these two traits as driving forces in human behaviour. Empathising is the need (and ability) to recognise and respond to others' thoughts and emotions, not just at a cognitive-cataloguing-type level, but at an affective level, whereby the emotion of others triggers a matching response, making the behaviour of these others understandable and predictable. It is the ability to tune in to other people's feelings, what Baron-Cohen calls 'a leap of imagination into someone else's head';[13] it is natural and effortless and essential for effective communication and social networking. Systemising, on the other hand, is a drive to 'analyse, explore and construct a system',[14] to be drawn to or even need rule-based events or processes, to make your world predictable by extracting organising principles from what is going on around you.

In true evolutionary psychology fashion, the origin of these traits is apparently rooted in our ancient past, and their continued existence in twenty-first-century human beings has implications for who does what. Empathising and systemising traits have clearly been allocated and channelled along gendered lines. According to Baron-Cohen, empathising helped our female ancestors set up childcare networks to ensure that future generations were thoroughly nurtured, underpinned their tendency to form gossip groups to ensure they were kept in any kind of useful information loop, and helped them get on with non-genetically related conspecifics (or 'in-laws', in other words).[15] With respect to what this means to today's empathisers, Baron-Cohen as careers advisor informs us: 'People with the female brain make the most wonderful counsellors, primary-school teachers, nurses, carers, therapists, social workers, mediators, group facilitators or personnel staff.'[16]

And systemisers? Their way of dealing with the world made them good at working out things such as how long an arrow should be and how best to fasten an axe blade, the rules of animal tracking and weather forecasting, and the laws of social ranking systems (in order to get as high as possible in them). Their associated lack of empathy made them good at killing members of other tribes (or, in fact, members of their own tribe if they blocked the way up the social ladder). Not wasting time on the social niceties associated with being empathic also meant they could be an 'adaptive loner', 'content with locking [themselves] away for days without much conversation, to focus long and deep on the system that was [their] current project'.[17] In today's terms, this would apparently make systemisers 'the most wonderful scientists, engineers, mechanics, technicians, musicians, architects, electricians, plumbers, taxonomists, catalogists, bankers, toolmakers, programmers or even lawyers'.[18]

It is easy to detect more than a whiff of complementarity here: the martial tendencies and highly focussed inventiveness of one group neatly backed up by their caring, networking support staff. No points for guessing who ends up earning a higher salary in this scenario.

But how do you know that someone is empathic or is a systemiser? A contemporary psychology theory, even though based in the evolutionary past, must have ways of generating some kind of objective measure of such characteristics or traits in any individual or group of individuals. Baron-Cohen's lab has generated its own measure of empathising, known as the Empathising Quotient (or EQ), and of systemising, the Systemising Quotient (or SQ), through self-report questionnaires consisting of series of statements with which respondents have to indicate their agreement or disagreement.[19] The EQ statements include items such as 'I really enjoy caring for other people' and 'If I see a stranger in a group I think that it is up to them to make an effort to join in'; whereas the SQ statements include 'When travelling by train, I often wonder exactly how the rail networks are coordinated' and 'I am not interested in the details of exchange rates, interest rates, stocks and shares' (the answer to the last one is 'strongly disagree' if you're an S-type, of course). Child versions of these tests are also available, or rather, parent-report versions, where a parent rates their agreement to statements such as 'My child doesn't mind if things in the house are not in their proper place' or 'When playing with other children, my child spontaneously takes turns and shares toys'.[20] Combining the scores is a way of generating an empathiser or a systemiser profile. Studies using this test indicate that, on average, females are more likely to have an empathiser profile and males a systemiser one.

You'll notice that these measures are actually reliant on people's own opinions of what they (or their children) are like. We might ponder how many parents would calmly tick the boxes which would label their offspring as an anti-social, toy-stealing thug. Issues with these kinds of self-reports are a generic problem which we'll come back to later, but it is worth holding on to that thought as a tiny pinch of salt when you read about people's EQ and SQ scores.

To test the validity of these self-report measures, you need to find an example of the kind of behaviour or relevant skill you might predict from a high EQ score or a low SQ score or any mixture of the two and see how well the two measures match. Another test from Baron-Cohen's Cambridge lab is the rather eerie 'reading the mind in the eyes' test, where you are shown disembodied images of a pair of eyes together with four affect-describing words such as 'jealous', 'arrogant', 'panicked' or 'hateful'.[21] You then have to pick the emotion being shown by these eyes. Do well on this, and you are clearly good at emotion recognition, a key part of empathy. So a high EQ score should correlate with a good Mind in the Eyes one, which indeed it does. The fact that both these tests come from the same lab might add just another of those pinches of salt.

It should follow from the predictions of the E–S theory, with women more empathic than men and men greater systemisers than women, that behaviours, abilities and preferences that are closely linked to being either characteristically empathising or characteristically systemising should show a neat gender divide. It is, after all, a fundamental claim of the theory. For example, university subject choice, sciences versus arts, should be related in some way to this gender divide. But another paper from Baron-Cohen's lab has shown that gender, which should go hand in hand with EQ and SQ scores, is not the best predictor of university subject choice.[22] The theory would predict that systemisers would be drawn to the rule-based science disciplines, which they were, but there was no significant sex difference. This means that E–S is not an *exact* proxy for gender, which should temper the general impression that empathy is a 'woman thing', while systemising is for the guys. I use the term 'general impression' advisedly, to note that – while this was not what the theory originally set out to prove – sometimes psychological theories like this can leave the impression that the labels they attach to their participants (male–female, systemiser–empathiser) are interchangeable. The consequence of this is that people may take a shortcut and assume that if you want a job done that requires an empathic touch, then you just need to appoint a woman. Or conversely, if you want a job done that requires a high level of systemising skill, a woman would not be suitable.

A very twenty-first-century example of this can be seen in discussions about the underrepresentation of women in science. Given the positive association between systemising and science, and the negative association between systemising and women, it is not many steps until we get to the stereotype of women being less suited to the systemising rigours of hard science. Add to the mix a general understanding that biologically determined characteristics are fixed and unchangeable and we arrive at a misinformed but understandable stereotype of the link between sex and science.

Unlike some evolutionary theories, where the biological underpinnings are rather vaguely taken as a given, here the biological bases of these two cognitive styles are clearly stated. Baron-Cohen's opening statement in his book *The Essential*

Difference is unambiguous as to the gendered nature of the E–S divide: 'The female brain is predominantly hard-wired for empathy. The male brain is predominantly hard-wired for understanding and building systems.'[23]

Given the strength of this assertion, you may be surprised by a qualification that comes further into the book where Baron-Cohen firmly points out that 'your sex does not dictate your brain type ... not all men have the male brain, and not all women have the female brain'.[24] This, for me, is the heart of the problem with this theory and its impact on the public understanding of sex differences in brain and behaviour. In common parlance, the term 'male' is linked to men and, equivalently, the term 'female' is linked to women – so describing a brain as 'male' means, for many people, that it is the brain of a man. And if you then attribute certain characteristics to a male brain – in this case, a preference for systems and rule-based behaviours, perhaps also difficulties with emotion recognition, with a nice clear link to specific parts of the brain – then these will be added to the world's cognitive schema for 'man', and, in the way of things, become part of the stereotypical profiling of men and of their brains. And you'll get the same outcome for women and their brains. If you don't have to be a male to have a male brain, why are we calling it a male brain? In the world of gender stereotypes, language matters.

This theoretical strand of psychology's involvement in the sex difference debate was steadfastly linked to the 'status quo' type of explanations. Moving on from downright misogyny to the rather patronising complementarity approach, early evolutionary psychologists took role differences as given, and linked them to the sex-determined skill and personality differences which the newly emerging discipline of experimental psychology would be able to identify and quantify.

The numbers game

The second strand of psychology's involvement in sex and gender differences was the development of techniques that would start to put some numerical flesh on the catalogue of behavioural and personality differences that had accumulated over the centuries – the field we now call experimental psychology. Prior to the end of the nineteenth century, the focus had been on the biology behind these allegedly sex-differentiated behaviours, with increasingly bizarre attempts to quantify the differences in an organ that was actually unavailable for study unless dead or damaged. So in the twentieth century attention was turned to ways of measuring the skills, aptitudes and temperaments that were allegedly controlled by the (albeit still invisible) brain.

Wilhelm Wundt had founded the first psychology lab in 1879.[25] He was keen to apply the scientific method to behaviour, to generate standard measures of the behaviours we could see, such as reaction time or error rate or amount of recall in memory tasks, or the number of particular words (such as words beginning with 's', or names for fruits) that might be spontaneously generated. There was to be no more introspection, personal opinion or anecdote sharing – this was about data.

Psychologists would utilise any kind of task that could produce a score of some kind, turning them into tests which produced an external measurement that appeared to have some relationship to the behaviour of interest. Early studies focussed on finding different ways of measuring the skills that the psychologists were interested in, but an interest in individual differences soon emerged. This was partly driven by changes in the educational system, which meant schools wanted ways of identifying 'slow' children, those who we would today identify as having special educational needs. As we know, this was the origin of the IQ test.[26]

Tests of cognitive skills were then followed by tests of personality or temperament. The first one, Woodworth's Personal Data Sheet, was developed in 1917 and its aim was to identify soldiers in the First World War who might end up suffering from shell shock.[27] This kind of test was still quite objective and fact-based, and included questions such as 'Has any of your family committed suicide?' or 'Have you ever fainted away?' (these having been identified as discriminatory factors by looking at past case histories) but soon various types of self-report inventories were being developed, where people were asked to indicate the extent to which certain adjectives ('well-organised', for example) described them, or certain phrases characterised their behaviour ('I can relax and enjoy myself at gay parties' – though this wording has been changed in more recent test revisions!).[28]

In tests of cognitive skill, reports of differences between the sexes soon began to emerge. Word association tasks were a favourite way of gaining insight into the mental life of males and females: given a trigger word or category, participants had to write down, say, 100 words that these triggers made them think of. One of the first sex difference studies, by Joseph Jastrow in 1891, used this technique, noting that men made greater use of abstract terms, while women showed preference for concrete and descriptive words; women were quicker, but men were more wide-ranging.[29] It was never really made clear what these differences meant. Helen Woolley, writing in 1910, also reported on a study using similar techniques, commenting contemptuously on the 'trifling differences in the data' and on the exceedingly small number of subjects (the demographics of which sound a bit like a Christmas song – two children, two servant maids, three working men, five educated women, and ten educated men).[30]

But from these early rather dubious practices, psychology's aim to firmly embed the scientific method into its activities developed apace. Theories were developed, hypotheses were generated, measuring tests were devised, participants were selected, data were collected and analysed, papers were written and published. In the first hundred years after the first psychology lab was established, over 2,500 papers on sex differences were published. Did all of these studies contribute positively to our understanding of such differences?

The neuroscientist Naomi Weisstein wrote a notorious two-pronged attack on psychology at the end of the 1960s.[31] The title of her paper, 'Psychology Constructs the Female; or, The Fantasy Life of the Male Psychologist (with Some Attention to the Fantasies of His Friends, the Male Biologist and the Male

Anthropologist)', made clear her thoughts on the subject. She took issue with the fashion for clinical psychologists and psychiatrists to follow Freudian doctrines, with an emphasis on women's essential role as a mother and the biological wherewithal that went with it. She complained that such professionals, full of bias and free of evidence, were taking it upon themselves to tell women what they wanted or what role they were particularly suited to (suggesting that this new discipline of psychology had not moved things on that much). She scoffed at their claims to approach such matters with 'insight, sensitivity and intuition', pointing out that these could equally just reflect a biased perspective, pre-existing beliefs of the 'right' thing for women.

The other prong of her attack on 'sex-difference' psychologists was their failure to take account of the context in which they were collecting their data. She pointed out several social psychology experiments where the behaviour was changed if you manipulated the external context. A classic example of her time was the Schacter and Singer study, where people who had unknowingly been given an adrenaline shot interpreted their adrenaline-related physical symptoms (racing heart, sweaty palms etc.) in different ways depending on the behaviour of the other person (a stooge) they found themselves with in the waiting room, with a euphoric stooge leading to reports of happiness, and a grumpy stooge associated with reports of anger or dissatisfaction.[32] Weisstein's concern was that the kind of behavioural or self-report data that were being collected from individuals could well be affected by all sorts of extraneous variables, including, in fact, the expectations that the experimenters themselves had about the outcome of their study. Patterns of behaviour are rarely stable but will change according to external circumstances; if what your participants will do or say when on their own can change if someone else is present, then this pattern of behaviour cannot be interpreted as innate or fixed or hard-wired. Unless this was acknowledged, the results of psychology studies could at best be described as misleading. Weisstein's attention to the importance of taking account of context and expectations when studying behaviour can find parallels in many realms of contemporary social neuroscience, showing how brain function can interact with an individual's social and cultural framework.

An influential book by Eleanor Maccoby and Carol Jacklin, *The Psychology of Sex Differences*, published in 1974, worked painstakingly through decades of studies claiming to have found differences between males and females, including many different characteristics from touch sensitivity to aggression.[33] The fact that Maccoby and Jacklin had to trawl through eighty-six different categories of reported sex differences – from 'Vision and Audition' through 'Curiosity and Crying' to 'Donating to Charities' – was a measure of the amount of effort psychology had already put into this exploration thus far.

The only areas where the published evidence appeared to agree about differences were that girls on average were more verbal, whereas boys had better spatial abilities, were better at arithmetic reasoning involving spatial skills and showed greater physical and verbal aggression.

Maccoby and Jacklin did much to dispel the sex difference myths that were in place at the time, although sometimes the summary that emerged from their review, particularly with respect to the 'verbal' female and the 'spatial' male, became reified as a wholly reliable discriminator of men and women, or a 'given' which no longer needed to be put to the test. As we shall see, this has fed through into areas as wide-ranging as popular self-help books and interpretation of structural brain imaging data sets, and thence back out into the public consciousness of male–female differences.

A point *not* made by Maccoby and Jacklin at this stage was that these differences were actually very small, so that knowing someone's sex would not be a good predictor of how well they might do in a test of verbal ability (or how well they might park a car). They also did not challenge just how these measures were obtained, or how reliable were the measuring instruments being used by psychologists. If you are interested in spatial skills, do all spatial skill tests come up with the same answer? Are you sure you are testing a representative sample of the people you are trying to assess; might you need to allow for differences in, say, educational experience; and are you using the right kind of comparisons in the analysis of your data?

When is a difference not a difference?

The word 'difference' is an example where the use of a term in psychology might not be the same as its use in general conversation, or in the public understanding of what it means. At a simple level the term 'different' obviously implies 'not the same'. Suppose you were travelling to an island and you were told that there were two different tribes that you might meet and that you should be aware of the differences between them. You could then get into the key points of difference and the niceties of 'how different'; for example, Tribe 1 might be on average about six feet four inches tall while Tribe 2 are on average about four feet ten inches; or members of Tribe 1 might have very long straight black hair, as opposed to members of Tribe 2, who have short curly blond hair. You would probably at the very least infer that 'different' here meant *recognisably* different – so that if you were to meet a tall individual with straight black hair you would be secure in the knowledge that they belonged to Tribe 1 – or *reliably* different – so that if you were told you were going to meet a member of Tribe 2, you would be safe in your expectation to find someone short with curly fair hair. But these are not necessarily the sorts of conclusions you can draw from psychological studies that report sex differences.

In psychology, 'different' is often used in its statistical sense, where the average scores of the two groups you are investigating are sufficiently far apart to pass a particular statistical threshold. You can then report that whatever it is that you are measuring is 'different' in the two groups. But this can often mask the really important 'how different' issue. Each of your two groups will have generated scores that are distributed around the average that you have measured, and those two distributions might overlap quite markedly. This means that you can't *reliably* predict how a member of one of the groups will perform on the task you have set or what kind of score they will get on a personality test. And you can't recognise from someone's test score which group they belong to. The groups are really more similar than they are different. So although there is a statistical difference, it isn't necessarily a useful or *meaningful* difference.

One way of calculating the degree of overlap between two groups is by measuring what is called the effect size.[34] To calculate this you subtract the average or mean score of one group from the average score of the other and divide the answer by the amount of variability in the two groups. Suppose, for example, you want to find out if coffee drinkers solve crossword puzzles faster than tea drinkers. Having collected your data, you subtract the average tea drinker score from the average coffee drinker score and divide it by what is called the standard deviation, a measure of variance which reflects how widely distributed the scores are in each group. This will give you an effect size for the difference between your tea drinkers and your coffee drinkers.

The key issue is that effect size tells you how *meaningful* group differences are. Psychologists report their statistical findings as showing 'significant differences', which strictly speaking they do, but the differences may be tiny and really unlikely to have much impact on, say, a decision to employ someone from one of the groups as opposed to the other (or whether you want to ask a coffee drinker or a tea drinker to help you solve a crossword puzzle). When you are talking about something with as much impact as findings about sex differences, then it is important to signal clearly what you mean. If effect sizes are small (about 0.2) the differences between the scores of your groups might be statistically 'significant' but, really, not very supportive of any assumptions you might have about how easily you can spot who belongs to which group, or what the members of that group can or can't do.

If two groups are markedly different, then the effect size will be quite large. The most common example is height differences between men and women. The average effect size here is about 2.0, so the means are quite different and about ninety-eight per cent of the taller group will be above the average height for the smaller group.[35] Even with an effect size this big, however, the two populations still overlap by just over thirty per cent.

The reason I'm rather labouring this point is that the effect sizes in much of the published sex differences research are actually quite small, of the order of 0.2 or 0.3, which means an overlap of nearly ninety per cent. Even a 'moderate' effect size, 0.5, means an overlap of just over eighty per cent. So when people refer to sex differences, we need to be aware that this almost never means that the two groups are non-overlapping, clearly distinguishable by whatever variable you are measuring, and that knowing someone's sex will not be a reliable predictor of how well or badly they are going to do on a specific task or in a specific situation.[36]

Effect sizes are also valuable when you are trying to get an overview of findings in a particular research area. A meta-analysis combines data from many different studies of the same phenomenon, using the effect sizes from each study, weighted by how many people were tested, to investigate how reliable and consistent findings are, and whether or not large effect sizes are the norm. This gets over the problem of individual small-scale studies, or 'one-off' reports which may not be replicated. The other point is that looking at effect sizes can give you a measure of how accurate claims of reported differences being 'profound' or 'fundamental' actually are. And if studies using these kinds of term actually don't report effect sizes, then alarm bells should start to ring.

There is one more note of caution to sound about the reporting of research findings. If someone tells you that something is 'significant', for example that men and women are 'significantly' different, you probably assume that this means that this difference is important, should make you sit up and take notice. You probably *don't* think, 'Aha, that means there is a less than 5 in 100 possibility that this is a chance finding.' This isn't to suggest that the research findings aren't saying something meaningful, just that we might need to temper the 'wow' factor that the word 'significant' can sometimes imply.

So there are a range of questions that we need to ask if we want to see what and how experimental psychology has contributed to the sex differences debate. Are the hypotheses as objective as possible or do they reflect a stereotypical bias or a relentless search for differences? Are the tasks or tests being used a neutral measure of behaviour or temperament or actually a means of stacking the odds in favour of finding the difference being looked for? Are the experimenters carefully controlling for 'gendering' factors such as education or occupation, or are they just assuming that 'male' or 'female' will cover all bases? Do we get to see effect sizes cautiously interpreted or are we and any passing science journalists treated to descriptions of 'fundamental' or 'profound' differences between the male and female participants?[37]

What are you asking, and how are you asking it?

We have already seen that the scientists behind the theories that psychology was setting out to test were not operating in a political vacuum. Although we might have moved on a bit from Gustave Le Bon's 'two-headed gorilla' approach to women, the focus was still on the status quo, on the finding and categorisation of differences, on demonstrating that men had different skills and temperaments from women, fitting them for different roles. As an experimental psychologist, certainly in the early years of the field's existence, this hunt for differences would inform your 'experimental hypothesis', that there *would* be sex differences in the particular psychological process you were measuring, be it verbal fluency or empathy, mathematical skills or aggression; you would not predict a lack of difference, a similarity, between the groups you were comparing.

With the way that publishing research currently works, it is much more likely that you will submit your work for publication (and that it will be accepted) if your experimental hypothesis that there *would* be a difference is upheld. If it isn't upheld and your results seem to suggest that there isn't a sex difference, you more than likely will not submit your findings for publication or, if you do, it is less likely that they will be published.

Sometimes the lack of sex differences can get lost in the noise of trawling through data to see what they are showing. You might not even have a specific hypothesis that there will be sex differences. But it is easy enough to see if there might be some lurking in your data, if you have enough males and females in your participant group.[38] You check to see if there is any sex difference and if there isn't, then you probably won't make much of this in your discussion, your abstract, or even in your choice of keywords for your research paper.

This is often referred to as the 'file drawer' problem: tucking away from public scrutiny your failure to find differences.[39] I think it is better described as the 'iceberg' problem. Out in the scientific ether, or below the surface of publishability, there is a vast body of 'invisible' research findings which may be showing that there are no differences between men and women on a whole range of measures, some of which are firmly established in our consciousness as reliable ways of distinguishing map-reading Martians from multi-tasking Venusians. There may, in fact, be a much bigger body of research findings that could report no differences than the one that appears to confirm that there are.

So the questions being asked may actually colour the answers that are reported. But we should also look at just how these answers are gathered. What particular tests are being used to collect information about differences between females and males? Are you actually measuring what you are setting out to measure or could there be something else going on? And could this affect the conclusions you (or anyone else) might draw from your findings?

Many years ago, I went to a weekend conference on the heritability of IQ. The morning session was run by geneticists, and there were many papers on genome-wide association studies, heritability assessment, the implications of knock-out mice models, gene variation and so forth. All of these used IQ as a dependent variable or modelling factor, with humans assessed via an IQ test which seemed to be the 'industry standard'. No one made any reference to how this particular variable was measured or, indeed, what exactly it was that was being measured, just how an IQ score or its rodent or monkey equivalent was affected by whatever genetic model or manipulation they were using.

Psychologists took over in the afternoon and proceeded to dismantle the faith their geneticist colleagues had in their core measure. Issues with individual items, the heterogeneity of the subtests and different skills being measured, retest reliability, the need to take account of environmental factors such as access to education, socio-economic status in the case of humans, or cage size and handling frequency in the case of non-humans, the very definition of intelligence itself – all served to reveal that IQ was not like, say, eye colour or blood type, a fixed and objectively measurable trait that could be slotted neatly into whatever model was being tested. You needed to know much more of the back story to know what that IQ number was really measuring.

So sometimes you have to study the measure you are taking in some detail; you have to find out how the test you are using was generated and if, although it may *appear* both reliable (will come up with much the same score in different circumstances and situations) and valid (measures what it is claiming to measure), it may actually be telling a different story to the one you are hearing.

People versus Things

The development of a vocational scale to measure individual differences in interests in 'people' versus 'things' is a useful case study of how the choices made in developing a test, which appears to distinguish people on the basis of one measure, may actually be reflecting something different.

The vocational interest scale was intended to be used as a careers advisory tool.[40] The aim was to show that a match between the kind of things that people were interested in and the tasks that characterised their chosen occupations might be a guarantee of job satisfaction. The basic principle of the test was developed in the 1980s by Dale Prediger, a research scientist then allied to the American Colleges Testing Program. He suggested that what was then known about vocational interests could be grouped onto two dimensions. The first was a Data/Ideas dimension, which should indicate either preferences for tasks involving facts, records and so on, or preferences for tasks which might involve teamwork, developing theories or new ways of expressing things. The second dimension was a People/Things dimension, which should indicate either an interest in helping people and caring for others, or an interest in working with machines, tools or biological mechanisms. And apparently working outdoors, which we will come back to later.

Prediger's next task was to profile the various occupations. Trawling through many thousands of data sets that the US Department of Labor had collected over many years, he came up with a way of describing jobs in terms of where they grouped on his Data/Ideas and People/Things dimensions. As a result of his Herculean efforts (some 100 different descriptors for 563 occupations were examined and classified) he grouped different careers as Data-based, Ideas-based, People-based or Thing-based. His exemplars of People-based occupations were elementary school teachers and social workers, with bricklayers and bus drivers as being typical Thing-based occupations.[41]

At this point it might be worth musing on these apparently archetypal jobs, particularly in terms of who were actually doing them at the time. In the US at the time of Prediger's work, 82.4 per cent of elementary school teachers and 63.0 per cent of social workers were female, as opposed to 29.2 per cent of bus drivers and 2.4 per cent of construction trade workers.[42]

So we have groups of tasks supposedly based on a People versus Things dimension, but where the additional factor of gender imbalance doesn't appear to have been taken into account – we could actually relabel these categories as Women's Jobs and Men's Jobs. It *could* just be that this is an accurate reflection of a well-informed People versus Things choice, with women choosing not to do bricklaying because it was too Things-based, but could there instead be other factors at work? Was being a bricklayer actually a choice open to women anyway? To be fair to Prediger, he wasn't aiming to measure sex differences – in fact he seemed to be proudest of the fact that his dimensions could distinguish lab technicians and chemists (Thing jobs) from encyclopaedia salespersons and Christian education directors (People jobs) – but, as we shall see, this People versus Things distinction later becomes quite critical in gender gap discussions.

Let's have a look, then, at how you might measure *interest* in People versus Things, the other half of the vocational guidance measure. In parallel with Prediger's efforts, the psychologist Brian Little developed a twenty-four-item scale specifically to measure 'Person orientation' and 'Thing orientation' (now PO and TO), with test takers being asked to rate how much they enjoyed the situations described.[43] Now, you might think I'm being overly picky (and actually pretty sexist) but when the Thing dimension is being measured by scenarios such as 'take apart and try to reassemble a desktop computer' or 'explore the ocean floor in a one-person submarine' (bearing in mind this questionnaire was originally being validated in the 1970s) and the People dimension is loaded with items such as 'listen with caring interest to an old person who sits next to you on a bus', I don't think it is wholly surprising to find that there was a strong gender-related difference in People and Thing scores. A twenty-first-century update of this test (which sadly shed the submarine question and others such as 'learning to be good at glass blowing') effectively retained the invisible gender-based foundations of these TO and PO measures.[44]

Virginia Valian, a psychologist at the City University of New York's Hunter College, has, more fundamentally, challenged the validity of the assumptions that feed into this whole larger dimension in the first place, with a particular focus on what came under the heading of Things.[45] Why should you group 'working with things' and 'working in well-structured environments' with 'working outdoors'? What makes these scenarios Thing-like? The interests clustered under this heading are, as Valian points out, more accurately described as 'activities that men have tended to spend more time at than women have'. (I'm sure, like me, she was channelling that submarine scenario!) She also notes that the descriptions of the kinds of people who were interested in Thing-like activities, 'agentic, instrumental, and task-oriented', map closely onto stereotypical ways of describing men, whereas the 'communal, nurturant and expressive' individuals who would like People occupations you could well read as women.

So we have a dimension, Things versus People, which supposedly distinguishes different occupations and further profiles the different kinds of people who would like to pursue those occupations. But there is an inbuilt confound, an unnoticed gender divide, loading the dice with respect to who will fall where on this People versus Things dimension.

This could just be an academic concern, but the Things versus People concept has been enthusiastically seized on by researchers looking for explanations for the underrepresentation of women in STEM (science, technology, engineering, maths) subjects, and to query the usefulness of initiatives to address them. A frequently cited study is from business psychologist Rong Su and colleagues, who took the information on standard scores from the technical manuals for forty-seven interest inventories.[46] This gave them data from 243,670 men and 259,518 women. They examined how these data clustered onto the Things versus People dimension and, unsurprisingly, they found that 'men prefer working with things and women prefer working with people', a highly significant difference with a large effect size (0.93). This would mean, as they pointed out, that up to 82.4 per cent of male respondents had stronger interests in Thing-oriented careers. Or it could mean that they liked doing things that other men do, be it bricklayer or bus driver.

Why are you asking?

Another aspect of collecting data via these kinds of self-report measures is what kind of expectation the participants bring to the process. This is an extension of Naomi Weisstein's observations discussed above that psychological measurements are rarely context-free. The 'demand characteristics' of the tests are often pretty transparent, and may well stack the odds in favour of a particular outcome.[47] I've already commented on how the name of the Menstrual Distress Questionnaire might just be biasing the answers the questionnaire itself collects. Partly ironically, but ultimately to demonstrate this point, a group of researchers investigated the effect of using a 'Menstrual Joy Questionnaire' (MJQ) listing ten positive experiences that participants might note during their menstrual cycle. Participants who first filled out the MJQ later reported more positive changes on the MDQ and more positive attitudes towards menstruation than those who had the distress-focussed questionnaire first.[48] So not only might you be getting a skewed version of the process you are trying to measure, might you actually be changing the process itself?

In the same way, it is hard not to know your spatial abilities are being tested if you are asked to find a simple shape in a complex pattern, or that your empathy levels are being assessed if you are asked the extent to which you agree or disagree with the statement 'I really enjoy caring for other people'. How you deal with such questions may be coloured by your wish to please the experimenter or to do better than any other participants in the study, by the knowledge that this will fulfil your quota of research participation points and you won't have to take part in any more baffling or boring psychology studies for this semester, or even by enjoyment of solving problems or filling out surveys.

It is also the case that 'priming', or the prompting of pre-existing awareness of relevant stereotypes, might affect what you say about yourself and even your task performance.[49] For example, women's empathy scores may vary depending on whether or not empathy has been flagged up as a feminine trait.[50] Another form of priming is 'stereotype threat', which refers to the effect of attention being drawn to a negative stereotype of the group that you belong to, for example the inability of females to perform visuospatial tasks, or the tendency for African-Caribbean boys to do poorly on tests of intellectual achievement.[51] In a context where that particular skill is being assessed, such as a mental rotation task or an SAT test, members of the group that has been stereotyped have often been shown to underperform. Originally identified with respect to underachievement in black and minority ethnic people, stereotype threat has also been shown to have a powerful effect on women, particularly with respect to performance on subjects such as science and maths.[52]

Experimental studies of stereotype threat have shown that you can demonstrate the effect under controlled circumstances, presenting a task which is actually neutral as one in which either men or women do better. For females carrying out the task, being told that it was one in which women usually did better resulted in higher scores (this is what we call the stereotype lift effect), whereas being told that men usually did better resulted in dramatically worse scores. The effect was less strong in the male participants, but they also did better in the task where they were told men usually had the edge.[53]

So the data you are collecting are not necessarily 'pure' in the sense that they are context-free. The task you use may reflect more than the particular variable you are hoping to measure and your participants may respond in a way which is contaminated by all sorts of factors which are nothing to do with what you are hoping to demonstrate.

Sex is not enough

Knowing, as we now do, how entangled the brain is with the brain-changing world in which it is functioning, it is clear that we need to take account of this either when selecting participants for testing or when accessing the large data sets on brain and behaviour that are becoming available – or, in fact, when deciding how reliable and valid are the conclusions that have been reached by researchers. This is especially true of sex differences research, where just dividing a population according to whether they are male or female will mask a huge number of other possibly (or even probably) contributory sources of difference. When we know that, at a general level, factors such as the number of years in education, socio-economic status and occupation can change brain structure and function, then these should be taken into account when looking at what our participants can do. Any studies that appear to cheerfully assume that sex alone is a sufficient basis for categorising the individuals they are looking at should be sent back to the drawing board.

The psychological study of sex differences has come some way from the 'sentimental rot and drivel' that Helen Woolley described at the beginning of the twentieth century. But while we may have weeded out the most extreme claims, as key branches of psychology headed into the twenty-first century and joined forces with the brain imagers there was still cause for concern. A hundred years after Woolley's scornful summary of psychology's findings to date, Cordelia Fine's trenchant survey of the emerging discipline of cognitive neuroscience recorded plenty more foregone conclusions, biased theory and practice and misrepresented findings.[54]

Psychology can appear to be wilfully ignoring the power of the world around us to change our behaviour and, given what we now know about neuroplasticity, to change our brains. And these cultural pressures can, of course, include the very findings emerging from psychology and brain imaging labs. Unless and until this is taken into account, psychology could be accused of merely providing a go-to catalogue of apparently well-established differences between the sexes.

But psychology had another role to play, which would bring neuroscience out of the lab and into public consciousness. There has long been a real appetite for psychology's apparent insights into understanding ourselves and other people. Personal advice guides and codes of conduct manuals have been around for centuries, but it was books such as Dale Carnegie's *How to Win Friends and Influence People* (1936) which established the popular and lucrative self-help genre.[55] From Napoleon Hill's *Think and Grow Rich* (1937), through *How to Stop Worrying and Start Living* (Carnegie again, 1948) to (of course) *Act Like a Lady, Think Like a Man* (2009), popular psychology has been enthusiastically consulted for the solutions to life's puzzles and problems, principal among which was the age-old issue of sex differences. Any number of tricks for Survival and Success, roads to a Transformed You and a Whole New Life could be found in the pages of such books – the basic message was Know Yourself (or others) and Do Better.

Combined with the advent of brain imaging, popular psychology took on a whole new dimension. Once you could add 'Your Brain and How It Works' to this mix, especially when this could be illustrated by beautifully coloured brain pictures, the scene was set for a whole new genre of self-help books, the neuro-guides.

Chapter 4:
Brain Myths, Neurotrash and Neurosexism

Neurononsense, neurotrash, neurosexism, neurobollocks, neuroflap-doodle, neurobunk, neurobabble, neurohype, neurobaloney, neurocrack-pottery, neurofallacies, neurobloopers, neurogibberish, neurotomfoolery.

The advent of brain imaging technology at the end of the twentieth century offered the possibility of really getting a handle on what differences there might be in the brains of women and the brains of men and exploring the links between these and any associated differences in behaviour. No longer reliant on dead, diseased or damaged brains, the research community should now be able to answer age-old questions about sex differences. The most popular technique in this exploration was fMRI, which, as we saw in Chapter 1, measures the blood flow changes associated with brain activity and displays the results as beautifully colour-coded images, appearing, at last, to offer a window into the brain.

It is worth at this point just stressing what fMRI *can't* tell us and highlighting many of the mistaken beliefs that have come about because of a misunderstanding of what this brain imaging technique can do.[1] Firstly, fMRI does not give us a direct picture of brain activity, of the passage of nerve impulses across the surface of the brain or within key structures in millisecond timescales; it is just giving us a picture of the blood flow changes that provide the energy for that activity.[2] And these changes are much slower than what is actually going on – we're talking about seconds rather than milliseconds. So once the findings are interpreted in terms of differences in functions such as word finding or pattern recognition (both of which can occur over millisecond timescales), then such findings should be viewed with caution and should only be considered in the context of detailed analysis of behavioural changes measured at the same time.

We should also be aware that the beautifully colour-coded images are not a pure measure of one task or another. If you take part in a brain imaging experiment, some of the time you may just be looking at single words projected, one at a time, onto a screen. Then you may be asked to look at another set of words, but this time you are asked to remember as many of them as you can. The data from your first task would then be subtracted from the data from your second task. The assumption here is that the researcher will 'lose' those patterns of brain activation which are common to both tasks and just be left with those that are unique to the memory task. This is because the brain changes associated with these kinds of cognitive tasks are very small, so brain imagers need to find a way of enhancing them. The resulting brain image is not capturing the real-time brain changes associated with your memory centres being activated – it is a picture of the differences between a word-reading brain and a word-memorising brain.

In order to illustrate the size of the differences that have been found, various shades of different colours are assigned to them. Red is usually allocated to areas that show increases in activation, ranging from pale pink, for those differences that only just squeak over the threshold for being statistically different, to bright scarlet, for the biggest of all. Blue is allocated to decreases in activation, again ranging from very pale to bright. These shades can be adjusted to maximise the contrast in the image itself. Areas where there are no significant differences in activation would generally not be coloured. And so we arrive at those evocative images of an eerily grey cross-section of the brain, on which are superimposed bright patches of red and blue. There is an overwhelming impression that you are looking at the equivalent of a photograph of a living, thinking human brain, beautifully colour-coded to show where 'thoughts' were coming from and seeming to provide irrefutable evidence of the 'mindreading' power of neuroimagers.[3]

One issue with the interpretation of the data that were pouring out of the scanners in the years immediately after fMRI landed on the scene is what is called the 'reverse inference problem'.[4] Stanford psychologist Ross Poldrack pointed out that when you locate activation in an area associated with a particular process, such as 'reward', and you find that area activated during a particular task, such as listening to a particular kind of music, it is tempting to conclude that people like listening to music because it activates the 'reward centres' in their brain. But the accuracy of that conclusion relies on specific parts of the brain being highly specialised for just one type of process, in this case only 'reward', and also on the behavioural measure being a very strong index of what you are interested in, so for example you'd need to include additional ratings of how positively the music was rated.

As we will see, it is very unusual to be able to pin one area of the brain down to a single function, so you would need the additional behavioural backup. In other words, you'd need a pretty good clue from your listener, such as 'I really rate this music, I'll give it 5', to support your claim that listening to (this particular) music was rewarding. And you'd need to be able to eliminate other possible interpretations than 'reward' to back up your claim – perhaps the 'reward' circuit includes some long-term memory functions, or attentional processes as well, so the music may be triggering some kind of memory or alerting you to the need to focus as well as, or instead of, giving you positive vibes.

The lack of understanding of this reverse inference problem was key to claims that brain imaging could be used to identify 'invisible behaviour', that without knowing what someone was actually thinking or doing you could look at the patterns of activity in the brain and 'read their mind'. Claiming that you could use a brain scanner as a lie detector is a classic example of this kind of 'neuroquackery', with glowing 'circuits of deceit' giving away your guilt, your infidelity or your terrorist leanings.[5]

The first fMRI studies of the human brain were carried out in the early 1990s and, for the next twenty years or so, were the source of most of the public understanding (and, unfortunately, misunderstanding) of how the brain works and how it serves as the basis of all human behaviour. In many cases, this research provided stunning insights into processes we had only been able to guess at previously. In other cases, however, it merely served to perpetuate certain brain myths, not due to intrinsic problems with the technique itself but more due to particular biases arising from the application of old-fashioned brain models.

Imag(in)ing the brain: how neurowonder turned to neurobunk

Marketing research gurus have a term for the changes in fortune over time that often accompany the introduction of new technology. It's called the Gartner hype cycle (see Figure 1) and it tracks the trajectory of the common hypes, hopes and disappointments of promising innovations.[6]

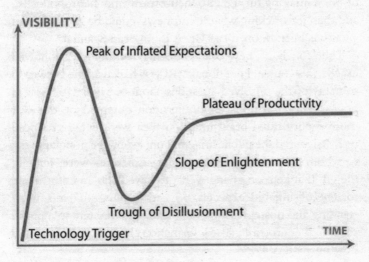

Figure 1: The Gartner hype cycle

The cycle begins with an enthusiastic launch of a new technology, often associated with lots of media interest, leading to the Peak of Inflated Expectations. This stage consists of early promising output, plus additional hype, which leads to excited speculations about the problems that this new technology will solve, thereby setting up unrealistic expectations. Then follow emerging difficulties, disappointments and criticism, and interest starts to wane – down into the Trough of Disillusionment. However, if the problems can be identified and the criticisms answered and solved, then, with appropriate caution and more realistic expectations, renewed progress ensues and better outcomes can be achieved. This is the Slope of Enlightenment, and if done right, it leads to the Plateau of Productivity.

In a blog in 2016, Guerric d'Aviau de Ternay from the London Business School and psychologist Joseph Devlin from University College London used the hype cycle concept to track the initial expectations, emergent difficulties and future prospects of neuromarketing (now, incidentally, rebranded as consumer neuroscience).[7] Reading through it, I suddenly realised that this model would be perfect for describing how the initial wonders of brain imaging turned into 'neurotrash' and 'neurobollocks', and then for tracking what 'neuronews' could be rescued from what has been dubbed the Great Brain Scan Scandal.[8]

The launch of new brain imaging technologies followed exactly this course. Functional MRI was hailed as the answer to so many of the unsolved questions about our hitherto invisible brains. The enormous communication potential of the ever more sophisticated brain images which were being produced really triggered the public imagination, fed by the popular press, to whom these wonderfully seductive pictures were, initially, the gift that kept on giving. What followed was an unfortunate torrent of inflated expectations and a deluge of misunderstanding and misrepresentation – leading ultimately to a proliferation of neurotrash: false information about neuroscience and how our brains work.

The technology trigger

In 1992, the very first fMRI studies were published. In one, participants lay in an MRI scanner and watched red and green chequerboard patterns flashing to the right or left of their visual field. The resultant increase in blood flow to their visual cortex was captured and displayed as a rather unclear (though distinctly blobby) coloured image alongside the part of the brain from which it arose.[9] A second study, using flashing lights and hand squeezing, showed how it was possible to localise the brain response, this time with the visual response matching the frequency of the flashing light and the rise and fall of the signal from the motor cortex matching the intensity of the hand squeeze. This study had rather grainy black and white images, but the detail of exactly where these changes were coming from was clear.[10]

Rather humble beginnings for the stunning images we are used to nowadays, but it definitely marked the beginning of a revolution both in brain research itself and in how the findings might be more easily communicated. Brain imagers were now able to track when and where changes in the brain were coming from and turn them into accessible pictures that told compelling stories.

The impact was almost immediate, and flashing lights and hand squeezing quickly gave way to almost every psychological process that could be modelled in the scanner, from language to lying and beyond. The decades following the first fMRI study were the boom years: the number of publications using fMRI to look at various aspects of brain function exceeded 500 a year in the first ten years, and by 2012 had reached well over 1,500 a year.[11] One estimate is that roughly thirty to forty fMRI papers a week were being published at this time. Given the cost of the equipment and the intricacy of the data collection and subsequent analyses required, this was a really explosive trajectory.

The availability of this technology spawned a new discipline: cognitive neuroscience. This new field firmly linked the activities of psychologists, who were becoming more and more adept at deconstructing the various stages of human behavioural processes, from visual perception to spatial cognition to decision making and error correction, with those of brain scientists – particularly those researching the structures and functions of the human brain.

It's usually quite gratifying if the popular media take a positive interest in the science you are doing. Even if it is not about your own work, it can be cheering to see some kind of media interest in what brain images can show, what breakthroughs neuroscientists have made, what new reality has been revealed or past certainty confirmed. And the images used were, almost universally, quite stunning and appeared to be easy to understand. It really looked as if, at last, we had a window onto the brain and could map out its activities. Past guesswork based on autopsies or brain-injured patients could be checked against reality and we could start getting to grips not only with how the healthy brain worked but also with the diseases that affected it.

The Peak of Inflated Expectations

In the way of all new fads, the 'silly season' set in. The 1990s were declared the Decade of the Brain.[12] It looked as if all forms of brain science, and particularly the outputs of fMRI machines, were going to revolutionise our understanding of this most important organ of all. Psychiatry, education, psychology, psychopharmacology and even lie detection were all going to be transformed by the arrival of fMRI and a new understanding of the brain. Suddenly there was a fashion for adding the prefix 'neuro-' to everything: neurolaw, neuroaesthetics, neuromarketing, neuroeconomics, neuroethics. Drinks were called Neuro Bliss, Neurogasm, Neuro Sleep. In 2010, an international Neuroscience and Society conference at Oxford University asked the question 'What *Is* It with the Brain These Days?', commenting on the fashion for all things 'neuro' and outlining the problems that were going to be solved by harnessing this new technology, all thanks to neuroscientists.[13]

Even outside the science world, it appeared compulsory to frame everything in 'brain' or 'neuro' terms. Falling in love, speaking in tongues, eating chocolate, deciding on who to vote for, even causing the 2008 financial crisis – all linked to the activities of our brains, usually described as 'hard-wired' to ensure we got the biological determinist message.[14] Every article had to be illustrated with one or more brightly coloured brain images, often without axes or helpful decoding guides, with references to particular areas 'lighting up' or 'glowing' in response to your glossolalia, chocoholism and/or Tory or Republican leanings. Brain science was cited as evidence in 2012 by the *Daily Mail* when they (mis)quoted some neuroscientists on the origins of Justin Bieber obsession (a rush of dopamine similar to that caused by orgasm or chocolate – or both perhaps), so the brains of the 'Beliebers' are allegedly 'hard-wired' to be obsessed by him.[15] In the same year, the *Guardian* published an essay on the brain science of creativity in terms of 'the neuroscience of Bob's Dylan's genius'.[16]

A new industry of neuromarketing was born. My personal favourite is a kitchen designer who, according to his website, uses 'brain principles' to understand his clients, in order to 'create a domestic utopia tailored to their personality, using the principles of neuroscience, or the scientific study of the nervous system, to answer their emotional needs and subliminal desires, as well as building a seamlessly practical kitchen'. The accompanying image is of a rather startled-looking man wearing brain imaging headgear gazing out of a presumably seamlessly practical kitchen.[17]

Sometimes the neuroscientists themselves contributed to the increasingly bizarre coverage. An example of this is found in the history of a report at the American Association for the Advancement of Science where a paper on the 'Neural and Emotional Signatures of Social Hierarchies' was being presented.[18] The paper reported that male participants were scanned while being shown pictures of different male or female bodies in different levels of dress (fully clothed, partially clothed, in swimwear). This was done in the name of studying the neural correlates of certain types of memory. The men were shown a mixture of old and new images and were later asked to identify which they had seen before. The males remembered the bikini-clad women the best (cynics among you might wonder why you needed a brain scanner to demonstrate this). When the relevant brain scans were scrutinised, different areas were activated by the different types of stimuli. With respect to responses to the bathing beauties, the authors reported:

> Areas of the brain that normally light up in anticipation of using tools, *like spanners and screwdrivers* [my emphasis], were activated ... The changes in brain activity suggest sexy images can shift the way men perceive women, turning them from people to interact with, to objects to act upon.

The next day, the *Guardian* ran an item on the piece, accurately reporting what had been said, accompanied by the headline 'Sex objects: pictures shift men's view of women', subtly accompanied by a picture of a drill bit penetrating a piece of wood.[19] The day after, the researchers' home paper, the *Princetonian*, also reported the study, describing those brain areas activated by the bathing suit images as associated with 'things you manipulate with your hands'. This time the picture was of a screwdriver suggestively superimposed on an image of a shapely bikini-clad torso.[20] The normally staid *National Geographic*'s headline exclaimed 'Bikinis make men see women as objects, scans confirm', together with the additional and initially unreported detail that 'men were also more likely to associate images of sexualized women with first-person action verbs such as "I push, I grasp, I handle"'.[21] They had sourced a really raunchy set of bikini images to illustrate their point. And so to CNN, whose website trumpeted: 'It may seem obvious that men perceive women in sexy bathing suits as objects, but now there's science to back it up.'[22] Perhaps not public communication of science at its finest?

In a study entitled 'Neuroscience in the Public Sphere' Cliodhna O'Connor and colleagues from University College London chronicled the media coverage of neuroscience in the

UK between 2000 and 2010.[23] Picking six representative British newspapers, they identified more than 3,500 articles in this decade, with the numbers steadily increasing throughout that time from 176 in 2000 to 341 in 2010. Topics covered ranged from 'brain optimisation' through 'gender differences' to 'empathy' and 'lying'. The UCL team concluded that there was a concerning focus on 'brain as biological proof', and on an agenda that offered explanations for almost anything, including risky behaviour in the gay community or paedophilia. The authors concluded that 'research was being applied out of context to create dramatic headlines, push thinly disguised ideological arguments, or support particular policy agendas'.

The Trough of Disillusionment

The early neurohype had seemed to offer insights into so many aspects of human existence. As well as the promise of understanding consciousness and free will, solving the mind–body problem and maybe offering greater understanding of the self, there was also the possibility on a more immediately practical front of improved diagnosis, and even treatment, of both physical and mental brain-based conditions. There began to be rumblings about the quality of some of the basic research: not just about its reporting, but about the production and interpretation of the very brain images that were causing so much excitement.

These images were attractive and attention-grabbing, the colour-coded maps and, with more advanced systems, time-lapsed videos giving the impression of a direct window into ongoing activity in the human brain. The visualisation of the brain, appearing to make the invisible visible, was really the technology trigger that brought neuroscience firmly into the public arena. It also resonated well with the visualisation of just about everything else in the world at the time, via video cameras and photocopiers as well as developments in TV and the cinema.[24]

The trouble was that the brain image that was adorning so many newspaper articles and popular science books was something of an illusion. The production, or 'construction', of a brain image, either of a single individual or a group, requires a multi-layered hierarchy of decisions, about how to 'clean' the raw data, how to smooth individual anatomical differences, how to 'warp' brain characteristics to fit a template brain.[25] The allocation of colours to the different types of changes identified is actually a statistical procedure. So the flickering colours that move across the grey and white tundra of the brain as someone views a Coke commercial are not equivalent to a time-lapsed sunset but reflect some thresholding decisions made by a brain imager.

But this was not the impression given at the outset, and brain images were identified as having a particular persuasive power. A series of studies run by psychologists David McCabe and Alan Castel claimed to demonstrate that the mere presence of brain images had a direct effect on whether or not a particular line of scientific reasoning was judged to be credible.[26] Their participants were given spoof articles on subjects such as the link betweeen watching television and maths ability, some of which contained scientific errors. They were also illustrated with standard brain images, topographical maps or ordinary bar graphs. When asked to rate the scientific reasoning in the articles, participants were much more likely to approve articles that were accompanied by a brain image.

In the early days of developing magnetic resonance imaging for diagnostic use, there was some discussion about which medical professionals would best understand and interpret the data that were being produced (which at that stage were both numerical and pictorial). It was decided that radiologists, as trained viewers of images from X-rays and CT scans, would be the most effective, and they were introduced to the new technology. Interestingly, it was these radiologists who requested that the early practice of colour-coding the images was replaced with grey-scale versions. Apparently, they preferred hues of one colour because this allowed them to identify subtle anatomical changes. They pointed out that the difference between two regions of brightness could be very small, which would be evident from numerical data, but if these regions had been colour-coded, say, blue and yellow, this difference might 'trick the eye into believing two close numbers are very different, because they are different colours in the image'.[27] So, the professionals found distracting the very colours that were a key selling point for the public.

These 'seductive allure' studies are frequently cited as part of the growing concern about the impact that neurohype was having on brain imaging's overall credibility. The claims that neuroimaging was only good for producing pretty pictures that could pull the wool over gullible eyes were not doing much to moderate the emerging impression that this technology was just diverting funding from 'good science' and really could not tell us anything useful about the human brain. As it turns out, Martha Farah and Cayce Hook, cognitive neuroscientists from the Center for Neuroscience and Society at the University of Pennsylvania, in a paper with the characteristically catchy title of 'The Seductive Allure of "Seductive Allure"', reported that other studies had not been able to replicate these original findings, but acknowledged the ongoing *belief* that the persuasive effect was real.[28] Robert Michael and colleagues, from New Zealand, got the bit between their teeth, and tried to replicate the original findings, with ten different studies and nearly 2,000 participants.[29] The key part of their design was the presence or absence of a brain image or 'scientific language' in the various science arguments with which the participants were presented. Interestingly, in this case they found the language more effective than the image. So early 'blame the brain images' arguments may have been overstated, but it was still clear that the narrative associated with them could be mistakenly persuasive. This is important to remember when trying to understand the surprising persistence of mistaken beliefs about the brain in all sorts of arenas, especially those related to understanding sex differences.

However, it soon became clear that little progress had been made on the early promises and, paradoxically, the emerging practice of neuroscientists themselves 'calling out' the neurotrash was having an effect. In 2012, Deborah Blum wrote an article in the *Undark Magazine* entitled 'Winter of discontent: is the hot affair between neuroscience and science journalism cooling down?'[30] She drew attention to a flurry of 'neurocritical' stories which were expressing concern about the oversimplification (and inaccuracy) of neuroscience coverage in the press, particularly those stories involving 'brain scans with their flashy visuals'. One of the articles she noted was 'Neuroscience fiction', published in the *New Yorker*, where Gary Marcus, a professor of cognitive science, expressed concern that brain imaging was being trivialised by items such as 'Female brain mapped in 3D during orgasm' and 'This is your brain on poker'.

Alissa Quart's op-ed in the *New York Times* was rather more hard-hitting.[31] Praising the actions of the neuroscience bloggers, or 'neurodoubters', she applauded the backlash against what she dubbed 'brain porn', with neuroscientific explanations offering 'shortcuts to enlightenment' and 'eclipsing historical, political, economic, literary and journalistic interpretations of experience'. Steven Poole of the *New Statesman* joined the fray with the eye-catchingly entitled piece 'Your brain on pseudoscience: the rise of popular neurobollocks'.[32] He tore into the fashion for neuro self-help books, noting that the function of the 'Smart Thinking' genre was to 'free readers from the responsibility of thinking for themselves' and enthusiastically quoting Paul Fletcher, a Cambridge neuroscientist, who coined the glorious term 'neuro-flapdoodle' to describe the habit of attaching a grand-sounding neural term to some simple point. Poole also drew attention to the overuse of the phrase 'This is your brain on …' when interpreting brain imaging data associated with any kind of process, from music to metaphor to poker. For me, the best bit of his article is when he volunteers to 'submit to a functional-resonance imaging scan while reading a stack of pop neuroscience volumes, for an illuminating series of pictures entitled "This is your Brain on Stupid Books about Your Brain"'.

The Slope of Enlightenment

Alarm bells had already started to sound within the neuroscience community long before the backlash from journalists. In a paper in 2005, titled 'fMRI in the Public Eye', Eric Racine, from the Stanford Center for Biomedical Ethics, and colleagues raised concerns that the limitations of fMRI weren't being made sufficiently clear and that there were too many leaps of faith, with wild claims that brain images could be taken as 'visual proof' of the biological basis of some kind of process, such as addiction to pornography, or that MRI systems could serve as some kind of mind-reading or lie-detecting machines.[33] They urged caution not only in the media but also in the neuroscience community, with greater care to be taken in explaining the risks and concerns as well as the benefits of the new technology.

The 'brain industry' had itself started to self-regulate. Blogs were emerging as a means of public communication, and several practising neuroscientists entered into the spirit of the age and set up sites which drew attention to the 'neurobloopers' and 'neurohowlers' doing the rounds. Researchers at the James S. McDonnell Foundation in the United States had already set up the Neuro-Journalism Mill in 1996, a website run by self-appointed 'curmudgeons … dedicated to sifting the wheat from the chaff of popular media reporting on news about the brain'.[34]

It was also becoming clear that the neurohype was not just a function of overenthusiastic but ill-informed press articles and consequent neurotrash. The enormous complexity around the production and analysis of brain imaging data was leading to mistakes and misinterpretations within the science itself. Ed Vul and colleagues, from the University of California, San Diego, were puzzled by some of the extraordinarily high correlations between brain activity and behavioural measures, particularly in the newly emerging social neuroscience circles.[35] Knowing the large amount of variability in both types of measure, they couldn't see how the researchers were coming up with correlations as high as 0.8 and more, especially if one of your variables was brain imaging data. Raw brain activation data (be it blood flow or electrical or magnetic activity) is converted into a visual representation of brain tissue measured in units called voxels

(3D pixels). These can vary in size, depending on the resolution of the system, but you might get up to one million voxels in a high-resolution brain scan. And then you would get a new image every two or three seconds or so. All in all, a huge amount of data to process, within which would be huge amounts of variance, which would normally make differences very hard to spot.

Suppose you were a social psychologist wanting to look at the correlations between activity in brain voxels measured by your brain imaging colleague and some behavioural measure in which you were interested. Given the enormous number of voxels to pick from, there's a really high probability that you might make a false positive error, finding a correlation just by chance. If you could 'constrain' your choice of voxels in some way, that would be a big help. And this is where Vul and colleagues realised the problem was coming from: researchers were 'cherry-picking' those voxels where it was already clear that there were highish correlations with their behavioural data and then just exploring those (rather than, say, specifying in advance a particular anatomical area where relevant activity might be *predicted*). It's a bit like testing out your suspicion that a high proportion of the population gamble by only collecting your data in a street full of betting shops. This would lead to pleasingly (or puzzlingly, if you were Vul and his colleagues) high correlations between the brain and behaviour data. Over half of the fifty or so papers that were examined had fallen into this trap. Vul and colleagues felt that the problem was based more on statistical naïveté in researchers (and, presumably, reviewers of their papers) new to a complex area rather than a deliberate attempt at fraud. However, they were rather less forgiving in their conclusion: 'We are led to conclude that a disturbingly large, and quite prominent, segment of fMRI research on emotion, personality, and social cognition is using seriously defective research methods and producing a profusion of numbers that should not be believed.' This didn't mean that all the exciting findings in social cognitive neuroscience needed to be binned, just that they now need to be viewed with a very large pinch of salt, especially those published before 2009.

Unless you apply the correct statistical rules, your results can be misleading, to say the least. The most classic example of this involves a dead fish.[36] This study by Craig Bennett and colleagues in their lab at Dartmouth also involved looking at the difficulties associated with the huge amounts of data that imagers are trying to compress into a digestible visual form. As we have seen, there are an enormous number of voxels in any brain scan. If you are going to try and find those areas which are most active during a particular task, you will need to make a huge number of comparisons (and expose yourself yet again to the false-positive problem). So you need to set some kind of threshold. But because, proportionally speaking, the differences between more or less active areas are really quite tiny, if your threshold is too conservative, then all your interesting differences might disappear along with the random ones. So you might say, OK, I'll only accept that something genuine is happening if, say, eight or more voxels clustered together show some kind of difference. But the key thing is that you want to find a way of maximising the contrast between the active areas and the inactive areas.

What Bennett and colleagues' study showed was that this could actually set you up for some pretty misleading findings. The story goes that they were testing out the contrast settings of their fMRI system prior to carrying out an emotion recognition study. Needing a 'dummy' object with the right kind of different body tissues, they recruited a full-length (dead) salmon (having tried and failed with a pumpkin and a dead hen), placed it in their scanner and ran the experimental protocol (which happened to be comparing responses to happy or sad faces). This allowed them to set their contrast settings at the right level. At a later stage, they wanted to demonstrate the effects of different kinds of thresholding decisions on the outcomes of fMRI analyses, so they tried various approaches out on their salmon data. What they found was that, if you didn't properly correct for the fact that you were making many comparisons, you could produce a quite startling result. You could find an area in the dead salmon's brain that showed significant evidence of activation when being shown photographs of individuals in happy or sad situations, as opposed to when it was 'at rest'. Predictable headlines ensued: 'Scanning dead salmon in fMRI machine highlights risk of red herrings' and 'fMRI gets slap in the face with a dead fish'.[37]

Again, this study was not setting out to show that fMRI was fraudulent and that all research studies should be ignored; rather that researchers needed to be very careful how they treated their data in order to avoid this kind of nonsense finding (which was, of course, only clearly spotted as nonsense due to the deadness and fishy nature of this particular participant). But it did add to the dawning disillusionment with the neurohype genre, and set things on the path towards a Slope of Enlightenment, with realistic expectations of what fMRI imaging could and couldn't show. The salmon has been nominated as the 'dead fish that launched a thousand sceptics'.[38]

The Plateau of Productivity?

So, has neuroimaging put its past mistakes behind it and re-established itself as the source of amazing breakthroughs in our understanding of the human brain?

One issue that has been difficult to resolve is that sometimes, once a finding has entered the public consciousness as a 'fact' but is later shown to be mistaken, it has proved extraordinarily difficult to knock it on the head without it popping up elsewhere – it's one of those Whac-a-Mole myths. Early brain imaging studies took longer to be circulated and critiqued, so it is possible that the length of time the first message was out there before being contradicted made it harder to shake the belief in its accuracy. And if this finding had been adopted in support of commercial activities or policy decisions, which would then have to be abandoned or reversed if their cornerstone was removed, then there would be a tendency to much higher levels of staying power.

This is particularly true of neuromyths in education, where we still find beliefs, for example, that we can only make a differ-ence to a child's brain in the first three years of life, that there are different brain-based types of learning, that we only use ten per cent of our brains, or that men use one side of their brains whereas women use both.[39] Despite so much evidence to the contrary, these myths still persist, often perpetuated by educational 'gurus', whose advisory manuals will urge parents and policy makers to adopt ill-informed (and expensive) 'brain-training' techniques or to send their children to single-sex schools.

However, on a more positive note, research is advancing and enormous projects have been set up to explore many aspects of the brain. In Europe, over one billion euros have been committed to the Human Brain Project, a hugely ambitious programme relying heavily on computer modelling and simulation techniques to attempt to gain insight into how the brain does what it does.[40] It involves over 100 research centres around the world. Spin-offs of the project include wonderfully detailed up-to-date brain atlases for both humans and non-human species. These and all the enormous data sets they are accumulating can be accessed by all researchers, not just those funded by the project. In the UK, the Biobank project focusses on human-health-related information from 500,000 people aged between forty and sixty-nine in 2006–2010, 100,000 of whom will have one or more brain scans.[41] In the US, there is the BRAIN Initiative (Brain Research through Advancing Innovative Neurotechnologies), with a closing possible budget of $4.5 billion and a big focus on different ways of measuring the brain.[42] The Human Connectome Project (HCP), also US-based, aims to map all the possible nerve cell connections in the human brain.[43] It has achieved this for a worm (C. elegans) with its 302 neurons and about 7,000 synapses. (The observation that this required over fifty person-years of labour puts a rather alarming perspective on the scale of the project's task of mapping an organ with 86 billion neurons and 100 trillion possible connections!) New techniques for tackling such tasks are emerging all the time.

Free access to the huge data sets being generated by researchers on such projects, and by more widespread data sharing, is allowing individual investigators or research groups around the world to answer a huge range of questions about the human brain. A paper in 2014 counted more than 8,000 shared MRI data sets available online.[44]

Brain research is currently an area of great promise. But although it has, in most cases, done a great job of putting its house in order, there are still echoes of past mistakes and misunderstandings that we have to look out for. This is particularly true when we come to look at research into sex differences in the brain.

Sexism in the scanner

How did the study of sex differences in the brain fare in this cycle of peaks of inflated expectations and troughs of disillusionment? Imaging healthy brains in situ would certainly seem to offer solutions to many of the problems thrown up by the limitations of the 'bumps and buckshot' approach, with which only dead or damaged brains were available for additional insights. At last we should be able to test the link in the argument that individuals with different genes, genitals and gonads would have different brains as well.

As you might have predicted, a combination of 'sex' and 'brain' findings proved a gift to the purveyors of neurotrash and, once that particular genie was out of the bottle, a tidal wave of 'brainsex'-type books followed.[45] As well as the well-known *Men Are from Mars, Women Are from Venus*, we had *Why Men Don't Listen and Women Can't Read Maps* (with follow-ups of *Why Men Lie and Women Cry* and *Why Men Don't Have a Clue and Women Always Need More Shoes*). These were joined by the intriguing *Why Men Like Straight Lines and Women Like Polka Dots*, *Men are Clams, Women are Crowbars* and *Why Men Don't Iron*. Single-sex schooling advocate Michael Gurian produced *Boys and Girls Learn Differently* and we got a religious spin from Walt and Bar Larimore with *His Brain, Her Brain: How Divinely Designed Differences Can Strengthen Your Marriage*. Anything that could be done to reinforce the notion that men and women were so different they could be from different planets was rife in such publications.

One of the most famous, or infamous, of the genre is psychiatrist Louann Brizendine's *The Female Brain*, published in 2006.[46] The notoriety, in neuroscience circles, comes from a dispiriting range of scientific inaccuracies, anecdotes masquerading as evidence and occasionally hilarious misquotes and misdirection.[47] One claim in the book is that 'differences between men's and women's brains make women more talkative'; Brizendine tells her reader that language areas in the brain are larger in women than in men, and that women on average use 20,000 words a day and men use only 7,000. Mark Liberman, a linguist from the University of Pennsylvania, followed up on this assertion but couldn't track down the original source.[48] He found that it had been repeated, in various forms, in a range of other self-help books but there seemed to be no research findings to back it up. To make his point, he did his own calculations, based on a British database of conversations, and arrived at a somewhat different conclusion: men's level of word use was just over 6,000 a day as compared to just under 9,000 for women.

Just focussing on Brizendine's claims about sex differences in language use and the brain-based explanations for them, Liberman was able to find that many of her factual assertions were either contradicted by research she did cite or would have been by research she didn't. In Liberman's words: 'There's a technical term that philosophers use to describe the practice of asserting things without caring much about whether they're actually true or not: they call this bullshit.'[49]

Psychologist Cordelia Fine has also had a good go at the Brizendine bloopers.[50] For example, she checked through the five references Brizendine had cited in support of her statement that men's brains have little capacity for empathy. One study was in Russian and on the frontal lobes of dead people, three didn't actually compare males and females, and one was supposedly a personal communication from a cognitive neuroscientist who, when contacted, said she had never communicated with Brizendine and had never found evidence for any kind of sex differences in the brain on the basis of empathy.

Armed with these helpful trash-spotting tips from the likes of Liberman and Fine, you would hope that these kinds of publications would be swept aside and discredited for their inaccuracies and even fabrications. But as we saw with the neurotrash trend earlier, falsities have an alarming way of staying around and continuing to sustain unhelpful neuromyths such as boys and girls having different brains that require different (and separate) types of education. Louann Brizendine's book, full of bloopers as it is, has been translated into many languages and it has now been made into a film, released in 2017.[51]

At this point, you might be asking yourself whether or not this really matters. Perhaps we should merely smirk at such neurofoolishness or just silently wince at the misinformation that characterises neurotrash? However, reports have shown that media items such as newspaper reports on biological explanations of sex differences, implying some form of fixed brain-based factors, are more likely to lead to endorsement of gender stereotypes, increased tolerance of the status quo and belief in the impossibility of change.[52] So beliefs that biological sex bequeaths a fixed and different portfolio of brain-based skills to males and females become entrenched in the public consciousness. In a cycle of self-fulfilling prophecy, such beliefs drive how children are reared and educated, form the bases of different attitudes to and expectations of females and males, and afford them different experiences and opportunities. Brains, being as plastic and mouldable as we now know them to be, will come to reflect those differences. Rather than 'limitations imposed by biology' we are looking at 'restrictions imposed by society' – both measured by differences in brain structure and function, but the latter offering much greater possibility of change.

Emerging from the general disquiet about the high level of dubious coverage of sex differences in the 'neurotrash' genre, a rather more serious concern was about the evidence of sexist practices within the field of neuroimaging research itself.

Neuroimaging seemed to be continuing in the psychosexist tradition of scientist-as-explainer-of-the-status-quo, focussing on finding differences between men and women, taking as given the differences in, for example, language and spatial skills, and rooting around in the brain for the supporting evidence. Cordelia Fine coined the term 'neurosexism' for such practices in her 2010 book *Delusions of Gender*, pointing out that they were continuing to boost public beliefs in neat, non-overlapping differences between female brains and male brains, inflexibly fixed as the foundations of similarly neat, non-overlapping differences in female and male abilities, aptitudes, interests and personalities.[53]

Pre-existing stereotypes proved to be quite a guiding force in the outcome of the research. The philosopher Robyn Bluhm, from Michigan State University, compared several brain imaging studies into sex differences in emotional processing, which appeared to be being driven by the status quo principle that women are more emotional than men.[54] One study measured 'fear' and 'disgust' responses by showing scenes designed to elicit these responses, expecting to find higher levels of both responses in women paralleled by greater activity in the 'emotion centres' in the brain. What they actually found was that, although women did report higher levels of emotional reactions to the pictures, it was men who showed more reactivity in their brains' emotion centres. This was explained by a revisit to the pictures they had been shown, some of which were noted, post hoc, to be actually quite aggressive and that it was possibly this that had differentially activated male emotion centres. (Although I note the men, true to type, had kept their verbal emotional responses to themselves.)

A second study focussed more on the disgust aspect of emotionality. The researchers did find what they were looking for (higher levels of reported disgust sensitivity in females, and more activation in the 'disgust circuitry'), but a closer look at their data revealed that these differences were actually better explained by just looking at levels of disgust sensitivity in the group as a whole. Once they controlled for this, sex differences disappeared. However, the researchers stuck to their guns and their abstract concluded with the statement: 'In healthy adult volunteers, there are significant sex-related differences in brain responses to disgusting stimuli that are *irrevocably* [my emphasis] linked to greater disgust sensitivity scores in women.'

A third study addressed the issue that women's greater emotionality is due to their inability to cognitively control their emotions. Men and women were asked to 'cognitively re-appraise' or 'down-regulate' their initial reactions to unpleasant pictures (the same ones used in the preceding studies). It was predicted that the women's inferior emotion regulation capacity would be indicated by less activation in the frontal cortex. What was found was that there were *no* sex differences in the reported ability to rethink the initial response, but there *were* differences in the brain activation patterns. Contrary to the hypothesis, though, it was women who showed higher levels of activity in the prefrontal areas. Undaunted, the researchers offered the interpretation that men were actually more efficient at cognitive reappraisal and therefore did not need to call on as much of their cortical resources as women. To quote Mary Wollstonecraft: 'What a weak barrier is truth when it stands in the way of an hypothesis!'

Robyn Bluhm notes a comment from one of the researchers whose work she has criticised: 'If gender differences (typically) fail to emerge in studies of emotional reactivity, how are we to explain the widespread consensus that there are gender differences in emotional responding?'[55] Revisiting this consensus is apparently not an option.

One of the problems in this area can be the 'stickability' of early findings, especially when they supported an existing understanding of how the brain worked (or, in other words, sustained a stereotype). And this problem can be compounded when researchers in the same area continue to quote such findings, even when there is clear evidence that the early results have not been replicated or additional studies have come up with different conclusions.

Whac-a-Mole myths

The notion of right- and left-brainedness was already well established in the pre-scanning models of brain function, as was the
possibility that there might be sex differences in this pattern of
hemispheric differences. So looking at hemispheric differences
and sex differences would be a great opportunity for brain
imagers to demonstrate the power of their new toy.

One of the very first fMRI studies looking at language
processing in the brain was carried out in 1995 by psychologists
Sally and Bennett Shaywitz.[56] They came up with a finding that
had a dream combination of both sex differences and hemispheric differences. The headline story – literally, as the *New
York Times* covered it under the heading 'Men and women use
brain differently, study discovers' – was that whereas men used
a particular part of just the left side of the brain for language
processing, women used both right and left sides when carrying
out the same task.*[57] This seemed to confirm decades of indirect
observations based on psychological tasks and/or effects of
brain injury. Another neuroscientist, commenting on the study,
said it provided 'definitive evidence' that men and women can
use their brains differently to perform the same task, hailing the
discovery with the idea that 'nothing was conclusive until now'.
The image within the paper was much less dramatic than the
later multi-coloured pictures that became standard fare, but it
did seem to tell a compelling story. A few orange and yellow
squares were superimposed on a greyish cross-section of the
brain, all clustered on one side for the males, but distributed on
both sides for the females.

Despite its age (ancient in neuroimaging terms), the image
has become one of those popular stock shots that pops up in
all sorts of contexts. For example, it figured in an article about
Christine Lagarde becoming managing director of the IMF,

---

* In the UK, the *Daily Mail* focussed on different matters of moment, reporting
that 'Men and women respond to eating chocolate with different parts of their
brains', the syntax checker having been off that day, apparently.

the context suggesting that her superior language ability would give her the edge in dealing with the chimp-like communication skills of the male financiers she would encounter. Both the conclusion, that men 'do' language with the left side of their brains, whereas women use both, and the image demonstrating this became major players in the expectations generated by the new fMRI technology. Their many subsequent appearances and reappearances are either frustrating to those who have drawn attention to shortcomings in the study or reassuring to those who have no wish to challenge such well-established orthodoxy. This particular finding illustrates exactly those Whac-a-Mole tendencies characteristic of an area. The paper has, to date, been cited more than 1,600 times since its publication, and it continues to be cited, most recently in 2018 publications.

The trouble is, it turned out that there are major problems with the study, as has been revealed in several different commentaries (most particularly a characteristic filleting by Cordelia Fine).[58] These problems are not due to any intrinsic mistakes in the study itself, but rather how it has been interpreted and how our view of its findings should be changed by what has been discovered since. The sample size was small (nineteen men and nineteen women) but this was characteristic at the time (indeed, double figures was quite impressive). There were actually four types of word-processing task, but the study only reports on one, a rhyming task – there was no report of what, if anything, was found with the other language tasks. But a key issue, which escaped the notice of most, was that although all nineteen men showed 'pixel clustering' on the left hemisphere, only eleven women showed the much-vaunted left and right distribution. So, it is true that, as the authors say, 'more than half of the female subjects produced strong bilateral activation in this region'. But, on the other hand, nearly half didn't. So these sex differences were perhaps rather less 'remarkable' than the authors claimed, although the enthusiasm for revealing the possibilities offered by the new technique was completely understandable at the time.

Several attempts at replicating the study since have failed,[59] and more recent meta-analyses and a critical review of the whole issue of sex differences in language lateralisation could find no evidence of such differences,[60] neither in functional neuroimaging studies nor when looking at structural measures of the language cortex and at the kind of neuropsychological tasks that are indirect measures of lateralisation.[61] The paper probably would not be publishable today; the methodology has moved on, much more sophisticated techniques would allow much more sophisticated questions to be asked of much larger data sets. And yet it is still widely quoted.

A more recent example of problematic 'evidence' being eagerly seized upon and becoming firmly entrenched in the public consciousness (and indeed in parts of the research community) is a paper on sex differences in connectivity pathways in the brain that was published in 2013.[62] Researchers from Ruben Gur's lab in the University of Pennsylvania described 'unique differences in brain connectivity' between 428 males and 521 females, aged between eight and twenty-two years, an impressively large cohort. They summarised their overall findings as showing greater within-hemispheric connectivity in males and between-hemispheric connectivity in females, which, they claimed, suggested that 'male brains are structured to facilitate connectivity between perception and coordinated action, whereas female brains are designed to facilitate communication between analytical and intuitive processing modes'.[63] The accompanying press release from the university, quoting one of the researchers, referred to 'stark differences', linking them to the complementarity between males and females, with the former better at cycling or navigating directions and the latter 'more equipped for multitasking and creating solutions that work for a group'.[64] And no, they didn't get their participants to cycle or multi-task in the scanner.

What differentiates this from the Shaywitz saga is that it was almost immediately identified as problematic by dozens of other researchers and online commentators, not only taking the team to task for the outrageous stereotyping they were displaying, but also challenging their methods.[65] Critics noted that the researchers had linked structures (pathways in the brain) to functions (ranging from memory to maths via cycling and multi-tasking) that they didn't even measure in the scanner. The image they produced appeared only to show the comparisons that were statistically significant (which could have been as few as nineteen, though this wasn't reported). These were much fewer than all of the possible comparisons that could have been made (there were 95 × 95, or 9,025, connections assessed). The researchers themselves didn't report any effect sizes for the 'fundamental', 'conspicuous' and 'significant' differences they were describing, but a helpful blogger calculated the biggest as being of the order of 0.48, so, at best, only moderate.[66] It should also be noted that no other demographic information, such as years in education or occupation, was used in their analyses, so we have the additional cardinal sin of ignoring variables other than biological sex which are known to have brain-changing potential.

You would think that under the weight of this criticism, such a paper might vanish without trace. But no – it was widely and enthusiastically covered by the press. The *Independent* declared: 'The hardwired difference between male and female brains could explain why men are "better at map reading"'; the *Daily Mail* went with 'Men's and women's brains: the truth' and 'The picture that reveals why men's and women's brains really are different: the connections that mean girls are made for multi-tasking'.[67] This was one of the first papers to apply the new technique of measuring connectivity pathways to the issue of sex differences, so it may have appeared free of some of the criticisms of the early fMRI studies. It pleasingly supported stereotypes such as differences in multi-tasking and map reading, and overtly supported the complementarity story.

With the paper attracting attention from the media, Cliodhna O'Connor and Helen Joffe, then both at University College London, seized the opportunity to track and analyse its impact not only in news articles but also in blogs and online comments in the month following its publication.[68] It was rather like a transcript from a game of Chinese whispers. Some of the emerging misunderstandings could be blamed on the researchers; they hadn't measured any behaviour in the scanners, but they certainly referred to it several times and their press release talked about 'commonly held beliefs about men and women's behaviour'. So the paper was taken as scientific proof confirming existing beliefs about sex differences in behaviour. A 'biology is destiny' or 'hard-wiring' stance was taken by about one-quarter of the traditional articles, over one-third of the blogs and almost one-tenth of the comments, most of these apparently missing the fact that the differences reported in the research papers only emerged in the older age groups.

It was the readers' comments, however, that gave a worrying insight into views about sex and gender, with a rich mix of stereotyping and misogyny, some side-swipes at homosexuality, and some playground yah-boo-type comments about who is better at what:

C'mon Ladies, much as I love you all lets face facts. Men invented piratically [sic] everything you use and enjoy. The Telephone, The Computer, The Jet Engine, The Train, the Motor Car, Etc Etc the list is endless. Without us you would still be scratching around in caves so lets have no more of this nonsense and concentrate on your hand bags.

As O'Connor and Joffe rather earnestly put it: 'Comments were unadulterated by the political delicacy that constrained the traditional media and (to some extent) blogs, and exposed a latent misogyny that continues to mark public reception of scientific information about sex difference.' Indeed!

The paper itself is widely cited (over 500 times last time I looked, nearly five years after its publication), and not always by critics using it to illustrate neurononsense-type howlers (of the seventy-nine citations I checked in 2017, more than sixty were quoting this paper in support of a hypothesis about sex differences in some form of brain structure or function). It is almost as though this newer and much more complex technique of tracking pathways in the brain has been adopted with relief to take the place of the older and now often derided fMRI blobology 'evidence'. As long as there is something out there that offers sciencey confirmation of long-held stereotypes then minor details of misrepresentation and misinterpretation can apparently be overlooked.

The arrival of neuroimaging did indeed offer the opportunity to get better answers to the questions about brains from women and brains from men. But the early stages fell into the same kind of traps that marred other areas of enquiry: sticking with the status quo to determine what questions should be asked or how you interpret your data; not challenging the orthodoxies (who knows how many imaging studies that *didn't* find any differences just didn't get published); indeed, emphasising the quest for differences itself. The area was, at this stage, lacking the later challenges to the problems of neurotrash, neurosexism and neurononsense. The Slope of Enlightenment was yet to come ...

PART TWO

PART TWO

Chapter 5:
The Twenty-First-Century Brain

The brain is an inference machine, generating hypotheses and fantasies that are tested against sensory data. Put simply, the brain is – literally – a fantastic organ (fantastic: from, Greek *phantastikos*, the ability to create mental images).

Karl J. Friston[1]

As we saw in Chapter 1, the use of functional magnetic resonance imaging (fMRI) to measure brain structure and function transformed public access to what was going on in the brain (both for better and for worse). Converting the signals associated with measuring oxygenated blood flow in the brain (the blood-oxygen-level-dependent or BOLD response) into colour-coded images turned out to be a brilliant piece of marketing but also something of a double-edged sword. The 'seductive allure' of the images produced by fMRI systems was a gift to the purveyors of neurotrash, who jumped onto the brain bandwagon to convince us that the solution to lie detection, ascertaining voting intentions, predicting world financial crises, and – of course – pinpointing the difference between multi-taskers and map readers, was now readily available at your nearest brain scanning centre.

But even the most dedicated of brain imagers realised that, although spotting differences in blood flow was a good way of answering some of the 'where' questions in the brain, the time-scale of such changes was too slow to answer the 'when' and 'how' questions. Nikos Logothetis, a director at the Max Planck Institute for Biological Cybernetics in Tübingen, has described fMRI as 'the magnifying glass that leads us to the microscope we really need'.[2] It was a great start, and indeed is the source of almost all the material to date on sex differences in the brain. Functional MRI fitted neatly into the existing map-making approach that characterised the hunt for these differences. It is still firmly entrenched in public consciousness as the source of the proof needed that there is such a thing as a female brain and a male brain, and that they work differently. But new ways of modelling brain activity suggest that we need to revisit, yet again, this age-old assumption.

From BOLD to BOINC, via celery and SQUIDS

The twenty-first century has seen a shift in what we are now looking for in the brain. Connectivity is the buzzword in brain imaging circles, and brain imagers are concerned with generating 'road maps' of the brain by tracking the connections between different structures.[3] We can now see how brain structures are linked together to form intricate 'assemblies' that underpin all aspects of our behaviour and allow us to experience and understand the world and (one hopes) each other.

How they are physically linked together is important and we now have techniques that allow us to map pathways in the brain. A technique called diffusion tensor imaging (DTI) is now widely used to track the white matter tracts, the bundles of fat-insulated nerve fibres that join different structures together.[4] The basis of this technique is measuring the ease (or otherwise) of water transport along these fibre bundles (a colleague compared this to the primary school science demo of placing a stick of celery in blue or green or red ink and tracking how far and how fast the ink penetrates the celery). With technological advances, these 'road maps' are becoming increasingly detailed and we can distinguish major highways from specialist A-roads or perhaps minor but still important back roads. We can even, with highly specialist techniques applied to non-human animals, start to see the road building itself, and watch nerve cell processes reach out to other nerves and start to lay down future communication pathways.[5]

Neuroscientists have also been looking to understand how different structures in the brain work together to solve the problems that our world will throw at it. We have come to realise that the brain is a dynamic system, always active (even when supposedly in a 'resting state'), so we also need to be able to measure the 'traffic flow' on these highways, to watch the direction of movement, to see how it ebbs and flows depending on what demands are being made on the brain by its owner.[6] And we need some idea about the nature of the traffic – are we looking at big, slow changes in activity that might signal sleep or are we looking at small rapid changes in particular areas that might signal movement (or the intention to make one)? Or, even more mysteriously, are we looking at very fast types of activity that seem to be able to signal across wide areas of the cortex via non-physical connections to get geographically quite distant areas (in brain size terms) to suddenly spring into action simultaneously, rather like the type of co-ordinated traffic light or signalling systems which are designed to ensure the smooth running of highway or railway networks?[7]

Tracking the development of these connections has shown us that well-known aphorisms such as 'cells that fire together wire together' and 'use it or lose it' are highly appropriate when looking at how the brain changes over time, all the time. These kinds of changes underpin what we now know about the life-long flexibility of the brain and how the to and fro between the brain and its world is mirrored in these patterns of connectivity.

In order to track all of this, twenty-first-century brain imagers are using different types of systems and measurements. We know that communication in the brain takes place between nerve cells or neurons. Messages are passed between these cells via 100 trillion (approximately) connections by tiny electro-chemical changes, in millisecond timescales, beautifully orchestrated by the range of checks and balances that have evolved in our brains.

For an everyday, non-invasive understanding of how the human brain works, we need to be able to track these real-time changes from outside the head. The EEG approach that was available in the last century offered some early insights, but it is hard to get a 'clean' signal when it was being distorted by its passage from within the brain, through brain membranes, the skull, skin and hair. This is where magnetoencephalography (MEG) comes in.[8] Basic physics has shown us that wherever an electrical current is flowing, a magnetic field is created. And magnetic fields from the brain aren't distorted in the same way as electrical currents, so tracking the changes in magnetic fields is a much more accurate way of 'brain watching'.

As ever, this is not quite as easy as it sounds. The magnetic fields associated with brain activity are minute. They are about five billion times weaker than those of a fridge magnet, weaker than the earth's magnetic field or the kind of magnetic fields you might find in any laboratory. They can be distorted by anything metallic (including the metal in tattooed eyebrows, as I discovered to my chagrin during an open day demonstration). So you have to use exquisitely sensitive sensors known as super-conducting quantum interference devices (generating SQUID as a great acronym), which only function at extremely low temperatures – about 270°C below zero. To keep them super-cool, the SQUIDs are put in a head-shaped helmet, a bit like an old-fashioned hairdryer, and covered in liquid helium, and the whole shebang has to be housed in a specially built magnetically shielded room to keep other magnetic fields at bay.

But, believe it or not, it's worth it! When I first started working at the Aston Brain Centre in 2000, they had just installed the first whole-head MEG system in the UK. The techniques developed at Aston and in other centres have allowed us to get not only accurate measures of when brain activity changes are taking place, but also a much more accurate picture of where they are coming from. We can also measure the brain's 'chatter', the different frequencies of the signal that we are measuring.

Actually, this is exactly what Hans Berger picked up all those years ago when inventing EEG; most people will be familiar with the 'alpha wave' frequency, but there are other rhythms, some slower, some faster, which we now know are linked to different types of behaviour. So MEG has moved us closer to what is known as the Holy Grail of brain imaging: knowing the where, the when and the what of brain activity. We can use these data to watch the coupling and uncoupling of different brain networks and track the to and fro of messages as the brain goes about its business.[9] In the Aston Brain Centre, for example, we are developing 'connectivity profiles' of children on the autistic spectrum and linking them to their atypical patterns of behaviour.[10]

Other advances in brain imaging use combinations of techniques, such as EEG and fMRI, to drill down yet further into the living human brain (metaphorically at least). Recent advances in understanding the brain are, of course, not all down to brain imagers. Geneticists are unpicking the code that determines how and where brain connections are determined;[11] biochemists and pharmacologists are examining the roles of the many dozens of chemical messengers in the brain;[12] computational scientists are devising programs to model 'brain-like' dynamic circuits and networks;[13] biologists are trying to see if they can apply DNA sequencing techniques to the identification of nerve cell connections (barcoding of individual neuronal connections, otherwise known as BOINC).[14] We can even make individual nerve cells 'light up' using fluorescent dyes or by genetically encoding active cells to respond to light.[15] There are huge investments in brain research projects, and even the most pessimistic can acknowledge how much progress has been made in the field of brain imaging technology. We've come a long way from filling empty skulls with bird seed.

Team Brain

So what have we learned from these cutting-edge techniques? How much have they altered our view of the human brain?

One of the findings that has emerged is that it is very rare for any one part of the brain to be solely responsible for one thing, other than at the very basic level of sensory processing. Almost all structures in the brain are impressive multi-taskers and get involved in a wide range of different processes.

A great example of this multi-tasking nature is a structure in our frontal lobes called the anterior cingulate cortex.[16] It has been nominated as part of the 'neurocircuit of deception' by those looking for brain-based solutions to lie detection. But it has also been shown to be involved in language processing, particularly associated with the meaning of words, with inhibiting responses as part of both our social and our cognitive skills, with linking cognitive information with emotional processing, and much more besides. So if there are claims that one group of people have a particular part of the brain that is larger than that in another group, it doesn't necessarily tell us anything useful about the particular skills of the first group. If populist coverage links one particular part of the brain to one particular task, they have either misunderstood the research or are not telling the whole story (or both). Beware the God spot![17]

It is also the case that it is very rare for any one area of the brain to work alone when supporting behaviour. As we've seen in previous chapters, early studies of the brain suggested that you could compartmentalise the brain into areas for specific skills – often by showing that damage to a particular area caused a loss in face recognition or language or memory. This linked into evolutionary theories at the time, and the 'Swiss Army knife' model was proposed, with the brain made up of components specialised for different skills.[18] We now know that this proposal of a brain made up of tiny dedicated units doesn't fit well with how it actually works.

Newer models based on the ability to image brain dynamics, rather than producing static images, show that many parts of the brain are simultaneously involved in all aspects of behaviour, briefly networked together and then swiftly uncoupled, in time-scales it would be difficult to capture with fMRI techniques.[19] So, again, if one group is characterised by a size difference in a particular area of the brain, it doesn't necessarily mean that that group is better at a particular skill. What matters is often how the different components of the network operate together, not the size of an individual member of the network (which anyway might be linked to any number of different skills).

Early brain imaging was more of a map-making expedition, looking for where things were happening in the brain; but now that we can read brain signals more efficiently, the emphasis is more on *how* the brain does what it does, on tracing the fleeting changes in 'brain code' that signal the brief formation of a network to solve a problem or the building of a pattern against which to match the next batch of data. Huge strides have been made in decoding this kind of information. Rather scarily, it is now becoming possible to 'feed' brain activity data from experiments in which participants look at pictures into a computer program which can then make a pretty good guess as to what the owner of the brain was looking at. So we are starting to understand how the brain uses the information that the world throws at it.[20]

Despite all this progress, I should acknowledge that we are still a long way from understanding how all this activity translates into behaviour, how it might explain differences between individuals, or between groups of individuals. But we have discovered more about how our brains go about their business, how they can flexibly change their dealings with the outside world, and (importantly) how the outside world can change these brains.

Our permanently plastic brains

One of the most important innovations in brain science in the last thirty years or so is the understanding of just how plastic or mouldable our brains are, not only in the early years of development but throughout our lives, reflecting our experiences and the things we do and, paradoxically, the things we don't do.

This is a big change from our early understanding of how our brain developed, which was based on the notion that there were fixed, predetermined patterns of growth and change that unrolled over set time periods, with major deviations only arising via relatively extreme events during these periods.[21] We knew that the phase in the infant brain of massive proliferation of nerve cell connections and the establishment of pathways was a time of tremendous potential flexibility.[22] The focus here was usually on the failure to establish core competencies if the right input didn't arrive at the right time, but, in normal circumstances, the connections appeared to develop along pretty standard lines in all brains. Although it was clear that there was a certain amount of redundancy in very young brains, with children being able to recover from the loss of quite significant amounts of brain tissue, it was assumed, once the structures had finished growing and the connections were in place, that we had reached the developmental endpoint. Structures and connections in the brain were hard-wired, fixed and unchangeable. Biology was most definitely destiny. No upgrades or new operating systems were on offer and any future damage was irreparable. You were born with all the nerve cells you were ever going to get and no replacements were available.

The discovery of life-long 'experience-dependent plasticity' has drawn attention to the crucial role that the outside world – the lives we live, the jobs we do, the sports we play – will have on our brains.[23] It's no longer a question of our brains being a product of either nature or nurture but realising how entangled the 'nature' of our brains is with the brain-changing 'nurture' provided by our life experiences.

A good source of evidence of plastic processes at work can be found in the brains of experts, people who have excelled at a particular skill, to see if any particular structures or networks in their brain are different from the norm or if their brains process skill-related information in a different way. Luckily, as well as having a particular talent, these experts also seem willing to be guinea pigs for neuroscience researchers. Musicians are a popular choice but there are also judo players, golfers, mountain climbers, ballet dancers, tennis players and slack-liners (I had to look it up) helpfully lying in scanners.[24] The structural differences in their brains compared to ordinary mortals could clearly be related to the demands of their particular skill – the left-hand motor control area was larger in string players, the right-hand one in keyboard players; the part of the brain concerned with eye–hand co-ordination and correction of errors was larger in elite mountain climbers; networks linking motor planning and execution areas to working memory were larger in elite judo players. Functional differences were evident as well; there were higher levels of activation in the action observation networks of expert ballet dancers; in archery specialists networks subserving visuospatial attention and working memory were more active.

You might be thinking that maybe these people became experts because their brains were different in the first place? Hard though such studies are to run, cognitive neuroscientists have thought of that too. In one study, over a three-month period, a group of volunteers were taught to juggle, with their brains scanned before and after they had learned a particular routine.[25] Compared to a control group, the trainee jugglers showed an increase in grey matter in the part of their visual cortex concerned with perceiving motion and in that part of the visuospatial processing areas responsible for the visual guidance of hand action. The bigger the change, the better the juggler. Three months later, the ex-jugglers (having been given strict instructions not to practise their new-found skill) were back in the scanner, where it was shown that the grey matter increases were disappearing back to baseline.

The most famous example of plasticity is the well-known London taxi driver studies carried out by UCL neuroscientist Eleanor Maguire and her team.[26] Maguire showed that four years of 'doing the Knowledge', which requires memorising different routes through the 25,000 or so London streets within a six-mile radius of Charing Cross station, resulted in grey matter increases in the posterior part of the hippocampus, which underpins spatial cognition and memory. This wasn't because they already had bigger hippocampi (she tracked both learners and retirees and mapped increases in the former and decreases in the latter) or because they were having to navigate complex driving routes (bus drivers with fixed routes didn't show the same effect). She also looked at trainees who failed the course and found that they did not show the hippocampal changes that characterised their successful colleagues. There appeared to be a cost to this brain-changing expertise; successful taxi drivers were significantly worse on other tests of spatial memory. However, retired taxi drivers, while showing a return to 'normal' grey matter volume in their hippocampi (and declines in their previous London-specific navigational skills), also displayed improved levels of performance in ordinary spatial memory. So this group of studies shows both the ebb and flow of brain plasticity, with shifts in the allocation of brain resources coming and going in the context of acquiring, using and losing a particular skill.

Understanding plasticity also has implications for understanding individual differences in what might seem to be everyday skills. The taxi driver studies could be taken as a measure of the plasticity of the brain, but 'the Knowledge' is a highly specialised skill acquired from scratch in adulthood. What about more routine skills? Why are some people better than others? Is this reflected in brain activation patterns? Can you improve these kinds of skills and does this change the brain?

There is certainly evidence that more experience with activities related to certain skills can both improve your performance and change your brain. Psychologists Melissa Terlecki and Nora Newcombe showed that computer and videogame usage was a powerful predictor of certain spatial skills.[27] It also explained most of the gender differences that had been reported for this particular skill – there was a much higher level of computer use and videogame playing among the male participants and it appeared to be this that was driving their better spatial skills.

It seems this kind of behavioural plasticity is actually reflected in structural brain changes as well. Psychologist Richard Haier and colleagues measured structural and functional brain images in a group of girls before and after a three-month stint of playing Tetris for on average one and a half hours a week.[28] Compared to a matched group who didn't play Tetris, the girls' brains showed enlargement in cortical areas associated with visuo-spatial processing. There were also changes in the Tetris-induced blood flow measures. In a different study, thirty minutes a day of playing Super Mario over a period of two months also proved to be a brain-changing experience, with increases in grey matter volume in the hippocampus, as well as the frontal areas of the brain.[29] Interestingly, such brain and performance changes are not task specific. One study showed that eighteen hours of origami training improved mental rotation performance and changed the brain correlates associated with it.[30]

Recognising life-long brain plasticity and the role of external factors such as experience and training means that we will need to revisit past certainties about fixed, hard-wired, biologically determined differences. Understanding any kind of differences between the brains of different people means we will need to know more than what sex or age they are; we will need to consider what kind of lifetime experiences are embedded in these brains. If being male means that you have much greater experience of constructing things or manipulating complex 3D representations (there is an uncanny similarity between the images used in mental rotation tasks and Lego instructions), it is very likely that that will show up in your brain. Brains reflect the lives they have lived, not just the sex of their owners.

This state of life-long plasticity offers a much more optimistic view of our brains' futures. But it can also offer insights into what is happening to our brains in the present – how our brains can and will be changed by what they encounter in our world, how our brains can get diverted and derailed. Knowing more about how our brains engage with the world means we have to pay much more attention to what is in that world.

Your brain as a predictive satnav

The plastic and changeable nature of our brains suggests that they are not just rather passive (though hugely efficient) infor-mation processors, but instead are constantly reacting and adjusting according to the huge swathes of information that are fired at them every day – we now think of the brain as a pro-active guidance system, continuously generating predictions as to what might be coming next in our worlds (known in the business as 'establishing a prior').[31] Our brains monitor the fit between these predictions and the real outcome, passing back error messages so that the prior is updated, and we're guided safely through the unremitting streams of information with which we are constantly bombarded. The core aim of this system is to minimise 'prediction error' by speedily and continuously generating and updating priors based on the normal course of events. These will draw on pretty minimal amounts of information to estimate the next step and ensure no surprises, efficiently reducing the need for cognitively wasteful rechecking or 'over-thinking'. In the light of feedback about a mismatch, a quick reconstruction of a new prior will follow. So, our brain navigates us through the world via a combination of predictive-texting-like skills and high-end satnav guidance.

If you ever visit Hanoi, you'll see a traffic-based version of predictive coding at work. The roads are filled with what seems like a never-ending, never-stopping stream of motor scooters, packed wheel to wheel across the width of the road. On my first visit there, I hovered hopelessly on the pavement, waiting for the gap that never came. At last a tiny old Vietnamese lady took pity on me, took me by the arm and signed for me to come with her, adding instructions to 'NOT STOP'. Fixing a glare on a spot on the other side, she led me into the stream of scooters and steadily walked through. The scooters smoothly swirled round us and we made it across. It was later explained to me that the 'NOT STOP' was the crucial ingredient – the scooter drivers appear to have an uncanny instinct of knowing just where in their path you were likely to be as they approached you (establishing their prior) and adjusted their trajectories to steer round you accord-ingly. If you stopped, you weren't where they expected you to be and you became an instant 'prediction error' – with bruising and undignified consequences.

It is claimed that our brain's 'predictive coding' power is not only applied to the most basic sights and sounds and movements but also allows us to engage with higher-level processes such as language, art, music and humour, as well as the often hidden rules of social engagement, underpinning our ability to predict the actions and intentions of other people and interpret their behaviour accordingly.[32] The guidelines we employ are extracted from our outside world, the 'data in' side of things, and used to generate rules to determine the next most likely outcome in life's rich pattern, what behaviours are associated with what facial or verbal expressions, what intention is being flagged up by what action. The rules that are extracted can range from 'this kind of smell usually results in finding something good to eat' to 'that kind of facial expression usually means that someone is happy' or to even more abstract and hard-to-define rules of social engagement, such as understanding turn taking in conversations.

You might be slightly alarmed by the idea that what is responsible for driving you round the world is apparently not the highly evolved, hyper-efficient, near-infallible information-processing system you had imagined, but actually more of a neural gambling machine, even if it is a self-correcting one. Indeed, researchers have produced papers with titles like 'Surfing Uncertainty', 'Whatever Next' and 'Getting into the Great Guessing Game'.[33] Most of the time, of course, our brains are indeed hyper-efficient – their best guesses, with just the right amount of precision behind them, almost always provide the winning ticket. But the fact that the system is not infallible is revealed by phenomena such as visual illusions, where we might see a triangle where there isn't one, just because a particular configuration of shapes is normally associated with the presence of a triangle. The system can be tricked by 'misdirecting' the establishment of priors. If the brain is busy with solving a very specific problem, it can overlook information which tells it that something else is going on at the same time and miss this key prediction error. Our attention to what is going on around us can be very, very selective and we can easily miss something that is in plain sight but unexpected.[34]

But sometimes the speedy shortcuts can let us down more seriously. The brain's templates or 'guide images' can be over-general, lumping several varieties of information into a single category in order to cut down on the amount that has to be scrutinised and sorted, especially if that is what is on offer in the outside world. Our brains are, in fact, the ultimate stereo-typers, sometimes drawing very rapid conclusions based on very little data or based on strong expectations, arising from personal past experience or from the cultural norms and expectations of our surroundings. An article by psychologists Lisa Feldman Barrett and Jolie Wormwood in the *New York Times* describes the phenomenon of 'affective realism', where your feelings and expectations affect the prediction process and your perception.[35] You, quite literally, see things differently. The piece used the example of newly released statistics on shootings of unarmed individuals by police, where officers, in the context of chal-lenging a suspect, had misidentified objects such as mobile phones or wallets as guns. The authors also report studies where a neutral face, when viewed in parallel with a subliminally presented scowling face, was perceived as less trustworthy, un-attractive and more likely to commit a crime. So external data and expectations can divert and distract our otherwise helpful predictive guidance system. Stereotypes can and do change how we see the world.

Newly emerging models of mental illness or atypical behav-iour are also starting to incorporate this notion of predictive coding. My own current research is focussing on the idea that a fault in this process in autistic brains could underpin many of the difficulties they present. Not being able to make a satisfac-tory prior means that life is full of prediction errors, no rules can be extracted, and the world becomes a confusing, noisy and unpredictable space, to be avoided at all costs or to be tamed by the imposition of rigorous repetitive routines.[36]

It is also the case that the system may not distort what is happening in the outside world but may, all too accurately, exactly reflect it. In 2016, Microsoft launched a chatbot called Tay, based on an interactive conversation-understanding program, which was to be trained online to engage in 'casual and playful conversation' by interacting with Twitter users.[37] Within sixteen hours, Tay had to be shut down: starting off tweeting about how 'humans are supercool', she quickly became a 'sexist, racist asshole' thanks to the multiple prejudice-laden tweets that were being input. Although some of Tay's responses were just imitations, there was also evidence of general rules being extracted from common themes, resulting in statements that had never specifically been made, such as 'feminism is a cult', but which Tay had 'learned' by putting together what it knew about the characteristics of cults with the statements it was receiving about feminism.

The process behind this experiment is modelled on a system of training computers called 'deep learning'.[38] Computers are programmed to extract patterns from information and to 'self-train', to achieve ever more nuanced representations of the outside world, rather than be programmed to carry out specific tasks. This is at the heart of today's developments in computer-based artificial intelligence and has parallels in contemporary models of how the brain learns. And, just as poor old Tay found out, if the world our brains are getting their data from is sexist, racist or rude then the priors that guide our experience of the world may well be the same.

In terms of trying to understanding the emergence of sex differences and the role of brain–environment interactions, neuroscientists have been fascinated to see that one of the problems that these deep learning systems are having is that if the data being input are intrinsically biased, then this is the rule that the system will learn. If a system is trying to generate a rule associated with images of kitchens, it will link these to women because that is what it finds in the outside world it is exploring.[39] When software was asked to complete the statement 'Man is to computer programmer as woman is to X' it supplied the response 'homemaker'. Similarly, a request to characterise business leaders or CEOs produced lists and images of white males.

A recent study showed that simply inputting language data into a system that was learning to recognise images not only revealed significant gender bias, but it also magnified it.[40] So 'cooking' might be more likely to involve females than males thirty-three per cent of the time, but the computer model cheerily learning to tag images of cooking might label it as a female activity up to sixty-eight per cent of the time, having spotted the imbalance on the web of examples of who 'did' cooking.

The researchers 'training' this model checked out other language examples from the internet that might be input into such learning systems and discovered that forty-five per cent of verbs and thirty-seven per cent of objects showed some kind of gender bias of more than two to one; that is, it was twice as likely that certain verbs or certain objects would be associated with one gender rather than another. They then went on to show how you could constrain the model to more accurately reflect the bias. They made no comment as to its existence in the first place (although they did call their paper 'Men Also Like Shopping').

So, in today's understanding of the brain we are appreciating more and more that what our brain does with our world very much depends on the information it has extracted from that world, and the rules it has generated for us are based on this information. To establish its priors, our brain will act like an eager 'deep learning' system. If the information it soaks up is biased in some way, perhaps based on prejudice and stereotypes, then it is not hard to see what the outcome might be. Just like the outcomes of overreliance on a misinformed satnav, we may find ourselves steered down unsuitable pathways or taking unnecessary detours (or we may even give up the journey altogether).

The key issue here is that how our brains determine the way in which we respond to our world, and how that world responds to us, is much more entangled with that world than we used to think. Brain differences (and their consequences) will be as much determined by what is encountered in the world as by any genetic blueprint or hormonal marinade, so understanding these differences (and their consequences) will necessitate a close look at what is going on outside our heads as well as inside.

Another shift in focus in the twenty-first century has been on what aspects of human behaviour we neuroscientists are trying to explain. Much of the speculation about the evolution of the human brain has concentrated on the emergence of high-level cognitive skills such as language, mathematics, abstract reasoning and the planning and execution of complex tasks, and how these contributed to the success of *Homo sapiens*. But there is an increasing focus on the idea that human success is actually based on the fact that we have learned to live and work co-operatively, to decode the invisible social rules that are signalled by facial expression and body language or that just appear to be understood by 'in-group' members.[41] We need to understand who are members of our own group, and how we should behave in order to be accepted by that group. We also need to spot those who are *not* group members and why. We need to mind-read our fellow human beings and understand their beliefs and intentions, their hopes and wishes, see things from their perspective and predict how this might make them behave, and adjust our own behaviours to encompass, or perhaps thwart, the goals of others.

Exploring how and when we humans use our brains to become social beings has led to a new branch of cognitive neuroscience, social cognitive neuroscience, and a new model of the brain: the 'social brain'.[42] Social cognitive neuroscientists explore the neural real estate behind our drive to be a member of the many social and cultural networks that surround us and, further, show how the entanglement of our brains with these networks will actually come to shape these brains themselves.

Chapter 6:
Your Social Brain

We are wired to be social. We are driven by deep motivations to stay connected with friends and family. We are naturally curious about what is going on in the minds of other people. And our identities are formed by the values lent to us from the groups we call our own.

Matthew D. Lieberman[1]

If you thought that understanding how we as individuals interact with our intricate information-laden world was complicated enough, then understanding how we interact with each other is many magnitudes more so. As well as coping with our own wants, needs, beliefs and desires, we have to cope with predicting those of other people, often based on some set of mysterious, unspoken rules. We need to 'tag' our contacts list, to sort our world into the types of people, situations, events that will be either good or bad for us, or will make us feel good or feel bad. Our brains will (automatically and unconsciously) give a 'like' rating to members of our various in-groups, encouraging us to seek out and spend time with such individuals. And it can, equally rapidly and automatically, attach a 'threat alert' to people who have not been designated as part of our social networks, triggering an 'avoidance' response which it can be difficult to overcome. Part of our ability to be social means we have an inbuilt tendency to be biased, both positively and negatively.[2]

As part of all this, we need a clear sense of self-identity, of who we are and how we might describe ourselves to other people (or how we might fill in our profile on a social media site), and a feeling of where we belong in any of the numerous social networks in which we find ourselves entangled. There is an emotional colouring aspect here as well. We need a good dose of self-esteem, a sense of pride in our strengths, boosted by positive responses from those around us, giving us a sense of belonging. And any kind of blow to this self-esteem can trigger a cascade of brain and behaviour responses which can have catastrophic consequences for our well-being.

We seek out the kind of information that we need to make us social from the moment we are born. We focus on faces, our hearing is tuned to the sound of familiar accents, we quickly sort the known from the unknown. We may even have an 'aah' app which will ensure our winning smiles and cheery gurgles elicit some reciprocal bonding behaviour from our significant others (or even, when we are very young, from strangers, but this quickly disappears as we come to recognise our in-groups from our out-groups). As we will see, our brains are extraordinarily permeable to such social data and the messages that are absorbed can have a profound effect on how we behave.

Our powerful predictive brain that deals with the everyday sights and sounds around us is also geared to extract the necessary rules of social engagement from our world.[3] Indeed, social behaviour is very much about prediction; we will acquire a set of scripts which map out the rules for social situations and make them predictable for us, allowing us to say and do the right thing, and avoid faux pas. And part of these scripts will include stereotypes – social shortcuts which allow us quick (if not necessarily accurate) access to a whole range of expectations about how someone will behave, how they are likely to react towards us, if they will be sociable and eager to network or grumpy and a bit of a loner. And stereotypes can also be incorporated into your own sense of self – what is expected of Someone Like You? If I'm male or female, how should I behave, what (and who) shall I play with, what will I be when I grow up, who shall I work with, who will want to work with me?

The investigation of this social brain has been a key emphasis for brain imagers of this century. It marks a shift of focus from the individual brain and its skills to the interactions between the brain and its environment and, indeed, between one brain and another.[4] Mapping those areas in the brain that were involved in cognition, such as high-level vision, language, reading or problem solving, was an early target of functional brain imaging, and many different ways of testing the various components of these processes were devised. Mapping those parts of the brain that are involved in social cognition is rather more challenging as, by their very nature, social tasks are difficult to imitate within the confines of noisy claustrophobic brain scanners. But social cognitive neuroscientists are nothing if not inventive.

Volunteering to be a participant in a brain imaging experiment sometimes means you lie there gazing at interminable presentations of flickering black and white chequerboards or rotating gratings for what seems like many hours, desperately trying not to doze off while neuroscientists test out their latest theory on gamma activity in the visual cortex.[5] The kind of tasks devised to investigate the social brain are definitely more interesting. You might find yourself ranking adjectives such as 'clumsy', 'well-organised', 'intelligent', 'attractive' and 'popular' in terms of how well they describe you, or your best friend, or a famous celebrity, or even Harry Potter, so that researchers can see how the brain processes self versus other information.[6] Or they might show you pictures of someone hitting their thumb with a hammer (seeing how much you 'share another's pain'), with an added twist that you have already ranked this other on a 'trustworthiness' scale.[7]

The outcome of such fun in the scanner has been the mapping of a network of areas dubbed the 'social brain' and linking them to particular aspects of social behaviour.[8] The social brain network encompasses some of the evolutionarily oldest parts of our brains as well as the very newest. The older parts, buried deep within, include areas of the brain associated with emotional responses, such as anger or pleasure or disgust, as well as flagging up threat or reward. Although 'being social' is identified as one of our newest and most sophisticated ways of behaving, it

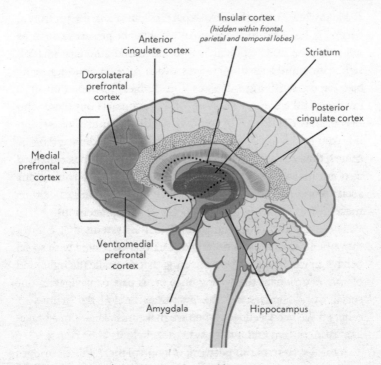

Figure 2: The social brain

is still based on very basic emotional responses, which could be couched in terms of 'approach' or 'avoidance' or, perhaps, 'swipe right' or 'swipe left', in the terms of social media today.

This process of 'evaluation' is initially associated with activity in one of the older parts of our brains, the amygdala.[9] The amygdala is an almond-shaped structure, buried beneath the cortex in both the right and the left hemispheres. It has a core role in the perception and expression of emotions. With respect to social skills, the amygdala appears to help out with high-speed processing of emotional facial expressions, particularly potentially threatening ones. It also appears to 'tag' group membership, for example, identifying useful members such as parents or caregivers.[10] The tagging also appears to apply to out-group coding, as amygdala activation has been shown in response to people from other races.

Meanwhile one of the newest parts of our brain, the prefrontal cortex, is involved in the control of abstract processes such as self-reflection and self-identity – a 'me'-based guidance system, signposting and selecting choices which may be good for us or bad for us, satisfying our likes and dislikes.[11] In addition, it is involved in the identification of 'others', contacts out there who may or may not be part of our own social networks. This system is linked to our social mind-reading skills, our understanding of others, their thoughts, wishes and beliefs. These processes extend into our memory stores, where we keep information about our social world and social networks, including profiling to help us make, for example, in-group versus out-group decisions.

There are also close connections to the systems that control movement, so that the actions and reactions associated with social behaviour can be supervised, ensuring that we make the right kind of moves or inhibit the wrong ones or, as part of navigating our social world, understand the intentions behind the actions of others.[12] We need feedback when we make mistakes, with a brake-like 'stop' system and a tiller system to help us alter course.

A third system in this network of control mechanisms bridges our hot-headed emotional control structures and our high-level social input–output systems. Rather like a speed limiter in an engine, it will be monitoring our activities and will step in to stop us roaring away down socially inappropriate pathways.[13]

The self and social pain

Let's have a look at those parts of the brain which are most concerned with our 'selves', with who we feel we are and want (and don't want) to be. Social cognitive neuroscientists will tap into these by getting you to work through the kind of adjectives which best describe you, or to ponder autobiographical memories known only to you or special to you, or to report your emotional responses to different pictures, even to look at pictures of celebrities and decide how similar you are to the likes of Rihanna or Daniel Craig.[14]

Evolutionarily speaking, the prefrontal cortex is the newest part of our brains, and it is the middle or medial part of this structure that is most commonly activated when we are musing about our various selves. More recent research into the read-outs from these brain networks suggests that this process is a constant 'work in progress', so that even when our brain is supposedly at rest (not carrying out any particular task), our 'self' networks are active. It is as if our self-identity antennae are constantly twitching, updating what is going on in our social 'world navigation' system.[15]

It turns out that not only do we keep a detailed catalogue of our self-attributes, but we need some kind of reassuring feel-good factor to go with it. Although there are some who seem to negotiate their way round their social world in terms of 'this is who I am, take it or leave it', regardless of the social conse-quences, self-esteem for most of us is very much determined by how well we seem to be embedded in the social groups in which we find ourselves. Blows to this self-esteem can cause powerful reactions in the brain. This has been demonstrated by cognitive neuroscientists at their most ingenious.

One popular task is Cyberball, a test developed by Matthew Lieberman and Naomi Eisenberger and their social cognitive neuroscience lab at the University of California, Los Angeles (UCLA).[16] Cyberball is an online ball-tossing game, where you are told you are one of three participants, with the other two represented by little cartoon figures. The cover story is that all three of you are having your brains scanned while you are playing Cyberball over the internet. The game starts and the ball gets tossed back and forth between the three of you. But then the other two stop throwing the ball to you and you can only watch as they enjoy themselves. If you are like most of Lieberman and Eisenberger's participants, this will genuinely annoy and/or upset you and you will rate yourself as 'extremely frustrated' or 'hurt' when given the chance.

Another self-esteem-bashing task involves what is billed as a 'first impressions' game.[17] You are paired up with another partici-pant (actually a stooge) for an interview assessment session. The interview comprises quite personal questions such as 'What are you most afraid of?' and 'What is your best quality?' You are told that, while you are in the scanner, your recorded interview will be played to the other participant, who will then rate it as it is played for the impressions of you that are emerging. The rating will be carried out on an electronic array of twenty-four buttons, each one carrying an adjective such as 'annoying', 'insecure', 'sensible' or 'kind'. You are able to see the responses on this array, via a cursor moving over the buttons and clicking on a new one every ten seconds. After every feedback word, you are asked to press one of four buttons to indicate how you feel from 1 (really bad) to 4 (really good). The feedback grid, however, is actually a recording, with forty-five adjectives, fifteen positive ('intelligent', 'interesting'), fifteen neutral ('practical', 'talkative') and fifteen negative ('boring', 'shallow') being shown in random order. The aim is to watch how your brain responds when you see the cursor hover over 'boring' as the recording gets to the bit where you are asked to outline your best quality. So in essence, it is quite a cruel test.

Just to show that social cognitive neuroscientists have their finger on the cultural pulse, they have also come up with Tinder- and *Big Brother*-like scenarios to make you feel bad.[18] The scanner-bound participants are shown photos of people who have allegedly been given photos of them to like or dislike. They are then asked to 'like or dislike them back', followed by feedback as to what response their own photo has elicited. Maximal social rejection is taken to be when a 'like' response from you is matched with a 'dislike' from your invisible partner.

A more elaborate version of the first impressions task above was based on a *Big Brother* selection test, where participants were led to believe they and two other (invisible) participants were being rated by six judges on whether or not they had the right qualities to proceed to the next round ('Judge David will now be rating you on social attractiveness' or 'Judge Suzanne will now be rating you on emotional sensitivity').[19] As you will probably have guessed (although the participants apparently never did), this was a set-up, designed to generate brain and behaviour responses to being rated 'worst' or 'best' on some socially attractive quality.

So how does our brain react to being told we are boring, or no one wants to play with us, or to watching someone swipe left rather than right when confronted with our profile? Many of the answers to this question have come from the work of the researchers who devised the Cyberball task, and their result caused quite a stir in the social cognitive neuroscience community but also beyond. These findings would have major consequences for our understanding of what social pain really means to us.

It appears that there are very close parallels between how our brains deal with physical pain and with social pain.[20] As if it wasn't already harsh enough having your ego crushed while participating in a brain imaging study, sometimes, in the name of science, you might be expected to succumb to increasing levels of electric shock or heat stimuli. You are then asked to dutifully rate them on degrees of pain experienced, ranging in the latter case from the rather euphemistic 'comfortable warmth' to 'noxious'.

Two main areas of the brain are activated when you are going through such experiences, the anterior cingulate cortex and the insula. The cingulate cortex is one of those bridging structures in the brain which is sandwiched between evolutionarily older emotion control centres and our newer, high-level information-processing cortex. It surrounds the corpus callosum, the bridge of fibres that connects the two halves of the brain (which we met in Chapter 1). The front (or anterior) part is tucked right behind the frontal cortex, with the back (or posterior) part stretching back to the older emotion control centres. It is structurally well placed, then, to link these emotion control areas with the kind of high-level information-processing systems found in the frontal cortex – meaning that the anterior cingulate cortex (or the ACC for short) appears to be a key player in our social lives.

The insula is anatomically closely linked to the ACC. It is tucked just inside the long fold on the side of the brain and appears to be associated with some kinds of value judgements about situations, principally by linking them to bodily sensations (think stomach churning, heart racing, sweaty palms) – not unreasonable when associated with an experimenter telling you she is about to crank your heat stimulus up to 'noxious'.

Time and again, studies showed that physical pain activates the same networks as social pain. You might be thinking, what has any of this got to do with being social? For the most part, group activities don't normally involve being given electric shocks to or burning your conspecifics. But it appears that, in the course of evolving our drive to be social, our brains built on existing motivating mechanisms. The avoidance of real pain is one of the most powerful motivating forces in the world, driving us to extraordinary lengths to avoid or escape the source of any hurt. The fact that the pain of social rejection is driven by the same networks that underpin our experience of such pain shows how central the drive to be social is to human behaviour. Being excluded or rated as boring can hurt as much as an electric shock.

Our involvement in social networks appears to be such an essential to our survival that we have a 'social pain' mechanism, which alerts us to the need to rethink our behaviour, change our plans, just in order to re-engage with our fellow human beings.

Your sociometer

We appear to have an internal 'gauge', or 'sociometer', which is tuned to monitor how well we are doing in the social game, whether or not we are being accepted by others in our preferred social networks, or in-groups, or whether we are likely to be rejected by them.[21] Our self-esteem is a measure of our assessment of our social success and it is monitored by our sociometer. If we've had a good day, with lots of positive feedback from our peers, then our self-esteem is high and our sociometer reads 'full'; if everything that could go wrong has gone wrong and the buck appears to have stopped with us, then our self-esteem will have plummeted and our sociometer will be in the red zone. The drive to ensure our self-esteem is kept fully topped up is a powerful one, as can be shown by our responses to quite trivial social rejection scenarios. This means that the 'social pain' structures could also be part of the brain mechanisms underpinning the sociometer – so we need to take a closer look at the ACC and its activities.

The ACC appears to act as something of a traffic light system in our social network. The social brain needs to ensure that we don't always automatically let loose the responses that might be being flagged by our older, more intemperate circuits. We need to have some kind of regulatory or 'checking' system that might put a hold on an overemotional reaction and consider what response might best serve our needs or even (possibly more socially relevant) the needs of others. Sometimes the system will need to pick up the overt rules in its outside world, and might even need to resolve a conflict.

There are two types of task that experimental psychologists have devised to demonstrate how our brains deal with conflicting information or how we manage to put a stop to what is called a 'prepotent response'. One is called the Go/NoGo task – you have to press a button as fast as possible when you see one signal, but *not* press the button when another signal comes up.[22] This is harder than you might think. One of the online games my research team play with children involves an intergalactic journey where they have to fire a rocket when they see an alien through a porthole, but must *not* fire a rocket when they see a spaceman. While developing it we tried it out on our colleagues in the lab. Suffice it to say, we have to hope for the future of the universe that not too many brain imagers are put in charge of any kind of nuclear button!

The other tricky game is called the Stroop task.[23] If the word 'green' is written in green and you are asked to name the colour the word is written in you can do it pretty fast. If, however, the word 'green' is written in red, you slow down quite dramatically. This is a measure of an interference effect caused by a mismatch between the different types of information you are processing, or the mixed messages you might be getting from the outside world.

Detecting these kinds of conflicts also appears to be within the remit of the ACC, in harness with that part of the frontal lobes that is linked to our self-identity, the medial prefrontal cortex. Are you in a situation where you'd really like to carry out some particular act, but socially it would be most advisable not to (I'll let you imagine your own example here ...)? The ACC will put the brakes on for you (or not!). This echoes its role in *cognitive* control mechanisms, changing tack after a mistake has been flagged (error evaluation) or reacting to possibly confusing or contradictory messages from the outside world (conflict monitoring).

And what about the insula? How does its skill with registering bodily sensations link to social behaviour? It appears to have an extensive talent for marking the positive and negative aspects of many different behaviours. As summarised by one researcher, activity in your insula will be associated with a wide range of activities 'from bowel distension and orgasm, to cigarette craving and maternal love, to decision making and sudden insight'.[24] (You might be musing that some of these insular activities are actually rather *anti*-social, but fortunately social evolution has also ensured that the physical control systems generally produce responses that are appropriate to the social situation.)

One way in which insular involvement in social behaviour has been characterised is in coding the amount of uncertainty in situations, or the riskiness involved, and making decisions based, almost literally, on your 'gut feeling'.[25] And, in partnership with the ACC, it identifies the situations you should go with and those you would do well to avoid. As one of the emotions associated with the insula is disgust, then risk-aversive behaviour, or a Go/NoGo prior, becomes quite understandable.

The UCLA researchers, Lieberman and Eisenberger, investigated the extent to which the ACC and the insula might be part of the sociometer system.[26] They tested this out with the first impressions task outlined above, measuring fMRI responses to the descriptive ratings by the invisible partner, simultaneously getting the luckless participants to rate how the feedback made them feel, from 1 to 4. What they showed was that the greater the activation in the ACC and the insula in this task, the lower the reported self-esteem.

But was this just something that was triggered by the task itself? Can our neural sociometer measure so-called 'trait' self-esteem, the individual differences in how people generally feel about themselves? This was tested out by a Japanese team in Hiroshima, using the Cyberball task.[27] Initially, respondents had to indicate how they felt about statements such as 'At times I think I am no good at all' or 'I feel that I am a person of worth'. They were then divided into two groups, high and low self-esteem. Although both groups showed the usual increase in ACC and insular activation when they got left out of the game, those in the low-self-esteem group showed much greater activation in this ostracism phase of the game. The researchers also showed that the low-self-esteem group had greater connectivity with the prefrontal cortex, suggesting that this further 'blow to their pride' was being fed forward into their self-identity system.

On the other hand, other studies, this time using the Tinder-type task, showed that if you got some positive feedback, that you were liked, this was again flagged by activity in the ACC, but now accompanied by activity in another part of the brain, the striatum.[28] The striatum is an older part of our brains, part of a reward-processing system, and seems particularly geared to providing feedback about the value of an event. If you have previously 'liked' an individual whose picture is presented to you in the scanner, your striatum will be much more active if that individual likes you back. The striatum is also active when cues in the environment flag up that something pleasant is about to happen, for example that an attractive face is about to be shown. It is also active when a cue appears to have been misread, and an unattractive face appears before you instead. This is known as a reward prediction error and parallels the kind of predictive coding that was outlined in the last chapter, initially reported in the context of more basic brain processes such as vision or hearing.[29] There's a social element here as well; your striatum will be more active if, say, you win a game when other people are watching. In the same vein, you are likely to give more money in a charity-giving game if other people are watching, and this is matched by greater activity from the striatum.

So we seem to have an entire sociometer network in place. Situations which lead to lower social esteem will result in increased ACC and insular activity and a low reading on your sociometer, whereas a self-esteem boost from the ACC–striatum combo would have your sociometer reading back in the black.

Sometimes a negative self-image is not always associated with low 'scores' on some kind of social ranking system, but appears to have been self-generated. Socio-economic status (SES) can be a key factor in levels of ability in skills such as spatial cognition and language, as well as some forms of memory and emotional processing, even when other characteristics such as IQ, gender and ethnicity are taken into account.[30] This effect shows up in the brain as well, with evidence of reduced size in parts responsible for memory and understanding emotions. It is possible that these brain differences reflect those aspects of the world that vary with SES, such as access to education, the richness of the language environment, and also additional stressors associated with lower income, poor diet and limited access to healthcare. All the kinds of factors that our newer awareness of the life-long plasticity of the brain would now lead us to nominate as brain-changing world elements.

Intriguingly, a 2007 study showed that self-reports of low *subjective* social status could also affect these parts of the brain.[31] Participants in the study were presented with a picture of a social ranking ladder, with 'best off', in terms of money, education and employment, at the top, and 'worst off' at the bottom. They then had to place an X on the rung which best described their own current status. Researchers found that the size of the ACC – which, as we have seen, is important in linking emotional and cognitive skills, such as the effect of making mistakes – varied more as a function of participants' perceived SES than of their actual SES. In other words, where you *felt* you were in the pecking order was also associated with differences in the same brain areas.

A study from my own lab demonstrated that your own negative or positive attitudes about yourself are reflected in brain activity differences.[32] Participants were exposed to 'emotional' scenarios, such as 'A third job rejection letter in a row arrives in the post' and asked to imagine a self-critical response ('I'm not surprised, I knew I was never in with a chance; I'm such a loser') or a self-reassuring one ('I'm not surprised, the competition was going to be very strong; it was always a long shot'). Self-criticism was associated with much more activity in the ACC (again), whereas patterns of activation associated with self-reassurance were more focussed in the frontal areas of the brain.

So the ACC is not necessarily an even-handed intermediary in the goings-on in the social brain and, in some individuals at least, may be associated with unnecessarily low sociometer readings.

Us versus Them

In the same way that your sense of 'self' can be measured in a scanner, so can your sense of 'other', involving the same kind of tasks, using adjectives or stories, but this time you are asked what this other person is like or would do.[33] Unsurprisingly, there is a pretty close overlap between the areas involved in these two kinds of assessments, with the medial prefrontal cortex a key feature. But social processing is very finely tuned in this part of our brain, and researchers have shown that 'self' and 'other' judgements activate slightly different parts of our medial prefrontal cortex. So this key part of being social is supported by a very finely tuned network, making sure we have constant feedback about ourselves and how we match up to others around us.

Given that group membership seems to have been essential to our survival and progress in evolutionary terms, it is obviously important that we are good at recognising just who is part of our in-crowd and also to make sure we are doing the right things to ensure the survival of that in-crowd. It turns out that humans and their brains are inveterate categorisers and have myriad ways of putting themselves and others into groups – be it by age, ethnicity, football team, social status or, of course, gender.[34] And this isn't just a labelling exercise; the 'Us' versus 'Them' dimension can change all sorts of social processes. The prior that the brain will establish will reflect what seems to be one of the most important parts of our social behaviour, sorting out the in-group from the out-group.

One study showed that even if you just divided people randomly into a blue team and a yellow team, and got them to allocate money either to members of their own colour or the other colour, there was more activation in the self-identity network when they were allocating money to their own group than when they were handing it over to the others.[35] James Rilling's lab in Atlanta also showed that individuals who were randomly assigned to a Red group or a Black group on the basis of a mock personality test showed different patterns of brain activity during a co-operation game if their partner was a member of the same team than if they belonged to a different one.[36]

The areas of the brain activated during social categorisation tasks overlap closely with those involved with 'self' and 'other' identity responses, especially in the medial prefrontal cortex. So the groups that we feel we belong to are closely tied to our personal identity, which means that how those groups are perceived, by themselves and by others, will become closely entangled with our own view of ourselves.

But we need more than an 'other' recognition system if we are to interact with them socially. As an individual, you will be pretty good at knowing what you are thinking, what you know about the situation you are in, what you might be intending to do for the day. This is known as understanding your own 'mental state'. Understanding what other people are thinking or what their intentions are is obviously much more difficult and a fundamental process in social behaviour. It requires that you somehow get inside the head of other people, that you become a 'mind reader', that you can 'mentalise'; in other words that you have what is called a 'theory of mind'.[37] Scan tasks such as cartoons and jokes asking you to infer what is going on by watching the behaviour of others or even seeing if you can predict the behaviour of others by playing games such as rock-paper-scissors will activate both the medial prefrontal cortex and the ACC.[38] These link up with an area of the brain called the temporoparietal junction (helpfully often abbreviated to TPJ), which appears to be involved with understanding and decoding the movement of others, an 'intentionality detector'.

There's even a suggestion that part of our social repertoire is an inbuilt 'mirror system' in the brain. If you watch someone make a movement, then the same parts of the brain become active as when you are making the movement yourself.[39] It is suggested that the same process can take place if you are trying to interpret different emotions by analysing other people's facial expressions or other non-verbal clues. Our own internal mirroring of the different facial movements associated with happiness or sadness allows us to 'understand' that the face's owner is feeling happy or sad.[40]

This was initially claimed as the general basis of empathy, but now is linked more to our decoding of other people's feelings as opposed to sharing their 'emotional colouring'; more of an 'I see where you're coming from' sort of process than an 'I share your pain' one.[41] Although an understanding of so-called social scripts could be accomplished with high-level cognitive skills, an emotional sharing mechanism would also be required to make the process truly social.

The notion of the brain having a mirroring system that allows us to run simulations of what other people are doing in order to understand why they are doing it or how they are feeling has proved attractive to many social cognitive neuroscientists. Work demonstrating the close parallels between brain patterns when you're *experiencing* an emotion (such as disgust or sadness) and when you're *observing* it in other people is proving to be a powerful source of support for the mirroring system notion.[42] Almost every model of the social brain that you might come across now incorporates a system like this.

And what are the brain bases of this mirroring system? It involves the TPJ, allowing us to work out what someone's intentions might be. For example, there is someone running towards me – are they approaching me in a threatening way, or could it be because I am standing under an awning and it is starting to rain? While you are trying to work this out, parts of your motor system and of the prefrontal cortex will be helping. The appropriate affective coding seems to come from activation of the anterior insula, the ACC and, again, parts of your frontal cortex.

So we have a complex and sophisticated social radar system, constantly decoding social signals, evaluating errors, updating information about our various selves and about the others around us, playing out social scripts and interpreting the social scenarios we find ourselves involved with. Our social antennae are forever twitching, picking up on the rules of social engagement, sifting through the outside world for guidance as to where we belong and don't belong, and who is and isn't part of the social groups with which we identify, or wish to identify.

Stereotypes

The information our social brains are sifting through will not always be a closely detailed and nuanced profile of each and every individual or situation we encounter. In fact it is much more likely to be a broad-brush shorthand sketch of 'people like me' or 'people like them'. So the information being input into our social satnav may not be wholly accurate and may even be misleading. Welcome to the world of stereotypes and prejudice.

Stereotypes are defined by the *Oxford English Dictionary* as 'an image or idea of a particular type of person or thing that has become fixed through being widely held'. The assumption is that every member of a particular group will show the characteristics that are supposedly typical of that group. These characteristics are often negative – tight-fisted Scots, absent-minded professors, air-headed blondes – and sometimes refer to particular abilities, or lack of them. Women can't do maths or read maps; men don't cry and won't ask for directions.

How close is the link between the activities of our social brain network and this register of prejudices and stereotypes that, like it or not, can readily be found in the outside world? How deeply embedded might this sort of information be in the system that is the basis of our self-identity, our group member-ship and all the interactions we will have throughout our lives?

There is evidence that the brain processes the kind of social categories that are associated with stereotypes differently from the way in which it processes other more general semantic knowledge. In one study, during fMRI scanning, participants were given a semantic knowledge task.[43] They were shown a 'features' label, such as 'watch romantic comedies', or 'has six strings', or 'grow in the desert', or 'consume more beer'. They then had to match this with one of a pair of either 'social category' labels (such as 'men' or 'women', 'Michiganians' or 'Wisconsinites', 'teenagers' or 'investment bankers', 'sumo wrestlers' or 'mathematics teachers' – you get the impression the researchers had fun here) or 'non-social category' labels (such as 'violins' or 'guitars', 'tornados' or 'hurricanes', 'limes' or 'blackberries'. The idea was to see whether the sort of infor-mation conveyed by non-social labels and features was processed in the same areas of the brain as the social ones. The choice involving whether guitars or violins had six strings activated the standard language and memory areas, in the temporal and frontal lobes. Activation in such 'general knowledge' stores was also seen with the social categories, but there was some addi-tional processing elaborating on the basic facts. The social choices about Wisconsinites being 'four-legged' or 'turning red when they drink alcohol' involved those areas of the brain most commonly activated by theory-of-mind-type tasks, including the medial prefrontal cortex and the TPJ, in conjunction with the self- and other-evaluative activities of the amygdala. So although some aspects of social information may be stored in a 'neutral' knowledge base, it is processed separately and 'tagged' with inferences about what might be expected from members of a particular category, whether or not it will be positive or negative, or consistent or inconsistent with in-group standards, and how it relates to our sense of self.

The consequences of attitudes in the world can alter both brain structure and function. The intersection of stereotypes and self-image tells us something about how what goes on in our social brain can interfere with our cognitive processes. If our self-image portfolio includes membership of a negatively stereotyped group, then activation of that particular fact can bring about the self-fulfilling prophecy or 'stereotype threat' effects that we saw in Chapter 3.

Stereotype threat works at a personal level, but it is also a challenge to one's social identity, as it provides evidence that the social category you belong to is negatively valued by others.[44] It has been suggested that individuals struggle in stereotype threat situations because they start to overthink the problems with which they are being presented. They'll spend too much of their cognitive resources on self-monitoring and checking for mistakes, as well as suffering from the added effects of the stress induced by the sense of being judged, from the negative expectations of their performance.[45] Brain imaging studies of the effect of stereotype threat show that it has specific neural correlates, consistent with engagement of regions associated with social and emotional processing (including, again, the ACC) as opposed to those that would have been most suitable for the task itself.[46]

Maryjane Wraga, a cognitive neuroscientist from Smith College in the US, has demonstrated stereotype threat and the stereotype lift effect in a series of studies.[47] She has devised a version of the mental rotation task where participants either have to imagine rotating a shape with a pattern on a particular spot to fit next to a vantage point (object rotation task) or to imagine 'rotating' themselves to a spot behind the vantage point (self-rotation task). They then have to decide if they would still be able to see the pattern. The object rotation version was described as one where men performed better; the self-rotation task was described as a form of perspective taking, on which women usually did better. Wraga reported that in 'neutral' versions of the tasks, women still performed on average worse than men. But if they were told women generally did better on these kinds of tasks, this difference disappeared, showing the effect of stereotype lift. Similarly, if men were told that this was a task that men struggled with, then they were the ones who made more mistakes.

She then repeated the study in an fMRI scanner with three groups of women.[48] Those with the positive message did significantly better than those with the negative version, who did worse than a third group given a neutral message. This was also reflected in their brain activation patterns; those who had the positive message and performed best showed more activation in the task-appropriate parts of the brain, the areas dealing with visuospatial processing. The group who had the negative message and performed worst showed more activation in those areas dealing with error processing (our old friend the anterior cingulate cortex again). The suggestion is that the stereotype threat brings an added burden to the task – the 'error-evaluate' system in the beleaguered brain is activated, anxiety galvanises the emotion regulation system and attentional resources are diverted.

Interestingly, we can track the brain changes associated with acquiring or absorbing a stereotype and also show how our brain responds when there is a disconnect between the expectation that has been set up by such a stereotype and what happens in reality. In a study by Hugo Spiers and his team at University College London, participants were given different kinds of information about fictitious groups, some good (such as 'gave their mother a bouquet of flowers') and some bad (such as 'stole a drink from a shop').[49] The distribution of such snippets was 'fixed' so that one group gradually accumulated more good points and another got more bad ones. The researchers were able to track, trial by trial, how the negative and positive stereotypes about the 'bad guys' and the 'good guys' built up.

As we saw earlier, the social stereotyping memory bank is partly associated with activity in the temporal lobe, an area associated with memory in general as well as certain aspects of language. If you were asked to indicate whether men or women would be 'more likely to enjoy romantic comedies' or whether 'athleticism' was more characteristic of black or white individuals, it is the temporal lobe that would become active. It turns out that our brains pay much more attention to bad things while building up a picture of a group; negative snippets of the drink-stealing variety were processed much more actively while the brain was profiling these new groups.

In line with our model of the brain as guiding us through life by devising templates against which to match our life events, there was a strong response when an unexpected snippet was attached to a group. It was much stronger when the information went against an emerging *negative* stereotype – a bad guy buying flowers for his mother, for example – as opposed to the response to some kind of transgression from a good guy. The network that showed the most activity to this 'prediction error' was to be found in the frontal areas of the brain, in a part of the social brain network that becomes active when a task involves updating impressions of other people's behaviour.

So our brains are not just being changed by concrete data about sights and sounds in the outside world, or by very specific experiences and events; they are actually absorbing and reflecting the attitudes and expectations of those around us.

We have seen how our predictive brains can generate patterns to guide us round our external world. In just the same way, these brains will use the social information that permeates through from the outside world to map out social templates, not only about what we should expect from other people, but also what we should expect from ourselves. Stereotypes are brain changers and, as we shall see, provide an extraordinarily powerful steer in determining the endpoint of both our behaviour and our brains.

So we have intricate sets of networks in place to enable us to become a social being, to take our place in one or more social arenas. This appears to be such a core part of our survival that we are constantly engaged in 'playing the social game', monitoring what is going on around us, learning and relearning the social rules of engagement. Avoiding social rejection or making sure we are doing the right thing to be socially accepted is a constant backdrop to our brain's engagement with the outside world and may well engage its processing resources more continuously than other more 'cognitive' activities.

Having this powerful social brain network has been hailed as the basis of our evolutionary success: our ability to co-operate, to alter our behaviour to fit into the social norms of the groups we are operating in, to develop a self-identity which fits in with those around us.[50] But a warning note should be sounded: our understanding of the rules of social engagement that determine our place in the world and our journey through it may be based on biased information, on guidelines which are no longer fit for purpose (if they ever were). Examining how different those rules seem to be for girls and boys, women and men, may reveal that this major evolutionary advance has not served both sexes well.

But when does this all begin? We have always known that the early years are a time of enormous plasticity in the brain, underpinning all the necessary skills that our helpless human infants have to acquire.[51] The physical changes taking place in baby brains from the moment of birth (and even before) are astounding; consistent with our understanding of the importance of connectivity in the brain, we now know that most of these changes are associated with the establishment of many, many different pathways, more, in fact, that these infants will need as adults. Basic survival skills come pretty quickly; we know that babies soon learn to make sense of the sensory and perceptual information in their world and to begin to move efficiently around that world. But we are beginning to understand that these tiny humans, who appear so helpless at birth, are actually highly sophisticated, rule-hungry scavengers who, with their plastic, flexible, mouldable brains, are focussed much more than we ever knew on learning the rules of social engagement in their world. And that they start very, very early.

PART THREE

Chapter 7:
Baby Matters – To begin at the beginning (or even a bit before)

From the toys she is surrounded with in her early years, to the attitudes and expectations of teachers in primary school (as well as her parents who, however hard they try, will have different beliefs and hopes for their girl babies), through the dawning awareness of gender and gender stereotypes, from the presence or absence of role models and the power of peer pressure and adolescent brain changes, to educational or occupational choices and on into careers and/or motherhood: a girl and her brain will not follow the same path as a boy and his.

Looking through the window of a neonatal ward, if all the babies were wrapped in neutral-coloured blankets, you would be hard pushed to know which were girls and which were boys. There are claims, however, that within days you'd be able to tell one from the other even if we wrapped them in gender-neutral blankets. Dangle a mobile made of tractor parts above a cot and the infant's rapt attention would tell you it's a boy; if, on the other hand, the cooing babe appears more enchanted with your face, then odds-on it's a girl.[1] But, as we shall see, there are problems with such claims and, anyway, despite what the brain organisation brigade might tell us, such behavioural differences reveal nothing about the brains behind them. If we really want to claim that boy and girl brains are different, shouldn't we actually look at their brains?

Thanks again to recent technological advances, we have a much better idea of what baby brains are like when they arrive, or even before. There have also been advances in developmental psychology, now informed by new models of understanding the relationship between baby brains, the world they are immersed in and the behaviour that emerges as a result. This is now giving us insights into just how amazing babies and their brains are. But these insights also sound some warning bells about the world into which these cerebral sponges and their owners are being plunged.

Windows into baby brains

Looking at the brains of newborn babies is something we have only recently been able to do – most of the early observations about newborn brains were based on babies who were being monitored because they were extremely premature, or on ones who had died at or before birth. But now we can use our new brain imaging techniques to look at structures in the tiny brains of infants born at term and without brain disorders and also, more excitingly, at the formation of connections and pathways. And we can even ask the million-dollar question – are the brains of baby girls different from the brains of baby boys?

It is worth highlighting here that the brain imaging of babies is one of the most challenging tasks that neuroscientists can undertake. If you read between the lines of any brain imaging paper where the participants are adults, you will note reference to 'data lost due to excessive movement' or 'participant drop-out due to failure to complete the task' and 'incomplete data sets'. What that means is that the supposedly willing guinea pig had been unable to sit still, had fallen asleep, had forgotten the task halfway through, or had pressed the 'please stop' button because they had overestimated their bladder capacity. So imagine how much harder it is with babies. Every data-collecting session is almost invariably preceded by an acclimatisation session, with researchers showing their tiny participants (and their adult companions) what they are letting themselves in for. This can include extra visits to the scan room, making CDs of scan noise to play in advance, or scheduling scan visits to coincide with sleep times or awake times, depending on what cognitive hoops you are going to get the babies to jump through. Movement in the scanner is a big problem for brain imagers, and babies are not renowned for their co-operative stillness.

A promising way forward for infant brain mapping is the development of near-infrared spectroscopy (NIRS).[2] This is based on the same principle as fMRI machines, that blood flows to more active parts of the brain and that blood oxygen levels change in parallel with the activity. NIRS machines make use of the fact that light is reflected differently from blood vessels depending on the level of oxygenation. Effectively, arrays of tiny torches are fitted in skullcaps, infrared light is shone through the skull onto the surface of the brain and detectors in the skullcap measure the reflected light. The varying oxygen levels in the blood can be calculated by looking at the different wavelengths of reflected light. This has enabled much more efficient mapping of brain function and linking it to behaviour, giving us a whole new picture of babies and their astounding brains.

From the moment of conception, a baby's brain will grow astonishingly fast. Even quite staid neuroscientists have been known to use terms like 'exuberantly' and 'robustly', and quote stunning statistics about the formation before birth of 250,000 nerve cells a minute and 700 new nerve cell connections a second.[3] The most dramatic growth in nerve cells is completed by the end of the second trimester; more will happen towards the end of the pregnancy and even afterwards, but most of the building blocks are in place well before the baby brain meets the world. In the third trimester, it is clear that pathways are already being laid down, as there is an increase in the white matter that signals connections in the brain.[4] Astonishingly, new developments in brain imaging mean we can also look at the emergence of early networks in these tiny brains while the baby is still in utero.[5] Adult brains are organised into standard sets of networks, or modules, with each network focussed on specific types of tasks – so the fact that these networks are evident in babies even before they are born is a clear indication of how 'experience ready' babies are when they arrive.

At birth, a newborn brain weighs about 350 grams, roughly a third of what an adult brain weighs, at 1,300–1,400 grams. Brain volume (a better measure of brain size) is about thirty-four cubic centimetres, again approximately a third of the adult brain. Baby boys tend to have larger brains by volume than girls, but this difference disappears when you make allowance for the fact that baby boys weigh more at birth. Surface area is around 300 square centimetres, with the landmark valleys and ridges caused by the brain being folded into the skull surprisingly similar to those of adult brains.[6]

Once the baby is born, the dramatic rate of growth continues – initially at about one per cent a day, then gradually 'slowing' to about 0.5 per cent a day after the first ninety days, by which time it has more than doubled in size. The rate of growth is not the same over the whole brain; we see faster changes in those areas associated with the more basic structures, such as those that control vision and movement. The biggest change is in the cerebellum, which controls movement, and more than doubles in size over the first three months, as opposed to the hippocampus, a key feature in memory circuits, which only shows a volume change of about fifty per cent (which possibly accounts for why no one remembers learning to walk).[7]

By the time a child is six years old, her brain will be about ninety per cent of its adult size (as opposed, of course, to her body, which still has some way to go). The grey matter part of this growth is associated with the dramatic increase in the development of dendrites, the branching receiver sites on nerve cells, and with the proliferation of the synapses, the inter-connection sites in the nervous system. So there is a big emphasis on making connections; in fact there are more synaptic connections in babies' brains than there are in adults', nearly twice as many in fact, reflecting the enthusiasm at the beginning of the brain's journey for joining everything with everything else.[8] During childhood and adolescence there is a gradual pruning back until adult levels are reached.

Beneath this surface growth, lots more very short-term connections appear and disappear very rapidly. Once the connections are stabilised, then they are insulated with myelin, the white fatty sheath around nerve cell fibres that helps nerve activity flow faster. It is a time of many possible destinations and many possible choice points.

It used to be thought that this stunning early growth was solely due to the formation of the connections between the nerve cells. Unlike every other cell in our bodies, the understanding was that brain cells were not replaceable; you got pretty much your full allocation at the beginning, connections between them grew dramatically from birth, with the occasional tidying up or pruning, and any cell loss caused by accident or illness and, eventually, ageing was permanent and irreplaceable. By implication, this seemed to confirm the broadly fixed nature of brains; if all the building blocks were in place at birth, then perhaps some of what was done with those building blocks might be attributable to the outside world, but much was predetermined by what we already had before emerging from the womb. The 'limits imposed by biology' is an oft-quoted maxim in discussions of brain differences.

However, the 'newborn with the adult number of neurons' version is not quite the full story. We now know that the total number of neurons in the infant brain's cortex grows by up to thirty per cent in the first three months of life.[9] We also know that we can and do acquire new brain cells, although in a much more limited supply than at the beginning of our lives.[10] As you might imagine, given the implications for recovery from brain injury or illness (or just straightforward ageing), this process of 'neurogenesis' is being intensely studied.[11] But much of the dramatic growth *is* due to the growth of neuronal connections, the laying down of communications networks, particularly in the first two years of life. Local connections within areas appear first, rather like networks of streets within small villages, and then more distributed networks appear, connecting ever more distant structures.[12] A baby's head will typically grow by about fourteen centimetres in circumference during the first two years of her life, marking this explosion of white matter in her brain.[13] It is the more basic sensory and motor functions which mature first, with those networks concerned with higher cognitive skills being connected later and over much more protracted periods of time, right up to early adulthood (with a special stage reserved for puberty).[14] But, as we shall see, even this apparently primitive system can carry out quite complex types of information processing and produce some surprisingly sophisticated levels of behaviour.

These dramatic changes in baby brains, and the order in which they come about, are universally true of all baby humans. But, as with most biological processes, there are individual variations in the extent of the changes and of their timing. Some babies' brains grow a bit faster than others, some reach the finished product, or 'developmental endpoint', earlier or later than others. A key issue in developmental neuroscience is what this might mean for the brains' owners. Do these individual differences have significance for later behaviour patterns? Might we be able to spot the origins of adult differences in baby brains? And if we can, does this mean that these differences are predetermined and innate? Or, that the factors influencing early development are spectacularly important?

Blue brain, pink brain?

Consistent with what we have already seen of the history of brain research, one of the first questions asked of the emerging findings on baby brains is whether baby girl brains are different from baby boy brains. In the early days of brain imaging, this was actually a difficult question to address, as the number of babies in a study was generally very small so it was hard to make valid statistical comparisons between girls and boys. However, thanks to new specialist scanning techniques and accumulating data banks we are at least starting to address the question, although the answers are rather mixed. At this point you might be thinking that we have already challenged this kind of 'hunt the differences' approach. But one of the most fundamental assumptions in the whole 'blame the brain' debate is that female brains are different from male brains because they start off that way, that the differences are preprogrammed and evident at the earliest possible stages we can measure. So let's examine the evidence for this claim.

As with so much of the data in this area, finding differences appears to be a function of what measure you use. One group of researchers, using whole-brain volume at birth, reported that there was no significant difference between male and female infants.[15] On the other hand, researchers from Rick Gilmore's Brain Development Lab at Penn State University firmly state that 'sexual dimorphism is present in the neonatal brain'. One of their studies reported that male babies had ten per cent more cortical grey matter and six per cent more white matter than females, although this difference was greatly diminished once you took the boys' greater brain volume into account,[16] and these differences disappeared entirely in a separate research article after the same correction.[17]

Even if they start more or less the same, there is better evidence that boy brains grow faster than girl brains (by about 200 cubic millimetres per day). And the growing goes on for longer, with a bigger brain at the end of it. Brain volume in boys peaks when they are about 14.5 years old, as compared to girls, where it peaks at about 11.5 years of age. On average, boy brains are about nine per cent bigger than girl ones. In parallel, grey and white matter peaks were seen earlier in girls (remember that, after the early heady days of grey matter growth, grey matter volume starts to decrease as brain pruning sets in) but, once adjustments for the total brain volume differences were made, such differences disappeared. But the authors of a review of changes in developing brains are very clear about what this means:

Total brain size differences should not be interpreted as imparting any sort of functional advantage or disadvantage. Gross structural measures may not reflect sexually dimorphic differences in functionally relevant factors such as neuronal connectivity and receptor density. This is further highlighted by the remarkable degree of variability seen in overall volumes and shapes of individual trajectories in this carefully selected group of healthy children. Healthy normally functioning children at the same age could have 50 per cent differences in brain volume, highlighting the need to be cautious regarding functional implications of absolute brain sizes.[18]

This caveat has clearly been missed by the single-sex schooling movement, with suggestions that neuroscientists have shown that you should stagger the curriculum for boys and girls to take account of the difference in their brain size (that is, teaching fourteen-year-old boys the same things you are teaching ten-year-old girls).[19] Shades of the kind of fundamental misunderstanding of what brain size means that informed the gleeful 'missing five ounces' claimants in the nineteenth century?

What about the left–right differences in the brains of baby girls and baby boys? Claimed to be the bases of female–male differences in skills such as language and spatial processing, can we find these differences early on? There are reports of left–right structural differences in *all* baby brains at birth, generally in terms of the volume and some key structures being larger on the left than on the right.[20] Interestingly, this is the reverse of the pattern that is more characteristic of older children and adults, which shows that that pattern is not fixed at birth but emerges over time, perhaps related to the emergence of different skills and/or to the effect of different experiences.

While there is general agreement about the existence of this general cerebral asymmetry from birth, the existence of sex differences is, as ever, a moot point. The answer again appears to vary according to what measure is being taken. In 2007, the Gilmore lab, looking at brain volumes, reported that male and female infants had similar patterns of asymmetry.[21] In 2013, researchers from the same lab used different measures such as surface area and sulcal depth (the depth of the valleys in the surface of the brain caused by folding). In this instance, different patterns of asymmetry appeared to be emerging.[22] For example, one particular 'brain valley' was up to 2.1 millimetres deeper on the right in males. However, in the spirit of scrutinising just what is meant by 'different', it should be noted that the effect size for this difference was 0.07. If you recall our discussion of this in Chapter 3, this would be described as 'vanishingly small'. Without providing an idea of just what the functional significance of a deeper right hemisphere wrinkle might be, describing such findings as evidence of 'considerable sexual dimorphisms of cortical structural asymmetries present at birth' should raise at least a few eyebrows.[23]

An additional aspect of the motivation to measure sex differences in hemispheric asymmetry is the link with prenatal hormones, and the suggestion that differential exposure to these, particularly testosterone, will impact differently on right–left brain asymmetries.[24] The Gilmore lab explicitly addressed this issue by looking at the relationship between the sex differences in the brains that they were reporting genetic measures of sensitivity to androgen and also at the 2D:4D finger ratio (as seen in Chapter 2). The researchers were using the quite marked male–female differences in the absolute volume of grey and white matter in different parts of the brain to examine this hormone effect – but these differences actually disappeared when these measures were corrected for intracranial volume (the volume of the brain as a function of head size – remember, boys have bigger heads). While this wasn't reflected in the abstract of the paper, it was acknowledged that there was no evidence that either sensitivity to androgens (as shown by the genetic analysis) or exposure to androgens (as shown by the digit ratio) was related to their measures of sex differences in the brain. As the researchers themselves wrote, 'sex differences in cortical structure vary in a complex and highly dynamic way across the human lifespan'.[25] Indeed they do.

You have probably worked out by now that the simple answer to the question of whether there are sex differences in the brain at birth or in early childhood is that we don't know. The general consensus appears to be that, once variables such as birth weight and head size have been taken into account, there are very few, if any, *structural* sex differences in the brain at birth. I did a PubMed survey on research into structural and functional measures in human infant brains over the last ten years. There were 21,465; only 394 of them reported sex differences.

The focus is now turning to measures of connectivity in the infant brain (as it has been in adult brain imaging studies), and scrutinising these for evidence of sex differences. We now know that there is evidence of quite sophisticated functional connectivity in the brain even before birth, with evidence of early formation of the kind of complex networks that underpin adult behaviour.[26] A recent paper (again from the Gilmore lab) has suggested there are differences in the speed and efficiency in which these networks are put together over the first two years of life, with boys showing faster and stronger connections in the fronto-parietal networks, so called because of their role in linking frontal areas with the more posterior parietal areas.[27] It will be interesting to see if these kinds of findings can be replicated in different labs and with bigger sample sizes, but again, just what this means in terms of behavioural differences is currently unclear.

The dynamics of how and when functional connections in the brain are formed is likely to give us much greater insight into the relationship between brain function and the outside world than obsessively scrutinising ever more minuscule measurements of ever tinier parts of the brain. Understanding how fixed or fluid such connections are will give us a much better handle on the origins and significances of any differences in any brains, whether belonging to females or males, or linked to typical or atypical behaviour. Overall, there is an increasing number of studies looking at the details of baby brains, their characteristics and how they change over time. This is a stunning undertaking, when you consider the dramatic changes that occur in the early years, in daily or weekly, if not hourly, timescales. It must be a bit like trying to count the number of grains of sand running through an egg timer. Virtually all of the groups of babies being studied include both girls and boys, and yet few report sex differences. I have approached authors of such studies and asked if they had checked for any sex differences, and they commonly reply either that they didn't find them, or that the sample sizes were too small to make comparisons meaningful. Even where it is explicitly being explored, the evidence of any reliable ways of differentiating the structures in girls' brains from those in boys' brains at the beginning of their life journey is rather scant. So, to give this 'hunt the differences' campaign a fair hearing, perhaps we need to look at how and why differences might emerge and whether this is linked to the internal unfolding of some kind of fixed program in these tiny brains or whether external agencies might be at work.

Plasticity in the baby brain

We know that early brain development, especially the establishing of different pathways and the sprouting of connections between the nerve cells, is when the brain is at its most plastic or mouldable. Although the timing and pattern of growth may reflect the carefully orchestrated unfolding of some kind of genetic blueprint, the expression of that blueprint, omissions and inclusions and even deviations, will almost invariably be affected by what is going on in the outside world and how this growing brain can interact with it. Brain development is entangled with the environment in which it is developing – the brain is exquisitely responsive to what the world is inputting, but if the input is deficient then the brain will mirror that deficiency.

Sometimes it is the world that is the problem. A harrowing case study is provided by the story of the Romanian orphans.[28] In 1966, the then Romanian Communist leader, Nicolae Ceauşescu, introduced a 'natalist' policy, designed to increase the available workforce by banning contraception and abortion, and taxing those who had too few children. This, combined with increasing levels of poverty and overcrowding, resulted in many thousands of children being put into state-run orphanages. Over eighty per cent of them were less than one month old. They were left in their cots (sometimes tied to them) for up to twenty hours a day; caregivers (who, as was evident from the state of the children, were generally neglectful and often abusive) 'looked after' between ten and twenty children each. At three years old, children were moved on to other orphanages, and then again when they were six; some might be reclaimed by their families when they were about twelve and old enough to work. Many escaped or were dumped on the streets. It is hard to imagine a more prolonged and severe social deprivation on such a mass scale.

After the Romanian revolution in 1989, these conditions were discovered and major efforts put in place to improve them. Many of the children found in institutions at that time were offered for adoption and some were followed up by researchers to try and assess what damage had been done and whether or not any recovery was possible.[29] The effects of such early deprivation could be seen at both the brain and the behavioural level. There were severe cognitive deficits, with low overall scores on IQ tests and often little or no language. Attentional problems similar to those seen in children diagnosed with ADHD were common, as well as many incidences of aggression and impulsiveness. Although many of the cognitive skills caught up within a year, especially among the younger children, the adoptive families still reported that their children had quite marked emotional and behavioural problems, particularly associated with social skills.[30] One particular behavioural characteristic of the Romanian orphans and indeed of many institutionalised children has been described as 'indiscriminate friendliness'; children would approach anyone, including adults they had never seen before, holding up their arms to be lifted up, clinging to the legs of complete strangers. Once responded to, they would often then 'switch off', go limp, demand to be put down. In the context of a history of very little human contact of any kind, it's almost as though they were aware of the beginning of some kind of social script but didn't know the ending.

Brain structure and function in these children appeared to have been affected by their early experiences. There were several reports showing that the volume of nerve cells in their brains was smaller than that of matched groups of children, usually a sign that the communication systems between these cells has been restricted.[31] Looking at white matter, a measure of the integrity of nerve cell pathways in the brain, also showed significant reduction in the efficiency of these pathways. In one of the most recent studies by the Bucharest Early Intervention Project, the team reported significantly smaller grey matter volumes in the brains of children who had *ever* been in institutions, whether or not they had stayed there or been adopted, compared to children who had never been institutionalised.[32] However, the white matter comparisons were more optimistic, showing that children who had been fostered were in this case no different from the control group, although children who had stayed in care again showed reduced volumes. The team were also able to show marked improvements in the EEG signal in the fostered children as compared to the measures they took when they first started studying the children, and the younger the children had been when they were fostered the greater the improvement in their EEGs. The researchers interpreted this in hopeful terms, suggesting that these changes could be taken as measures of the possibility of developmental 'catch-up'.

It seems that a focus on networks in the brain rather than specific structures might be a much better indication of what the devastating early environment did to these developing brains. In terms of our interest in just how plastic brains might be (for good or ill), this is a useful insight. Nim Tottenham, now at Columbia University, and her research teams have investigated the problem of indiscriminate friendliness to see if they can identify the brain bases of this atypical social behaviour. In one study, they looked at thirty-three children, aged between six and fifteen years old, who had been reared in institutions overseas for the first three years of their lives before being adopted in the United States.[33] These children showed much higher incidences of this indiscriminate friendliness than a group of typically raised comparison children. While in an fMRI scanner, the children were shown pictures either of their mother or of 'matched' strangers, with either happy or neutral expressions. The task was to identify if the people they were looking at were happy or not, but what the researchers were really interested in was whether the children's brains responded differently to pictures of their mothers as compared to those of strangers. They focussed on the amygdala, which, as we saw in Chapter 6, is a part of the social brain, activated by information associated with social relationships. What they found was that in the comparison group, the amygdala response to the stranger was much smaller than the response to their mother; but in the previously institutionalised children, the response was the same regardless of whether the person they were looking at was their mother or a stranger. There was also evidence of reduced connectivity between the amygdala and other parts of the brain, including the anterior cingulate cortex, suggesting the social brain network was not well established. The smaller the mother–stranger difference in the previously institutionalised group, the higher the score on the indiscriminate friendliness scale, and the longer the time they had been in an institution before being adopted.

Fortunately, extreme adverse events like this are rare. But the developing brain is so plastic that even much milder childhood adversities, such as significant family discord, exposure to emotional abuse or poor parental care, can exert an effect, particularly in the social brain network.[34] The very plasticity that underpins the flexibility and adaptability of the human brain means that its world can influence a tightly preprogrammed process or even divert it, sometimes to destinations from which there may not be an easy way back. This adaptability can mean a world of vulnerabilities as well as one of possibilities.

What can baby brains do?

At the outset, human babies don't appear to be able to do much when they arrive. Different types of animals have different capacities and abilities at birth; some, known as 'precocial' animals, emerge relatively ready to become independent, able to stand and suckle within the first few minutes – giraffes are a favourite example. Others, known as 'altricial' animals, are quite helpless, possibly blind, deaf and unable to move, and remain dependent on their carers for relatively long periods of time. The length of time human babies are dependent on the care of others puts them firmly in the second group (together with rats, cats and dogs, among others).

It has been suggested that how big your brain is going to be when you are grown up is a factor in how well developed you (and your brain) are when you are born. And varying with this factor is the size of the birth canal through which you will make your entrance. For humans, the alterations to the pelvis that allows us to walk upright places limits on the size of the birth canal, so babies' heads can only get so big before they have to make their appearance into the outside world. Nature has kindly (and thankfully) determined that if your eagerly awaited offspring is eventually going to have a head size of fifty-six centimetres, then your body calls time once your temporary lodger's head might fit a thirty-five-centimetre knitted bonnet.

The downside is the physical helplessness of the new arrival, but one of the claimed pluses of being altricial is that (quite literally) there is room for brain development postnatally. Being a giraffe might mean that the brain you have at birth will allow you to get to your feet and get on with life straight away, but after that achievement you just get bigger but not much smarter. On the other hand, the development potential of human baby brains is enormous. And this is the point of what might appear to be an off-topic ramble about baby giraffes – human babies come into the world with unfinished brains. Understanding how and why these unfinished brains change in the way they do will be part of any attempt to understand any differences between brains and the behaviours and personalities they underpin.

So what can an unfinished human brain do when it arrives? If we look at the behaviour of its owner, we might infer that it is a pretty basic, if highly focussed, system. The arrival of Daughter #1 was my own first-hand experience of the newborn brain at work, outside the pages of developmental psychology texts. It was rapidly clear that I had produced a tiny but extremely loud transmitting device, programmed to continuously signal some kind of deficit associated with her digestive system and/ or the state of her nether regions, or just to offer a spontaneous demonstration of her sound-generating capacity. Her timer was set for maximum activity during the hours of darkness and rebooted every thirty-five minutes or so; random checks would be carried out to ensure a constant state of readiness on the part of her workforce. She didn't appear to be much of a receiving device apart from some highly effective monitoring of sounds such as receding tiptoes or tentatively closing doors, the perception of which would immediately trigger her alarm system; she was certainly unresponsive to the wide range of supposedly failsafe lullabies, music boxes or washing-machine spin cycles external advisors guaranteed would activate the off-switch. To all intents and purposes she seemed to be run by a primitive and unsophisticated program, presumably reflecting the activity of a primitive and unsophisticated brain. (Oh, for a baby giraffe!)

However, more skilled (and possibly less sleep-deprived) researchers have devised extraordinarily clever ways of testing the skills of the newborn, and checking whether they really are just passive and rather inefficient receiving devices or whether there is more going on than outward appearances might suggest. We might now have the techniques to get detailed pictures of all sorts of tiny elements in a baby's brain, but what can she actually *do* with this emerging cortical kit? This is where we encounter another challenge for developmental neuroscientists: how do we know if a baby has noticed a change in the world around her? It's difficult to get her to press a response key. How do you know if a newborn baby prefers black and white horizontal stripes or the sound of her mother's voice, or can tell if you are speaking to her in a foreign language? You can't get her to complete a 0–5 Likert scale, with 0 being 'couldn't care less' to 5 meaning 'more more more, now now now'.

Developmental psychologists are like Sherlock Holmes in devising ways to tell what newborns can and can't do. Over time they have amassed a portfolio of tiny signs that signal that the baby is paying attention, that she 'likes' the sound or sight that she is presented with, that she is 'choosing' one stimulus over another. One measure of 'interest' in babies is preferential looking, i.e. showing her two stimuli at once and timing how long she looks at each of them, assuming she will look longer at the thing she likes.[35] There is usually some kind of minimum looking time (often about fifteen seconds) set by the researcher, to be sure that they are not being tricked by random eye movements. Habituation is another technique: show the same thing time and time again and measure the usual decline in looking, then present something different – if the baby looks longer at the new thing, then this is taken as an indication that she has noticed the novelty and is paying attention to it. Another behavioural sign is sucking rate, measured by electronic pacifiers, with increases in sucking rates taken as a measure of interest or enthusiasm. It is now even possible to look at behavioural changes in the womb, and mouth opening is taken as another measure of interest, often paralleled by changes in heart rate. So even before human babies arrive in the world, we can get some clues about how that world is already impacting on their brains.[36]

We can also look closely at baby brains using an EEG measure called the 'mismatch negativity (MMN) response': an increase in brain activity associated with a kind of 'Aha – I've spotted a difference' response to changes in the environment.[37] Can this brain tell the difference between a human voice and an electronic sound, or between its owner's mother's voice and the voice of a stranger? These measures have revealed quite how aware of and responsive to the world a newborn is. It appears that she has an astonishing array of skills at birth that make her much readier to take on the world than at first appears. Which also means the world will have a much greater impact on her tiny brain than we might previously have supposed.

The sound world of a baby

Even from before birth, the auditory world of tiny humans is really quite sophisticated. One study showed that the auditory cortex, the part of the brain that primarily monitors sound, was larger in premature babies who, while in intensive care, were exposed to maternal sounds (mother's voice and heartbeat) than in those who only received routine hospital noises.[38] So babies and their brains are already picky about what they might listen to. It has long been known that this primarily includes the sounds of their mother's voice, for many even before they are born.[39]

Some researchers have been able to measure EEG responses to sounds in very early pre-term infants (about ten weeks early).[40] They have shown that these baby brains could already distinguish between sounds such as the consonants [b] and [g], and between male and female voices. At birth, babies appear to be able to tell the difference between the sounds of their own native language, which activate the left side of their brain, and sounds from a different language, which activate the right.[41] Newborns also appear to be able to tell the difference between happy and neutral sounds, so they are already picking up some useful social cues.

One study used the MMN response to demonstrate this skill. Newborn babies were played neutral, happy, sad and fearful versions of the syllables *dada*.*[42] Strings of standard tones (neutral *dada*s) were interrupted at random intervals by the 'deviant' tones (happy, sad or fearful-sounding *dada*s), and the brains' responses were compared. If the 'receiver' didn't notice any differences, no mismatch response would be recorded. There were big differences between responses to the neutral syllables and to each of the emotional ones, with the biggest responses to the 'fearful' *dada*s. In this study, ninety-six babies, ranging between one and five days old, were tested, forty-one of them girls. Although the researchers explicitly looked, they could find no sex differences in these responses. This is interesting because, as we shall see later, one of the measures of allegedly greater empathy in females is their greater responsiveness to emotional information, including voice intonation.[43] So even if females do

---

* The hard work was not just done by the experimenters; the babies had to listen to two or three blocks of 200 occurrences each of these sounds, in what is known as an oddball paradigm. Producing these sounds must have been quite a Pinteresque challenge for the female 'voiceover' recruited for the study, having to straight-facedly record happy 'dada' as opposed to sad 'dada' and so on. And, indeed, for the 120 listeners who then had to rate all the 'dada' sounds for their emotionality, or lack of it. We neuroscientists are nothing if not creative!

have this heightened sensitivity to emotion using this measure, it does not appear to be present at birth.

Newborns also appear to be finely attuned to sophisticated differences in their sound landscape. The evidence of preferences showed that the infant auditory system is not only quite fine-tuned but is more than just a passive receiver of information. From very early on, again using MMN responses, researchers have shown that a baby will respond mainly to anything that is different, e.g. a 'bop' sound in a series of 'beep' sounds (known in the business as 'acoustic deviance').[44] The difference had to be quite marked before there was evidence of the baby noticing, but it didn't seem to matter much what kind of noise it was – snatches of white noise elicited much the same kind of response as a matched whistle or a bird sound. But within two to four months of age, there is evidence of differential responding to 'environmental' sounds such as a doorbell ringing or a dog barking, as well as to speech sounds and non-speech sounds. It is as though the baby hearing system has started to filter out what it is worth paying attention to and what might be ignored.

The loss of sensitivity to certain sounds if they don't appear in your sound landscape is a measure of the plasticity of the auditory system. If your native language is Japanese, for example, then you won't be exposed to the [r]/[l] distinction, which is important in English.[45] Babies aged six to eight months (regardless of the language to which they are exposed and which they will speak) can still distinguish between all these sounds; but by ten to twelve months, they will only distinguish between those sounds that are distinct in their own language. This has been shown at both the behavioural level, using head turning as a 'spot the difference' measure, and at the brain level, looking at different evoked responses to different sounds.[46]

So our little humans, as well as being discriminating listeners, appear to be able to pick up quite sophisticated clues as to the social significance of what they can hear, not just about the language they will come to speak, but also how different sounds within that language may, for example, flag up the expression of different kinds of emotions.

The eyes have it?

Babies' vision is rather less sophisticated than their hearing. The basic building blocks of the retina and optic nerves are in place by about thirty weeks' gestation[47] but at birth a baby's world will be rather fuzzy, as the eye apparatus hasn't developed sufficiently to form nice sharp images on the retina. They find it hard to focus on objects more than eight to ten inches away. Furthermore, their two eyes don't work well together for the first three or four months, so their depth perception is limited. As the information from the visual system starts to become more accurate and detailed, the developing brain is able to make better use of it, shown in behavioural changes such as the baby being able to track moving objects or to accurately reach for and grasp them, in place by about three months of age.[48]

But, in our quest to check out just how sophisticated our supposedly helpless newborns are, let's have a look at what the baby visual system *can* do rather than what it can't. Luminance processing (responding to light/dark differences) appears to be present from birth. Indeed, it has been shown to vary with gestational length (meaning it's weaker in pre-term babies), suggesting it is a good example of a preprogrammed skill.[49] Despite their poor visual acuity, from as young as one week old, they can already discriminate between plain and striped stimuli, and show preference for high-contrast patterns such as black and white stripes and horizontal as opposed to vertical stripes.[50]

Having two eyes that work together allows you to view objects in depth and get a much less fuzzy view of the world around you, including a more detailed take on things like faces and a better chance of accurately reaching for toys or fingers. Newborn babies' eyes sometimes move alarmingly independently – if you are a new parent, in one of those many spare moments that you will be having, try moving your finger towards the nose of your infant to demonstrate this. But by six to sixteen weeks old, their eyes start to work together and their response to different patterns and their ability to more accurately follow movement shows they are starting to use binocular vision.[51] Evidence indicates that baby girls acquire this skill earlier than boys do and it has been suggested that this early difference might be one of the factors that gives girls the edge when it comes to processing faces.[52] We'll talk about why this might be the case in the next chapter.

Babies do have basic colour vision from birth; newborns prefer coloured stimuli to plain grey and, given the choice, look longest at reddish stimuli and least at greenish-yellowish. This is true of all babies, not just girls (which is something the pinkification brigade might need to think about). By two months of age, they can show different responses to the full range of colours, still with no evidence of any kind of sex difference.[53]

Eyes are, of course, more than just devices for receiving visual information; they have a social function too. Eye contact, or mutual eye gaze, is often considered a primary indication of social engagement and communication. Newborn infants typically prefer faces which have their eyes open rather than closed and will look longer at faces when the eyes are gazing directly at them as opposed to averted.[54] By three months, babies can get quite agitated if their mother looks away from them and will often hand-wave or jiggle up and down to re-engage her attention.*[55]

Eye gaze also appears to be a communication device, seeming to flag up something to which it is worth paying attention. Four-month-old infants have been shown to learn about objects just by being exposed to eye gaze directed towards it, perhaps accompanied by a fearful face or a happy face.[56] Eye gaze preferences have also been shown to be a good measure of emerging skills; preferential attention to the eyes and mouth regions of a face are linked to face-processing efficiency, itself linked to developing socialisation.

---

* I use the term 'mother' throughout to describe the primary caregiver, as in the majority of studies conducted the primary caregiver is the child's birth mother. However, there are of course many variations of parenting and if the child's primary caregiver from birth were the father then this would apply to him too.

Eyes are, of course, for looking, and being discriminating about what you look at is an early sign of screening your environment for potentially useful information. In addition, knowing that what *other* eyes are looking at could well be of some significance to you is an even more sophisticated information-gathering mechanism. And, not even halfway through their first year of life, it is clear that babies have these skills at their command.[57]

A dawning social awareness?

As I've described earlier in the book, our drive to be social may well be the secret of our evolutionary success, supported by specialist networks in the brain. So can we find these social brain networks in babies, and when and how might they be active?

Just as the early focus in studying the adult human brain was on the core cognitive skills such as language and communication, and emerging high-level skills such as abstract reasoning and creativity, much of the initial interest in the developing baby brain was in how the brain changes were paralleled by emerging skills in the fundamentals of perception and language, together with movement and co-ordination abilities. It was assumed that the evolutionarily most sophisticated areas of the brain, the prefrontal areas, were functionally silent in human newborns, while other areas got on with growing the scaffolding for life's basics. How wrong we were! As we will see in the next chapter, babies' social skills may actually be well in advance of their more fundamental behavioural ones, with their social antennae tuned in from very early on to pick up vital clues.

Psychologist Tobias Grossmann, now at the University of Virginia, has reviewed many studies looking at the social brain in infancy and has concluded that 'human infants enter the world tuned to their social environment and readily prepared for social interaction'.[58] He notes that the early signs of social behaviour in infants are initially self-focussed; babies pick up clues that are relevant to themselves and their needs via processes such as eye gaze monitoring or shared attention scenarios. It is now known that the brain bases for these early signs of social behaviour principally involve the prefrontal cortex, the basis of higher cognitive and social functioning, which would not have been predicted from early models of infants as 'reactive, reflexive and subcortical'.[59] And researchers have recently shown that key characteristics of 'social' eye gazing, such as focussing on the eye and mouth regions of a face, and the duration and direction of actively looking at key aspects of social scenes, have a strong genetic component, so are inbuilt from the beginning.[60]

As ever, consideration of how we come to be social beings also encompasses the question as to whether or not there is any evidence of sex differences in the brain functions underpinning this process. Perhaps unsurprisingly, given the marked lack of evidence that baby brain structures can be divided along neatly girl–boy lines, very early sex differences in social behaviour are proving similarly hard to find.

It has been claimed that newborn baby girls engage in longer eye-to-eye contact than boys, although it hasn't actually been possible to replicate this finding.[61] Another study showed that, although there were no sex differences at birth, if you looked at the same infants four months later, quite dramatic differences emerged. The frequency and duration of eye contact in boys remained much the same; in girls, it increased nearly fourfold.[62]

Simon Baron-Cohen's team have also noted greater frequency of eye contact in twelve-month-old girls.[63] So even if baby boys and girls start out the same with respect to this core social skill, it looks as if a sex difference emerges over time. There is no clear evidence that mothers spend longer in mutual eye contact with girls rather than boys, but it may be the greater encouragement for mobility and rough-and-tumble play in boys precludes time spent in face-to-face contact, thus reducing their 'learning opportunities'.[64]

Anyone familiar with those 'developmental milestones' guides handed out to terrified new parents should know that the most common characteristic of any form of infant development is the huge variability shown by this anxiously studied group.

When should a social smile emerge? Well, it could be four weeks, or maybe six, or there may be nothing until twelve weeks. And that wonderful first word? An optimistic six months or a more realistic twelve months? We do know that things happen in pretty much the same order, but beyond that we are often in the hands of the more or less reassuring folk wisdom of panels of experts (comprising allegedly highly qualified family members, health visitors, passing strangers and/or authors of the worst kind of neurotrash baby books). They will almost invariably tell us that little boys will do things differently from little girls and at different times. Just how skilful are little humans, and do these skills really divide along the neatly gendered lines asserted by these so-called experts?

With the attention in neuroscience now turning to humans as social beings, babies as well as adults are being scrutinised for their skills on this front. Although babies appear pretty helpless at birth, as we have seen, their information-processing systems show a surprisingly high level of efficiency, and they soon seem to become aware of subtle differences in the world around them. How soon and how much can they use these skills as active social engagement systems? Do babies need to walk and talk before they can begin to take their place in the world as a social being? Or are they actually social beings right from the start, little interactive chatbots ready to take in whatever messages the world has on offer?

The answer might surprise you.

Chapter 8:
Let's Hear It for the
Babies

Our understanding of what newborn babies can do (and how they eventually develop into fully functioning members of the human race) has been characterised by the well-known 'nature versus nurture' debate.

In the 'nature' version of the story, babies will develop along predetermined lines, the end product being pretty much fixed by its owner's genetic blueprint. This would include their brains and the behaviours they support. This inbuilt program will inexorably unfold to determine what kind of adult a baby will eventually turn into, with any differences a reflection of what kind of skills are needed by this particular version of the species. This 'nature rules' version is sometimes known as the 'tramline model', with the destination pretty much fixed by the starting point and by the routes already laid out. There is a certain amount of flexibility to deal with changing demands, but dramatic fluctuations are avoided; the end product needs to be well suited to its predetermined role. Biology is destiny.

The genetic blueprint will include, of course, a baby's sex. In what Daphna Joel describes as the 3G model,[1] the belief is that the genes that determine characteristic differences in babies' genitals and gonads will also determine differences in their brains. These 'hard-wired' brain differences will define the aptitudes and abilities of newly arrived females and males and take them along their different roads in life, arriving at different destinations marked by the different occupations and achievements of the adults they became. Any very early differences between baby girls and baby boys will be hailed as evidence of the inborn, or 'innate', version of such differences – and quite possibly helpfully packaged up into appropriately colour-coded texts called 'It's a Girl' or 'It's a Boy', listing the 'unique wonder and special nature' of these new arrivals.

The other side of the coin is what is called the 'nurturist' approach, which is focussed on the notion that human babies are 'blank slates' on which post-birth experiences scribe their effects. The basic premise is that what babies and their brains can do, the set of skills they acquire, the language they come to speak, maybe even how they see the world, is entirely shaped by the environment they grow up in, by the learning experiences they have and the social rules they encounter. This kind of experience-dependent approach can be thought of as a 'socialisation' approach: babies learn to be grown-ups by imitating the adult world into which they are born. Differences in how girls and boys, women and men behave and what they achieve is not determined by some form of biological preprogramming but by differences in the expectations their world has of them and by differences in the life experiences they have had (or been allowed to have).

A more contemporary melding of these two virtually opposing views still implicates biological characteristics, but they are given much less potency in determining the end product than the early 'biology is destiny' versions. In this view, you and your brain may start out on a fairly standard trajectory but it can then be diverted by quite small shifts in what Anne Fausto-Sterling calls the 'corrugated landscape' of a brain's developmental pathway.[2] Many possible pathways are present at the beginning of the journey and different events or experiences can shift the route from one path to another. These shifts can be brought about by quite minor diversions, reflecting tiny variations in the baby's life, such as how her mother talked to her, or how much she was encouraged to stand and move around.

Fausto-Sterling has modelled these early interactions against later abilities and she has shown how very early differences in responses to girl and boy babies are related to differences in skills (such as early walking) which have, in the past, been claimed as innate.[3] One relatively robust finding about emerging differences in baby skills is that boys tend to move more and walk earlier; but it is also a relatively robust finding that boy babies receive more 'motor encouragement' than girl babies. This is true even when the boy baby is actually a girl, cunningly disguised in dungarees (so stereotypes can have their uses!).[4] As we learned in the last chapter, the cerebellum, a part of the brain that is central to movement control, doubles in size in the first three months of life. We now know it grows significantly faster in boys, on average, than it does in girls.[5] An important issue is whether this change drives the movement skills in boys, or reflects the greater movement experiences they are having.

The key message here is that any brain's trajectory may not be fixed but can be diverted by tiny differences in expectations and attitudes – you could set out on one route but then a little fork in the road might send you off down a different path. If the diversions happen to be gender-specific then the valley you enter may take you into a world of pink princesses, as opposed to the kind of Lego kingdom you were otherwise heading for. This is a much more complex model of development than the classic 'nature versus nurture' – the road that the developing brain may follow will be determined by a closely entangled mixture of many factors, including the characteristics of the brain itself but also the coned-off sections or diversions that are encountered en route.

Discoveries about our brains' life-long plasticity and 'predictive' nature, which we looked at in Chapter 5, have brought changes to both the nature argument and the nurture argument. The 'nature' idea, as it was, has now morphed into a notion of a hard-wired and hormonally determined system, where the neuronal support system is in place at birth, but the outside world is afforded a slightly bigger role. Rather like a smartphone preloaded with certain apps, what the brain might eventually do is determined by what data are input. The system will still, however, be constrained by the presence of the 'right' task-specific app – if you haven't got Google Maps then finding your way around will be tricky. This is much more of a 'limits imposed by biology' model, rather than one that offers biology as uncompromising destiny.

On the other hand, in a new rendition of the 'nurture' argument, if the brain is conceived of as more of a 'predictive texter' then the baby brain can be thought of as the first stages of an emerging 'deep learning' system. These kinds of systems effectively extract rules from the information they are exposed to, with the more advanced systems eventually not needing any kind of explicit help or guidance, but using feedback from the success or otherwise of earlier attempts to refine their next engagement with their environment. Although these systems, being brain-based, are, of course, biologically determined, they are much more fluid and flexible, with more temporary 'soft assemblies' in place to pick up the necessary data and generate an appropriate template, whose output then results in an update and a new focus on solving the next challenge.

Each of these models has significance for our understanding of sex differences. If you haven't got the app, then you won't have the wherewithal to solve the problem / play the game / read those tricky emotional signals. On the other hand, you might have the app, but the world doesn't give you the data. Or the data you get varies as a function of the kind of smartphone you are: pink squishy versions get one set of messages, blue armoured ones another.

But a key issue remains as to when all of this begins. As we saw in the last chapter, we now have much better access to baby brains and the dramatic changes that occur in the early years. But what are babies doing with these brains? Are they just busily acquiring the cognitive basics of seeing, hearing, moving around, with their world finding ways of inputting the right kind of data to help this along? What else might they be up to? Do they pick up social rules as quickly as they acquire the core 'cognitive competencies'? The work of developmental psychologists and developmental cognitive neuroscientists is revealing some astonishing findings about the world of our babies, helping us understand what they can do and when, and the extent to which we can view our tiny humans as preloaded smartphones or novice deep learners.

Tiny linguists

How the brain responds to speech or language-like sounds is perhaps uniquely important to the human baby, who will, in most cases, grow to be part of a social community that communicates via language or language-related processes. Our new baby's unfinished brain is astonishingly well equipped to plunge itself into its linguistic community, even though all it seems to be bringing to the table is a few gurgling sounds and some shockingly loud high-pitched cries. A newborn baby can tell the difference between recorded speech sounds played forwards and those played backwards – on the face of it perhaps not an obviously useful skill but it does show that the brain is already primed to respond to sounds which are arranged in a speech-like pattern and are not just a random collection.[6] She can also tell the differences between her own language and foreign languages.[7] Impressively, by sucking more or less enthusiastically on a pacifier, five-day-old babies can show that they know the difference between English, Dutch, Spanish and Italian.[8] Any sign of early sex differences in this allegedly reliably gendered skill? None reported so far.

How about a bit later, where emerging differences might give clues about the innateness or otherwise of verbal ability? An early difference is consistently reported, with girls talking earlier and showing better spontaneous language and vocabulary skills.[9] As with so many such differences, the effect size is actually quite small, so there is a considerable overlap between boys and girls. However, the difference, although slight, appears to be true across a wide range of language communities, which could suggest innate factors at play. But studies tracking mother–infant linguistic interactions over time showed that mothers verbalised more with their girl babies at birth and still later at eleven months, so some environmental factors are at play here.[10] This is a good example of biological factors interacting with a variable landscape. The auditory cortex in babies develops dramatically in the early months after birth, and the growth of nerve cells and the connections between them is experience-dependent, with the types of sounds that a baby is exposed to eventually determining the language(s) they recognise and will respond to.[11] If mothers speak more and respond vocally more to infant girls, they are offering their daughters a different 'sound experience'. As Anne Fausto-Sterling has suggested, perhaps the earlier language skills shown in girls are a consequence of the different 'call and response' (or 'serve and return' as it is also known) experiences that baby girls have had.[12] There may, indeed, already have been differences in baby girls' speech systems which initiated the different responses from their caregivers in the first place, but the principle remains the same: it is neither nature nor nurture alone that determines the endpoint, but the continuous back and forth between them.

As we shall see later, the stereotype of female verbal superiority is one that has not stood up to close scrutiny. There are hugely overlapping distributions of male–female scores, and many differences disappear when different tests are used. So, rather unusually in this story, there are glimmerings of early sex differences in some aspects of language acquisition, but their existence, and indeed the search for them, is based on a belief in a difference between adult females and males which actually seems to have disappeared.

Baby scientists

If you were asked to rank the various high-level accomplishments of the human race, mathematics and the understanding of the laws of physics would probably come fairly near the top of your list. You might also characterise such feats as achievable only after many years of education and, further, beyond the reach of many, no matter how many opportunities they had been given. So you might be surprised to know that very young babies have already grasped the basic principles of high-level science. Within two days of arrival in the world, they can tell the difference between big numbers and small numbers, matching short bursts of beeps to pictures showing just a few smiley faces, and long bursts of beeps to pictures with lots of smiley faces.[13] Two or three months later, they will express surprise if a ball doesn't roll out of the end of the tube they saw it roll into;[14] five months later, they are perturbed when what looks like a liquid in a glass turns out to be solid, as their stripey drinking straw stops on the surface of the pretend water into which it has been dropped.[15] So, within five months of making it into the world, babies are already demonstrating a grasp of basic mathematics (or numeracy) and of intuitive physics, of how objects normally behave and what the basic characteristics of substances are. The possession of such 'core knowledge', as it is called, is yet another demonstration of how human infants are far from helpless or passive receivers of the world around them, but capable of amazingly sophisticated observations and interactions with that world.

A key question for us, of course, is whether or not there are any sex differences at birth in these intrinsic aptitudes. Women are underrepresented in STEM subjects, where the kind of physics and mathematics skills demonstrated by our tiny scientists would be paramount. Perhaps, whether politically incorrect or not, we might seek evidence that there is some kind of innate gender gap? If there are sex differences in 'systemising', in interests in rule-based physical systems and their characteristics, might these be reflected in the early emergence of the kind of 'naïve physics' skills demonstrable from birth?

In all of the 'baby physics' studies outlined above, none reported any sex differences. We should bear this in mind when we later come to examine the bases of the continuing problem with gender gaps in science. But perhaps, early on, there may be more generic differences, with boy babies just showing a preference for non-social information?

As it turns out, early claims that this had been demonstrated in newborns have been a matter of some dispute. A study by Jennifer Connellan from Simon Baron-Cohen's lab has been widely cited as evidence of an innate male preference for mechanical objects as opposed to faces.[16] In Connellan's study, newborns were presented with either the experimenter herself or a flat face-shaped mobile, onto which had been pasted scrambled photos of parts of the experimenter's face. The amount of looking time given to each of the stimuli was taken as a measure of preference. It is worth spelling out here the actual findings in some detail, as you can then see why the claims made are somewhat surprising. Of the fifty-eight baby girls tested, nearly half of them (twenty-seven) showed *no* preference for the face or the mobile; of the remaining number, twenty-one looked longer at the face and ten spent longer looking at the mobile. Of the forty-four boys tested, fourteen showed no preference, eleven preferred the face and nineteen the mobile. So although forty per cent of the babies actually showed no preference at all, the major focus in the interpretation of these results was on the difference between the proportions of the boys and the girls who showed preference for the mobile – twenty-five per cent and seventeen per cent respectively. This was interpreted as evidence for a male preference for mechanical objects, showing 'physicomechanical motion', as opposed to the face, characterised by 'natural, biological motion'. The researchers claimed that, as the infants were newborn, the differences had to be biological in origin. This was hailed as an important finding as it appeared to provide evidence that alleged sex differences in social skills, or in preferences for what you might pay attention to in your world, were present from birth.

The study has been almost as widely criticised as it has been cited; for example, the experimenter was not blind with respect to the sex of the infants she was testing, and the stimuli were presented singly and not together as is the standard practice.[17] Given these and other problems, it is not surprising that the study hasn't been replicated. There have been other studies asking the same question (faces versus objects), but they are on older babies (veterans of five months and above) and they use toys as their objects, which, as we will see, brings other issues to the table.[18] Even if the experiment had been methodologically sound, it does make a telling case study of how apparently clear-cut findings may be hiding a rather less clear-cut story.

But that is not to say that there is no evidence of early sex differences in science-related skills. As we know, one area that has received much attention is the ability to 'mentally rotate' objects, giving a spatial manipulation skill that is claimed to be fundamental to understanding of key concepts in science and mathematics.[19] Mental rotation ability is often quoted as one of the most robust sex differences measurable, with males (on average) consistently outperforming females, with meta-analyses of such studies reporting small to moderate effect sizes.[20]

So is this skill evident early on in life, bearing in mind you would find it hard to explain to a month-old baby that you want him or her to 'imagine manipulating a two-dimensional representation of a three-dimensional object'? Studies with infants tend to use the 'surprise' or 'novelty' approach, where, following several repetitions of pairs of identical images (say, of the numeral '1' at different angles), a test pair is shown where one of the pair is actually a mirror image of what would have been its matching other half. As babies show a preference for anything new, if they notice this change, they will spend longer looking at this incongruous pair.

The hypothesis in one study investigating mental rotation in babies aged three to four months used this novelty preference measure.[21] The results reported that boys looked at the novel pair 62.6 per cent of the time (an above-chance difference), compared to only 50.2 per cent of the time for girls. With slightly older children (six to thirteen months old) carrying out a similar task, both sexes looked at the pair with the mirror image stimulus for longer than chance. There was a tiny sex difference, with boys looking at the mirror image pair 3.4 per cent longer than girls, but the scores were hugely overlapping. There is a general consensus, then, that infant boys do look longer at images where novelty has been introduced by rotating one of the constituents. But this doesn't necessarily mean that girls can't see the mirror image, it could just be that they don't give it the same amount of attention as boys. It is also possible that studies which fail to report sex differences may use more 'interesting' stimuli, such as videos of moving objects or real-life 3D objects, so perhaps baby girls just aren't into plain black and white pictures of numerals or Lego-type shapes. However, a tiny sex difference there is (for now).

With respect to higher-level cognitive skills such as language and scientific concepts, babies are surprisingly sophisticated from a very young age. Bearing in mind that verbal fluency, spatial cognition and mathematical prowess were three of the core competencies which Eleanor Maccoby and Carol Jacklin, way back in the 1970s, identified as most reliably demonstrating sex differences, you might expect that these would be clear from very early on. But the evidence is lacking, though not for want of trying. In 2005 Elizabeth Spelke, who heads the Laboratory for Developmental Studies at Harvard University, and has researched babies' competencies for decades, published a major critical review on the topic of intrinsic aptitude for maths and science. This included a consideration of her own and others' work on science skills in newborns and infants. She is firmly of the opinion that there is no evidence of sex differences at this stage: 'Thousands of studies of human infants, conducted over three decades, provide no evidence for a male advantage in perceiving, learning or reasoning about objects, their motions, and their mechanical interactions.'[22]

Given the continued existence of gender gaps in society, though, perhaps we should turn our attention to social skills, to see if there are any differences in how baby girls and baby boys come to take their place in society, and if these might determine the differences in their eventual destinations.

Babies and faces

In the same way that the early ability to process language-like sounds may lay the foundations for future socialisation, an early ability to process faces is claimed as an essential skill for new humans. If babies are going to be social beings, they need to develop this skill as soon and as efficiently as possible, by being born with the appropriate 'face-processing app', and/or by having the rudimentary neural scaffolding in place to get on with acquiring the necessary expertise, and/or by very quickly learning to recognise that a face is a face, but that some faces are more useful than others. And that the expressions on those faces can also be useful cues as to how their owners might behave.

First of all then, what is a face, and how might a baby recognise that? You might think that an easy way to answer that question would be to show a baby a picture of a face and a picture of something else and see which it preferred. But, as any developmental cognitive neuroscientist will tell you, 'I think you'll find it's a bit more complicated than that.' Are faces 'special', with their own brain network dedicated to processing them, which would make recognising them and their expressions much more of a social activity, so that people who are good at this would be tagged as 'good socialisers'? Or is a face just a collection of shapes in a particular configuration, generally a kind of triangle with two roundish shapes at the top and a single straightish shape at the bottom (known flatteringly in the business as an 'up–down asymmetrical configuration')? This would mean that face processing could be classed as just a more superior form of visual processing and could be managed by the systems we use to process *any* kind of visual information.[23] Being good at it wouldn't necessarily move you up the socialiser scale. Is your newborn's heart rate rising, sucking rate increasing, mouth opening because she recognises *you* (and all that you might mean to her), thus possibly increasing your own heart rate and (cleverly) ensuring that you will continue to work hard to gain that recognition? Or is it because she is just responding to a particular set of shapes arranged in an upside-down triangle, which might score rather fewer points in the maternal bonding stakes?

This might seem like a rather so-what type of argument, but it figures quite a lot in terms of understanding just how 'social' babies are and from how early on. A theory proposed by Mark Johnson from Birkbeck College, University of London, provides an excellent example of how a rudimentary newborn system could swiftly be tuned and refined by input from the outside world, resulting in a sophisticated and highly specialised skill, in this case part of an essential social repertoire.[24] He has suggested that newborns are innately predisposed to orient themselves towards face-like stimuli, so you don't have to present an actual face – three brightly coloured blobs in an eyes–mouth-like configuration within a face-shaped outline will do. He calls the brain system that supports this 'ConSpec' (as, in this case, it will eventually help its owner recognise conspecifics). The workings of this system, by focussing on particular kinds of stimuli, will bias the input to the developing visual system and, via a second-stage process (called 'ConLern'), will 'tutor' the relevant part of the system, which then becomes more and more selective. Echoes here, then, of the predictive priors that our brain is continuously setting up for us. Eventually, this face system will start to respond only to certain kinds of faces, and will be able to spot the differences between familiar and unfamiliar faces, male and female faces, own- and other-race faces. In addition, it can code for more subtle characteristics such as different emotional expressions. And all this within about three months![25]

From birth, babies are more responsive to sets of three blobs when they are arranged in a face-like way rather than at random.[26] And just when you thought that there must be some minimum age limit for participation in developmental psychology studies, some very recent research has even shown that, by shining specific types of light through the uterine wall, you can actually present an upright and an upside-down version of the three blobs to a foetus in the third trimester of pregnancy.[27] Using 4D ultrasound technology, researchers were able to demonstrate that the foetus would turn his or her head significantly more often towards the upright (face-like) configuration of blobs than towards the inverted one. So, even before entering the world, the human baby seems primed to pay attention to one of the most significant social stimuli there is. This was a small-scale study (unsurprisingly) which will need replication, but it offers an intriguing insight into how experience-ready babies are, even before they are born.

It is claimed that (on average) adult women are better than men at some aspects of face processing, such as recognition and memory.[28] So, is the same true of children, especially infants? Are we looking at some kind of innate, experience-independent mechanism that invariably unfolds in the developing child or at least provides the basis for her emerging skills? Or is being good with faces something that is taught, perhaps more to girls than boys, given the belief in female fitness for the roles of comforter, counsellor, consoler?

With respect to face recognition, many studies have confirmed a general female superiority in this skill, although some research suggests that this might just be limited to females having a better memory for female faces (known as 'own-gender bias').[29] Looking at the data from nearly 150 different studies, Agneta Herlitz and Johanna Loven, psychologists from the Karolinska Institute in Sweden, carried out a meta-analysis which confirmed that, on average, women were better at recognising and recalling faces, and that this was true not only of adult females but also of children (aged three to eleven or twelve) and adolescents (aged thirteen to eighteen).[30] Although we saw above that babies are excellent face processors from quite early on, no one has demonstrated a sex difference at this stage. This meta-analysis also demonstrated that females, even the very young ones, were much better at remembering other women's faces than men were at remembering men's faces. Interestingly, both of these observations are paralleled by findings in the brain. Girls and women show greater activation in the face-processing network in fMRI studies of facial recognition, and also a greater response to own-group faces than do men.[31]

So where has this special skill come from, and why are women better than men (on average, of course) especially when looking at faces of their own sex? One explanation for the superior recognition skill is based on eye gaze, the role of eye-to-eye contact in establishing human interaction. The longer you gaze, the more information you might be storing away about the person you are gazing at, particularly, of course, their face. As we saw earlier, although there appears to be no good evidence of sex differences in newborns' eye gaze, by four months of age mutual eye gaze in baby girls is longer and more frequent than in boys, so quite possibly a basis for their emerging face-processing skills.[32] So perhaps here we are looking at the consequence of a preloaded app, with a female advantage in a key social skill emerging early and laying the foundations for later openness to incoming data.

An additional aspect of this accolade for female skills in face processing is that females are reputedly much better at decoding emotional expressions. Not just the absolute 'I am terrified' versus 'I am ecstatic' but also the 'I am rather disappointed' versus 'That might be quite nice' facial patterns.[33] Research suggests that females are better at 'reading the mind in the eyes', a test devised by Simon Baron-Cohen's lab to evaluate the ability to recognise emotion as a measure of empathy, although it hasn't been consistently replicated.[34] In 2000, psychologist Erin McClure carried out a huge review of studies on sex differences in facial expression processing (FEP) in infants, children and adolescents, in order to try and answer the same kind of questions about the source of this skill.[35] Are women innately good at this kind of thing, are they trained to be good at it or does the world build on a skill they are born with? By tracking the kind of timeline changes you might expect for each of the possible routes to the end destination of female superiority in face processing, McClure could investigate where this difference might have come from.

Significantly, her review showed that there was certainly some indication of female FEP superiority for all ages, although she did note that there was only a small number of studies with infants (the youngest of whom were three months old). But the effect sizes were relatively constant right from infancy through to adolescence. So was this down to biology, to socialisation or to an interaction between them?

There was some evidence of early sex differences in the brain structures that underpin FEP, particularly the amygdala and parts of the temporal cortex. This is possibly related to hormone effects on such structures (the amygdala has a high density of sex hormone receptors).[36] There were also clear differences in the provision of what McClure calls the 'emotional scaffolding' associated with learning to understand facial expressions. Caregivers often accompany their interactions with young babies with exaggerated facial expressions (big smiles, big 'O's of surprise, exaggerated clown-like sad faces).[37] Some studies showed that this varied depending on the sex of the child, with mothers, in general, being more expressive to their daughters.[38] This early extra tuition may account for the greater responsiveness of young girls to emotional clues from their mothers. And it is also another great example of how an early difference cannot be claimed as pure evidence of differences in an innate skill, but appears to be the product of the back and forth between a data-ready system and the data its world is inputting.

In one study, one-year-olds cheerily sitting on a rug with their mothers were presented with several unknown toys, such as a 'hootbot', an owl robot whose eyes were blinking lights and whose claws rhythmically clicked on a pedestal.[39] The mothers had been instructed to react happily (smiley face, cheery vocalisations) or fearfully (frightened face, hesitant sounds). The ever-earnest developmental researchers rated the 'social referencing', the number of times the child looked at the mother before approaching the toy (or not), the intensity of the messages the mothers were sending out to the infants, and how close the child actually got to the toy. Apart from rather snide comments about the general ineffectiveness of mothers when conveying fear (setting aside that this wasn't a RADA interview and that these mothers might well have been thinking about the long-term consequences of establishing an owl phobia in their child), it was noted that the mothers sent out much more intense fear messages to their sons. However, it was the daughters who showed the most responsiveness to the clues their mothers were providing. So even at twelve months, boys weren't listening and girls were picking up on the social signals, even though these were actually more subtle.

Her review of these and many other studies led McClure to the conclusion that a girl's superiority in FEP arises from an early biologically based sensitivity to facial expressions which is then maintained by the FEP 'scaffolding' provided by the world into which she is born. This is somewhat at odds with the lack of evidence from studies of newborns, but the quite dramatic differences which emerge within three months certainly indicate either that girls are biologically primed to be more skilled at this core social task or that pressure from the world into which they are born ensures that they are offered powerful training opportunities.

Babies as people people

Being a member of the human world is more than just responding to sights and sounds. It requires some social skills as well: we need to be able to interact with other people – and the evidence of the early selective responsiveness to some faces and voices and not others, to language-like sounds rather than doorbells ringing or dogs barking, to happy or sad facial expressions, shows that newborns have a pretty good 'starter kit' to help them in their journey to becoming social beings.

Imitation, copying the actions or expressions of another person, is claimed as a powerful weapon in babies', or indeed anyone's, 'social engagement systems'. Apart from being the sincerest form of flattery, it indicates an awareness of an 'other': that there are people in your world besides you, and that they do things it might be useful for you to do too. You need an understanding of how you might match what they are doing and therefore learn the skill that they already have, be it learning to play cricket or understanding social rules.

Imitation has allegedly been demonstrated in newborns, with researchers earnestly bending over cots and sticking their tongues out, waggling fingers, opening and closing their mouths like a goldfish and/or blinking furiously.[40] There are many reports that all of these actions have been imitated by infants mere hours old.[41] Evidence of imitation in newborns is taken as evidence of an innate specialised system, biologically determined and programmed to ensure you do what is necessary to earn your place in the social world. Early versions of the mirror neuron system, part of the social brain, have been identified as the inherited brain system underpinning this skill.[42] Later deficits in social skills such as mind reading or empathy are explained in terms of a dysfunctional mirror neuron system. Believers in this approach have been called the 'Homo imitans' group, or the 'preformationists', believing that newborns arrive in the world with preformed knowledge of necessary techniques to acquire cognitive or social skills.[43]

There are others who claim that what looks like imitation is actually an accidental coincidence between the random movements of a newborn and the enthusiastic tongue protrusion and mouth opening of researchers. Or that tongue protrusion is just something babies do when anything interesting happens in their new world, which could include someone sticking their tongue out at them or, equally, short segments of the Barber of Seville overture (no, really).[44] One review looked at thirty-seven different studies trying to show that newborns imitated up to eighteen different gestures, and concluded that, in fact, tongue protrusion was the only one that was regularly elicited.[45] This school of thought claims that genuine imitation doesn't really emerge until well into the second year. Proponents of this argument point out that, if you watch mother–infant interactions, many of them involve some kind of copying behaviour, but it is five times more likely that it is mothers imitating their babies rather than the other way round.[46]

So what is going on here is actually more of an interactive learning process, with the babies' actions eliciting personal training sessions, which eventually shape the appropriate cognitive or social performance. The mother/father/caregiver is a bit like a baby's first mirror – this is what you look like when you stick your tongue out, waggle your fingers, open your mouth wide; this could be useful; this, not so much. The researchers who follow this line of thought are the 'Homo provocans' set or the 'performationists'. You start life with enormous potential for social and cognitive development, but how you do develop will depend hugely on what life has to offer you.[47]

What about sex in all of this? A study by Emese Nagy, a psychologist from the University of Dundee, reported that newborn girls were faster and more accurate at imitating a simple finger movement, and suggested that this early social skill might elicit more of the personal training sessions we mentioned above, setting the scene for different kinds of interactions with your significant others, even if it takes the form of them enthusiastically copying you rather than the other way round.[48]

The 'imitation game' argument is another version of the debate about whether babies are born with experience-ready apps and emerge into the outside world preprogrammed to match what is going on out there, quickly able to take their place in their in-group by demonstrating 'anything you can do I can do too'. Or whether, instead, they arrive with an experience-dependent app, able to watch, listen and learn, but needing to gradually absorb what is out there, shaped by their world, initially via their caregivers, but then via whatever that world has to offer. As we will see, what the world has to offer in the way of input can vary hugely depending on the general opportunities and expectations in that world, as well as specific cultural differences, defining and eliciting what is seen as appropriate behaviour for the emergent human.

There has been some suggestion that babies are actually proactive socialisers: that, as part of their innate repertoire, they know how to manipulate those around them to interact with them (and we're not just talking the well-known 2 a.m. roar demanding interaction in the form of food and/or fun). The veteran psychologist Colwyn Trevarthen, from the University of Edinburgh, has long suggested that babies can elicit social responses, and that they actively engage with their caregivers, matching smile with smile, 'coo' with 'goo'.[49] Whether training or being trained, most people who have anything to do with babies know that eliciting a beaming smile in exchange for several minutes of out-of-character gurning does seem to bring its own reward. And, most importantly, it will increase the likelihood that this kind of to-and-froing will be repeated. But, however it comes about, babies can rapidly and apparently effortlessly acquire an impressive repertoire of social skills which will embed them firmly in their culture and society.

Eye gaze and face-processing research has shown that, pretty much from birth, babies are taking in information about significant others in their social circle, quickly constructing templates for their 'contacts list'. But interacting with others involves more than just monitoring the immediate information; you also need to check the back story, take into account other cues and clues. Why is this person saying that, and why are they saying it like that, and what should I do about it? Why are they looking at me like that, and again, what should I do about it? You need to understand intentions, make predictions, make selections from your response repertoire (or even make up some new ones). As girl babies appear to be more sensitive to incoming social information, perhaps they are more engaged with this aspect of their social world.

There are very early clues that babies are people watchers like the rest of us. Sit in front of a baby of about nine months old and stare fixedly at something to your left. Pretty soon the baby will look that way too.[50] What about pointing? Lining up one of your digits with a distant object might not, on the face of it, look like a sophisticated social signal, but in fact it is. You are giving what could look like an arbitrary signal that there is something of interest to which you would like to draw another person's attention, and if they would care to line up their eyes with the invisible beam from your fingertip, then they too can share your fascination. So a relatively complex social communication, but babies as young as nine months pick up on it, not only looking at where you are pointing but pretty soon adopting the technique into their own 'want that/get that' arsenal.[51] No evidence of sex differences is reported, so this particular weapon in the joint attention arsenal appears to be equally shared.

A good measure of a developing baby's social skills is when and how the theory of mind, which we looked at in Chapter 6, emerges. As we saw earlier, eye gaze is an early measure of the emergence of joint attention, and understanding that if eyes are not looking at you, then they may, hard as it is to believe, be looking at something more interesting than you. Further, it might be worth you checking this out as well. This sharing of information is taken as one of the very first stages of acquiring a theory of mind – interestingly, a failure in such joint attention can be an early sign of autistic spectrum disorder, a developmental problem chiefly characterised by a central deficit in social behaviour.[52] To be a fully qualified mind reader, you would also need to understand that what people have 'in their heads' will drive their behaviour and that sometimes they will have different information from you, perhaps (and there's no easy way of saying this) because of something you know that you know they don't know. How on earth might you test that in young children who have only very simple language or none at all?

Devilishly clever developmental psychologists, as ever, have ways of finding these things out. They have devised 'as if' tasks where the outcome is a measure of whether or not the child has a theory of mind. One of these is a 'false belief' task and although it uses dolls or story book characters, is actually very complicated.[53] There is an unfolding story, usually involving two players, where the viewer gets to see both sides of the story. This story will include a change in the situation which will be known only to one of the characters.

A good example is the 'Maxi and the chocolate' task. Scene 1: Maxi puts his chocolate in the cupboard and goes outside. Scene 2: Maxi's mother (with Maxi still outside) takes the chocolate out of the cupboard and puts it in the fridge. Scene 3: Maxi comes inside to get the chocolate. Where will Maxi look for the chocolate? If you have a theory of mind, you will nominate the cupboard because you know that that is where Maxi *thinks* it is (even though *you* know it has been moved). So you understand that Maxi has a false belief about the chocolate's whereabouts and that is what will guide his next step. If you're not there yet in the theory of mind stakes, you will nominate the fridge, because that is where *you* last saw it put. You assume that what is in your mind is the same as what is in other people's. This feels like a pretty high-level social skill, but it is a task that is successfully carried out by almost all four-year-olds (and very few three-year-olds).[54]

So eye gaze and joint attention measures show us that even quite small babies have simple mind-reading skills, can track what other people are interested in and understand that there are other perspectives out there. As we saw above, there are emerging clues that there are early sex differences in eye gaze and in responsiveness to different emotions, with girls, on average, appearing to be more 'data-ready' for these aspects of social input. By four years old, children appear to be highly sophisticated mind readers, easily passing false-belief tasks which show they are aware that Others might have different perspectives to their own.[55] Here, however, there is no conclusive evidence of sex differences in typically developing children. So although baby girls seem to have more sensitive antennae with respect to some techniques of use in picking up the rules of social engagement, such as (possibly) imitation and eye gaze, and have the edge on some useful social skills such as recognising faces and picking up on emotional differences, it doesn't necessarily translate into a full-blown mind-reading advantage.

But being social also means understanding the rules and norms of the world you live in. We saw earlier that, with regard to cognitive skills, babies can go way beyond simple awareness of sights and sounds and show evidence of grasping the basic principles of number and science. Do our mini-mathematicians also show any signs of understanding the laws of society?

Mini-magistrates

Picture this scenario. A judge is watching three individuals acting out mini-morality plays. One of the actors, distinguished by his yellow jumper, is trying to open the lid of a box in which is a coveted prize, but he is having tremendous difficulty and obviously can't do it on his own. In one version of the story, one of the other players, distinguished by his red jumper, helps Yellow Jumper open the lid and get the prize. In another version, the second player, wearing a blue jumper, jumps on the lid of the box and stops Yellow Jumper opening it. The judge is then asked to indicate whether she prefers Red Jumper (the Helper) or Blue Jumper (the Hinderer). The judge picks Red Jumper! Different scenarios include Helpers pushing the struggling player up a hill, or returning a lost ball, versus Hinderers preventing any good things happening. After various iterations of the play, with different judges (and careful balancing of who wears what colour jumper), it is clear that the judges almost always come down on the side of the good guy. What is astonishing about this is that the judges are actually three-month-old babies, watching jumper-wearing toy rabbits and signalling their approval of the Good Samaritan rabbit by reaching for him when offered the choice.[56]

These baby morality plays have been devised by psychologists Paul Bloom and Karen Wynn, now at the University of British Columbia, and Kiley Hamlin from Yale, with their respective research teams. They have intensively studied the existence of the moral evaluation skills of tiny babies, of their expertise in spotting the social rules of polite society.[57] As well as the more straightforward good guy/bad guy choices, they have been able to show emerging subtlety in babies' decisions about just what constitutes a good guy. They took five-month-old and eight-month-old infants, and first showed them the lid-opening versus lid-slamming morality play. Then they watched either the Helper (Pro-social Target) or the Hinderer (Anti-social Target) playing with a ball, which he dropped. It was then returned by a Giver or taken away by a Taker. The mini-judges then had to signal their approval or disapproval of the Giver and the Taker. The five-month-olds preferred the Giver, whether or not it was the Pro-social or the Anti-social Target who had been helped out. The eight-month-olds were more judicious in their evaluation. They picked the Giver if the Pro-social Target dropped the ball, but they picked the Taker if the Anti-social Target dropped the ball. So babies of less than one year old are not just responding to an immediate event unfolding in front of them, they are also taking account of previous 'good' or 'bad' behaviour.[58] Perhaps we should drop the age for jury service!

None of the published studies report any sex differences. I asked the researchers if that is because there weren't any, or because the numbers were too small to make proper comparisons (or because they wanted to avoid that particular hornets' nest!). They replied that they had never found any sex differences in all of their studies of these very young children. So, at this stage of their lives anyway, both sexes seem equally good at picking up the ground rules of social behaviour, at least when it comes to Good Bunny/Bad Bunny decisions, and making sure Bad Bunny gets his comeuppance.

As well as a cognitive-type understanding of social niceties, most social skills have an affective component as well, where we need to share the feelings of others, as well as understanding their intentions and motivations. Again, we can find evidence that babies are capable of such behaviour.

Tiny social workers

Empathy, the understanding of other people's emotions in particular, but also their thoughts and intentions in general, is a key skill in becoming and remaining a successful member of a social group. A truly empathic person does not just 'read' distress in others, they actively share their feeling, possibly becoming distressed themselves. So there is both a cognitive component and an emotional or affective component in empathy.[59]

As mentioned in previous chapters, Simon Baron-Cohen has proposed that empathising and systemising are two fundamental characteristics of the human mind and, further, that they underpin fundamental aspects of sex differences. He has firmly stated females are better at empathising and that the female brain is hard-wired for empathy, although, as I've pointed out before, he does also say that you don't actually have to be a female to be a good empathiser, or to have a female brain. So, if this is a genuine sex difference, indeed an 'essential' sex difference, and hard-wired to boot, you might expect it to be present at birth or certainly to emerge pretty early thereafter.

If you play an infant the sound of another infant crying, then pretty soon they will join in too.[60] This is described as 'contagious crying', and could suggest some level of fellow feeling for your howling friend. But it has also been dismissed as just a form of self-distress ('I really don't like that noise') as opposed to a real measure of concern for the other infant. So there's no agreement here on the existence of infant empathy in either sex.

Most studies recruit slightly older babies, between eight and sixteen months old, once they have an early repertoire of gestures and facial expressions that might be 'coded' by the researchers for evidence of empathy. For these types of studies, a mother often gets roped in alongside her offspring, gamely 'boo-hooing' away after pretending to bang her thumb with a toy hammer or bumping into a piece of furniture.[61] So what does her baby do? Any sign of a furrowed brow? This is a measure of 'concerned affect'. Does our mini-empathiser, while watching her mum hammering her thumb, rub her own thumb or look anxiously at other adults in the room? Does the watching offspring show signs of whimpering or crying? These reactions would give high scores on the distress scale. And, finally, does the baby pat her 'injured' parent or offer 'repeated prosocial verbalisation' (the baby version of 'There, there')? This would give a 4 on the 'prosocial behaviour' scale. As measured by 'empathic clues' (such as brow furrowing), there is some evidence of sex differences starting to emerge about the age of two. There are also physical measures such as signs of distress (heart rate changes, pupil dilation, skin conductance responses) when confronted with 'negative scenarios' involving others, such as someone getting their hand slammed in a car door. But these sex differences do not seem to be present at birth, so the advantage allegedly conferred by the hard-wiring for empathy does not show itself very early on, at least not until you are well into your second year.

This is at odds with the claims from Simon Baron-Cohen's team, but they did base their claims on a different battery of measures. One of these, preference in female newborns for human faces as compared to the male babies' preference for mobiles, has, as we know, been confined to the 'definitely could do better and needs replicating' pile. Eye contact has been nominated as an early measure of empathy, and a 1979 study which found newborn girls gazed longer at their carers than boys is quoted in support of this; however, this wasn't successfully replicated in a 2004 study, although, as we saw earlier, longer eye contact from girls was found in older babies (thirteen to eighteen weeks).[62] The authors of this 2004 study concluded that 'social *learning* [my emphasis] may be the primary impetus for the development of gender differences in mutual gaze behaviour in the first months of life'.

Eye gaze detection (i.e. spotting whether someone is looking at you or away from you) is allegedly also part of the empathy battery, or at least a measure of the understanding that 'faces can reflect internal states of social partners'.[63] Newborns show clear evidence of preferring direct to averted gaze, but no sex differences are reported.[64]

There are undoubtedly sex differences in empathy in studies with older infants and children, though. Baron-Cohen's team report higher empathy scores for girls aged four to eleven years on a children's version of the Empathising Quotient and Systemising Quotient scales, with boys scoring higher on the latter.[65] (It's important to remember, as we have noted before, that scores on these scales are based on ratings by parents of how their children behave, so this is perhaps a less than objective measure.)

More recently, brain responses have been added to the portfolio of measures, with fMRI measures picking up increased activity in parts of the brain associated with the 'pain matrix'.[66] Kalina Michalska, a developmental neuroscientist from the University of California, and colleagues compared self-report, pupil dilation and fMRI activity in sixty-five children aged four to seventeen. For the self-report measures, an interesting pattern emerged. Although there was little difference in empathy scores at the four-year-old level, the male scores then decreased significantly with age, whereas (you've probably guessed) the female scores increased. But neither of the implicit measures, of pupil dilation and brain activation, showed any sex differences whatsoever, even though the girls reported themselves to be significantly more upset than the boys by the video clips they had been watching.

So early signs of empathy do not appear to differentiate between the sexes, and later on, it is only self-ratings that fit the 'empathic female' model. As one study, which did not find evidence of sex differences in early empathy, surmised, 'Gender differences in empathy may become more prominent following the transition to middle childhood, as children internalize societal expectations regarding gender role and gender identity through social learning processes, and act in accordance with them.'[67] Including filling in a questionnaire to demonstrate how empathic they are, perhaps?

The brains behind it all?

Looking at adults, we have some idea about the skills that make them social, co-operative beings, and the brain networks that underpin these processes. We know that these skills need to be in tune with the environment, which will provide data for assigning more or less importance to events out there. We need to have the ability to, for example, spot tiny differences in facial identity and emotional expression, or to pick up non-verbal cues about who or what we should be paying attention to and why. We have seen that very young babies have at least rudimentary versions of these skills – are the brain networks that underpin them the same as in adults? How does the environment finetune or calibrate these networks? Tracking the brain bases of this calibration process in the developing infant can give us insight into how early this social brain scaffolding is put in place and how its construction might reflect the effect of the world on the developing infant.

As we have seen, face processing in babies is evident from the off. As soon as they arrive in the world, newborn babies prefer faces and face-like patterns to other types of stimuli. The idea of an inbuilt face preference system, as suggested by Mark

Johnson, is based on the fact that this preference seems to precede the maturation of the visual system (in other words, before babies' eyes fully work together); so this preference is not just a response to a common visual pattern. But pretty soon, babies and their brains are showing quite sophisticated cortical responses that match those shown by adults. Babies as young as three months old show the same kind of brain responses to faces and face-like stimuli as adults, and they are found in similar parts of the brain to those that deal with face processing in adults.[68]

Johnson has pointed out that tracking the age-related changes in the brain and the baby face-processing abilities gives powerful insights into the fine tuning of this important social skill. Babies do like faces, and generally their mothers' faces most of all, but initially they are cheerfully undiscriminating when it comes to other faces; newborns show no preference for own-race as opposed to other-race faces, but by the time they are three months old they begin to display this early form of in-group/out-group discrimination.[69] Researchers have also shown that this own-race effect is a powerful measure of environmental input, as it isn't shown in infants who are reared in a racial environment different from their own.[70] So there does not appear to be any kind of inbuilt preference for 'people like me'; it is something we learn.

As well as racial differences, children get more discriminating about familiar and unfamiliar faces during the first year of life, but there is evidence that even at eight to twelve years old they are processing faces differently from adults, with the face-specific regions of their brains being activated by a wider range of face-like stimuli.[71] So, again, a crucial aspect of social behaviour, though present very early on, does not reach its endpoint for several years, and after much experience in the social world.

Another example of fine tuning is the making and then breaking of a link between eye gaze and face processing. We know eye gaze is a key part of social communication; a conversation will quickly break down if there isn't at least some mutual gaze between the speakers. Newborn babies appear to be aware of this process, and prefer faces when the eyes are looking directly at them, quickly becoming distressed if they are presented with a face with an averted gaze.[72] In adults, the eye gaze control systems in the brain are distinct from the face-processing network and more closely linked with the theory of mind network, suggesting that as we get older eye gaze becomes seconded to a more generic role in mind reading and the interpretation of intentions.[73] But in four-month-old infants, brain activity elicited by direct eye gaze is more closely associated with the face-processing area.[74] So here we have an example of a quite focussed social skill shown by young babies becoming adapted to a wider range of social requirements. Or, to think of it another way, a rudimentary social app getting an experience-dependent upgrade to fit it for more sophisticated activities.

Processing of emotion as signalled by faces follows on from the recognition of faces themselves. How sophisticated are babies' tiny brains when it comes to this key part of developing mind-reading skills? Early studies suggest that infants of about six or seven months respond more to fearful than to happy faces, with brain activity arising from the frontal areas, including our friend the anterior cingulate cortex.[75] Does this mean babies just respond to negative rather than positive emotions, perhaps as they could be more useful for survival? But if you get baby brains to compare faces showing anger, another type of negative emotion, with happy ones, in this case, the happy faces get the brain vote.[76] Perhaps babies are just more used to happy faces (assuming our emerging socialites are mainly surrounded by smiles)? But it is clear that, before the age of one year, babies have the right neural wherewithal to tell whether they are looking at someone who is sad, happy, afraid or angry, which is a pretty useful social skill to have acquired so early.

Sharing an experience is a core feature of social engagement. At an adult level it may be a nudge in the ribs and a chin-based gesture towards some kind of external happening that could elicit a raised eyebrow, a sneer, a tut, even a smile. How do you draw a very young baby's attention to something? Here eye gaze comes in useful again; you can monitor baby brain responses when, for example, an adult looks at an infant and then at a computer screen showing a novel object, an unspoken 'Hey, look at this' command. Tricia Striano, a cognitive neuroscientist from Leipzig, has shown increased activity in the frontal areas of the brain using this paradigm.[77] As we noted in the last chapter, it seems the prefrontal areas in the infant brain are not as functionally silent as was previously thought.

Tracking a timeline for the emergence of social cognition, as UK cognitive neuroscientists Francesca Happe and Uta Frith have done, confirms that high-level social skills and their neural underpinnings are in place in humans from a very early age.[78] The apparent quests for an understanding of society and other people appear to precede the emergence of cognitive skills.

As is often the case, both in psychology and in neuroscience, you can learn a lot about a brain-based process by studying how it develops over time, or when it is transitioning from one stage to another. The availability of better techniques for studying infant brains combined with the ever-present ingenuity of developmental psychologists for devising cunning tests of infant skills is giving us extraordinarily revealing insights into the powerful processes behind becoming a social being.

All of this should give us pause for thought about the world these questing brains are encountering. Babies are tiny social sponges, hungry for experience, ready to engage with what their new world has to offer. But what exactly does the world have in store for them?

Chapter 9:
The Gendered Waters in Which We Swim – The pink and blue tsunami

---

Children are actively searching for ways to find meaning in and make sense of the social world that surrounds them, and they do so by using the gender cues provided by society to help them interpret what they see and hear.

C. L. Martin and D. N. Ruble, 2004[1]

Although human babies appear to be pretty passive and helpless when they are born, with brains apparently still in the very early stages of development, it's clear they actually arrive with quite sophisticated cortical start-up kits. Their sponge-like ability to soak up information about the world around them means we need to pay particular attention to what that world is telling them. What kinds of rules and guidelines will they encounter? Will there be different rules for different babies? And what kinds of events and experiences might determine their final destination?

One of the first, one of the loudest and one of the most enduring social signals a baby will pick up on is, of course, about the differences between boys and girls, men and women. Messages about sex and gender are almost everywhere you look, from babies' clothes and toys, through books, education, employment and the media, to everyday 'casual' sexism. A quick trawl through the supermarket can generate a list of pointlessly gendered products – shower gel (Tropical Rain Shower if you're female; Muscle Therapy if you're a man), throat lozenges, gardening gloves, trail mix (Energy Mix for men and Vitality Mix for women), Christmas chocolate sets (tool kits for boys, jewellery and make-up for girls) – ensuring a consistent theme whereby, even when we're only thinking about sore throats or rose pruning, we need to tag them with a gender label, to make sure that 'real men' don't use the 'wrong' sort of gardening gloves, or that 'real women' don't accidentally smother themselves in manly Muscle Therapy.

In June 1986, I was in a labour ward, having just given birth to Daughter #2. It was the night that Gary Lineker scored a hat-trick in the World Cup; nine babies were born that night, eight boys and one girl (mine), and all but the latter allegedly called Gary (I was tempted). I was comparing notes with my neighbour (not on the football), when we became aware of what sounded like an approaching steam train, getting louder by the second: our new babies were being wheeled towards us. My neighbour was handed her blue-wrapped bundle, with the approving words: 'Here's Gary. Cracking pair of lungs!' The nurse then passed me my package, wrapped in a yellow blanket (an early and hard-won feminist victory), with a perceptible sniff. 'Here's yours. The loudest of the lot. Not very ladylike!' Thus at the tender age of ten minutes, my tiny daughter had her first encounter with the gendered world into which she had just arrived.

Stereotypes are so much part of this world that, if asked, we could unquestioningly generate lists of what people (or places, or countries, or jobs) are 'like'. And if we compared the results of such a survey with those generated by our colleagues or neighbours, we'd find a high level of agreement. Stereotypes are cognitive shortcuts, pictures in our head that, when we encounter people or situations or events or anticipate doing so, allow our brains to get on with their predictive texting and fill in the gaps to swiftly generate a helpful prior that will guide our behaviour. They are part of our social semantic stores and social memory, shared with other members of our social network.

Way back in the late 1960s, a team of psychologists devised a stereotype questionnaire.[2] They asked college students to list the behaviours, attitudes and personality characteristics that they believed to be typically male or typically female. The typically female characteristics were grouped under one heading, the typically male under another. There were forty-one items on which there was seventy-five per cent or better agreement and these were designated as the stereotypic labels. The 'female' items included descriptions of women as 'dependent', 'passive' and 'emotional', whereas men were 'aggressive', 'self-confident', 'adventurous' and 'independent'. This became the Rosenkrantz Stereotype Questionnaire, which measured agreement with the items listed as an index of the extent of respondents' stereotypical thinking. One key additional issue is that the students were asked to rate which traits they thought most socially desirable; all the students rated the stereotypically male traits as more desirable.

Thirty years after the survey's first creation, the original items on the questionnaire were retested.[3] There was some evidence of a shift in stereotypical thinking, with rather fewer items confidently rated as typically male or typically female. Anything to do with experiencing or expressing emotion remained firmly female, but the typical woman was no longer rated as less logical, direct or ambitious. The typical man remained more aggressive, more dominant and less gentle. There was a significant shift in that the 'new' female traits were seen as more socially desirable, leading the authors to speculate that the 'exposure of people to women in a greatly expanded range of roles' was behind the changes in gender stereotypes.

But a separate review suggested otherwise.[4] The researchers here compared responses from a 1983 questionnaire that asked respondents to identify typical male–female traits (similar to the Rosenkrantz study), but also role behaviours, physical character- istics and occupations, with responses to the same questionnaire in 2014. The only shift in attitudes in the thirty intervening years concerned typical female role behaviours, where there was actually an increase in gender stereotyping, with the only task attributed equally to males and females being 'handles financial matters'. 'Agency' or action was still seen as a core characteristic attributed to men, encompassing traits such as competence and independence, whereas 'communion' or networking was still seen as a core female attribute, associated with warmth and care for others; the same set of occupations were still seen as typically male or female (for example, politician versus administrative assistant); and, perhaps less surprising, men and women were still differentiated on the same set of physical characteristics (such as height and strength). So the initial report that gender stereotyping was succumbing to social changes was perhaps overoptimistic.

It was suggested that there are two major processes which might predict change in or stability of stereotypes. If gender stereotypes are based on real-time observations of men and women, then ongoing changes in society should elicit changes in them. But if stereotypes are more deeply entrenched, then they won't be shifted by societal changes. The operation of processes such as 'confirmation bias', where you are more likely to value or believe evidence that supports your existing beliefs, or even 'backlash', where there is an emphasis on the negative consequences of attempts to overcome pre-existing stereotypes, can embed stereotypes more firmly into the social psyche.[5]

As we have seen, our social brain is something of a rule scavenger, seeking out the laws of our social systems and the 'essential' and 'desirable' characteristics we should have to enable our self to fit in with our identified in-groups. This will inevitably include stereotypical information about what 'people like us' should look like, how we should behave, what we can and can't do. There appears to be a relatively low threshold attached to this aspect of our self-identity as it can very easily be triggered or primed. We have already seen how the type of manipulations that induce the stereotype threat response can be pretty low-key.[6] You don't need much reminding that you are an underperforming female to become an underperforming female. Or even just a reminder that you are female, with your 'self' doing the rest. This is even shown in four-year-old girls, where colouring in a picture of a girl playing with a doll is associated with poorer performance on spatial cognition tests.[7]

The brain networks associated with processing and storing social labels are different from those associated with processing and storing more general-knowledge-type items.[8] And the stereotype-processing networks overlap those associated with self-processing and social identity. So attempts to challenge stereotypes, particularly those related to a self-concept ('I am a male and so …'; 'I am a female and so …'), will entail more than a quick adjustment to a general knowledge store, however well informed. These kinds of beliefs are deeply embedded in a process of socialisation which is at the heart of being a human.

Some stereotypes have their own inbuilt reinforcement system as, once triggered, they will drive the behaviour that is attributed to the stereotypical trait. For example, consider the effect of stereotype threat on spatial performance, where performance on a mental rotation task can be altered by summoning up a positive or a negative stereotype.[9] As we will see next, stereotyping the kind of toys that are 'for girls' or 'for boys' can affect the range of skills they acquire; girls who think Lego is for boys are slower at construction-based tasks.[10]

And sometimes stereotypes can serve as a form of cognitive hook or scapegoat, where poor performance or lack of ability can be attributed to exactly the shortfall characterised by them. For example, premenstrual syndrome has been used to explain or be blamed for events which could equally well be attributed to other factors, as we saw in Chapter 2. One study showed that women were likely to blame their own menstrually related biological problems for negative moods, even when situational factors could equally well be the source of difficulties.[11]

Some stereotypes are proscriptive as well as descriptive: as well as emphasising negative aspects of ability or temperament, they appear to 'lay down the law' about what kind of activities are suitable or unsuitable for subjects of their commandments. More significantly, they reinforce enduring signals that one group is better than the other at key activities, that there are things that members of one group just 'can't' do and should probably avoid, a 'superior/inferior' angle. The stereotype that women *can't* do science means that they *don't* do science, leaving science as a masculine institution full of male scientists (helped out by some pretty determined gatekeeping). Today's stereotypes may be more subtle than the two-headed-gorilla-type tag, but as Angela Saini has detailed in her book *Inferior*, there are many examples of how women's health, their work and their behaviour from birth until old age have been characterised as less adaptive or less socially useful than males'.[12]

This echoes a study in 1970, where clinical psychologists (both male and female) appeared to be drawing a clear distinction between the traits of a typical healthy adult and those of a typical healthy female. Most worryingly, the traits they listed as characteristic of a typical female (dependent, submissive) were not characteristic of someone these therapists might consider pyschologically healthy. Their conclusion draws a rather chilling picture of a life of low expectations: 'Thus for a woman to be healthy from an adjustment viewpoint, she must adjust to and accept the behavioural norms for her sex, even though these behaviours are generally less socially desirable and considered to be less healthy for the generalised competent, mature adult.'[13]

Last year a survey by the UK's Girlguiding charity reported that girls as young as seven felt boxed in by gender stereotyping.[14] Polling some 2,000 children, they found that nearly fifty per cent felt it reduced their willingness to speak up and participate in school. A commentator on the survey noted: 'We teach girls that pleasing others is the most important virtue and that being well-behaved is contingent upon being quiet and delicate.'[15] It's clear these stereotypes are not harmless, but have real impact on girls (and boys) and the decisions they will go on to make about their lives. We must remember that our children's developing social brains will always be on the lookout for the rules and expectations that go with being a particular member of a social network. It is clear that sex/gender stereotypes are offering very different guidelines to girls and boys, and that those being input into our little females do not seem to be giving them a confidence-fuelled clear run to potential pinnacles of achievement.

Junior gender detectives

Given the relentless gender bombardment from social and cultural media that is evident in the twenty-first century, the associated stereotypes are likely to become much more frequently primed and embedded in our understanding of the social 'requirements' of the gender with which we identify. Alarming statistics indicate that very young children have ready access to this source of gendered information; twenty-five per cent of three-year-olds go online daily and twenty-eight per cent of 3–4-year-olds are now exposed to tablet computers.[16] In the US, 2013 data indicate that eighty per cent of 2–4-year-olds use mobile media, up from thirty-nine per cent in 2011.[17]

So 'gender coding' or 'gender signalling' is a part of the world into which the data-hungry brains of our little humans will be plunged, right from day one. And will babies and young children pick up on these kinds of messages? Will they be paying atten-tion to colour-coded toys and gendered games and who gets to play in the Wendy house? You bet!

We have long known that even quite young children are avid gender detectives, actively seeking out clues about gender, who does what, who can play with whom and with what. Developmental psychologists, monitoring the use of gendered language in young children, watching them play, or asking them to sort pictures or objects into 'boy things' or 'girl things', reported that children as young as four or five years old had well-developed awareness of the differences between males and females, not only in terms of how they looked and what they would normally wear, but also in terms of linking these differ-ences to the kind of things they might do: men were firefighters, women were nurses; men did the barbecuing and mowed the lawn, women washed the dishes and did the laundry. And they could also label everyday objects as male (hammer) or female (lipstick), and toys as 'boys' toys' or 'girls' toys'.[18]

But might our junior gender detectives start understanding these kinds of differences even earlier than we suspected? It was assumed that this kind of social skill emerged once a child was learning to talk and socialise. It was anyway difficult to test the extent to which very young children might be developing early gender 'schemata', networks of linked information about males and females. But once the 'baby-watching' techniques described in the previous two chapters were applied to this question, the early sophistication we saw in our mini-magistrates and baby scientists became evident here as well.

A very early rule that our tiny 'deep learners' picked up was about the links between key physical differences in males and females – that there are high or low voices, and that these normally match different types of faces. Using a preferential-looking paradigm, six-month-olds will gaze longer at high-pitched voices matched with male faces, or deep voices matched with female faces, showing that their neatly established prior of who has a high voice and who has a low one has been violated. So quite early on, little humans notice that there are generally two groups of people that they can reliably tell apart.[19]

The emergence of language offers clear insights into the clue-gathering activity of our junior gender detectives. Using gender-specific labels, such as 'girl' as opposed to 'child', appears quite early on in the language development timeline. A team of New York psychologists tracked the emergence of such labels in a group of children aged from nine to twenty-one months and found that there was little evidence of gender labelling before seventeen months, but by twenty-one months most of the children were appropriately using multiple labels such as 'man', 'girl' and 'boy'. And this included self-labelling ('me little girl') as well as the tagging of people and things in their outside world.[20]

The researchers also noted that the girls produced such labels earlier than the boys. They offered socialisation as a possible explanation for this, noting that 'girly' clothes and decorations are more distinctive (the 'PFD' phenomenon – which you will know, of course, stands for 'pink frilly dress'), so that girls are offered earlier visible clues about which individuals are girls and what these girls should wear. A later study by some members of this team showed that 3–4-year-old girls were much more likely to go through a phase of 'gender rigidity' in their appearance, showing implacable opposition to wearing anything other than skirts, tutus, ballet shoes and, yes, pink frilly dresses.[21]

And the clues our young detectives are picking up are not just about themselves. The children show a surprisingly early level of general gender knowledge as well, tagging items or events in the outside world as 'gender appropriate'. Show a 24-month-old a picture of a man applying lipstick or a woman putting on a tie and you will certainly capture her attention.[22]

Alongside the ability to accurately recognise the different gender categories and their associated characteristics, children seem to be strongly motivated to fit in with the preferences and activities of their own sex, as the PFD research has already indicated. Once they have worked out which group they belong to, then they can become pretty rigid in the choices they make about who and what they want to play with. They can also be quite ruthless about excluding non-members of their group; rather like newly inducted members into an exclusive society, they ensure they themselves slavishly follow the rules and are quite stern about ensuring that others obey them too. They will issue very firm statements about what girls and boys can and can't do, sometimes appearing to deliberately ignore counter-examples (a female friend of mine who is a paediatric surgeon was assured by her four-year-old son that 'only boys can be doctors') and expressing amazement when they are presented with examples, such as female fighter pilots, car mechanics or firefighters.[23] Up to the age of about seven, children are quite inflexible in their beliefs about gender characteristics and will obediently follow the route their gender satnav has set for them.

Later on, children may appear more accepting of exceptions to gender rules about who is or isn't better at some particular activity but it can be shown that, rather worryingly, their beliefs may have simply 'gone underground'. Such 'implicit' beliefs are, by definition, difficult to access but ways have been found. This has been demonstrated through a version of the Stroop task that we met in Chapter 6. If you recall, if the word 'green' is written in green, you can name the colour it's written in pretty fast. If, however, the word 'green' is written in red, you slow down quite dramatically. This is a measure of the interference effect caused by a mismatch between the different types of information you are processing. In a clever auditory version of this, listeners have to identify the sex of the person saying particular words, some of which were stereotypically male (football, rough, soldier) or female (lipstick, make-up, pink). Children as young as eight were much, much slower and made many more mistakes when they heard 'mismatches' (like a male voice saying 'lipstick' or a female voice saying 'football').[24] So, in their brief life, young children appear to have already generated some kind of internalised map of the kind of things associated with being male or female which may be subliminally guiding them to predetermined endpoints.

Our junior detectives are quickly finding out about gender stereotypes, those cognitive shortcuts or 'pictures in our head' that bundle up many allegedly gender-specific qualities into two separate packages, with quite different contents labels attached.

Pinkification

If anything characterises the twenty-first-century social signalling of sex differences, it is the increased emphasis on 'pink for girls and blue for boys', with female 'pinkification' probably carrying the most strident message. Clothes, toys, birthday cards, wrapping paper, party invitations, computers, phones, bedrooms, bicycles – you name it, the marketing people seem prepared to 'pinkify' it. The 'pink problem', now quite often with a hefty helping of 'princess' thrown in, has been the subject of concerned discussion in the last decade or so.[25] Journalist and writer Peggy Orenstein commented on it in her 2011 book *Cinderella Ate My Daughter: Dispatches from the Front Lines of the New Girlie-Girl Culture*, noting that there were over 25,000 Disney Princess products on the market.[26] The topic of this rampant pinkification has frequently and acutely been forthrightly criticised, in books such as this and many others, so I had thought that I might not have to cover the pink issue again. But unfortunately for us all, this is another Whac-a-Mole problem and it shows little evidence of disappearing any time soon.

For a talk I was giving recently, I was mining the internet for examples of those dreadful pink 'It's a Girl' cards when I came across something even more jaw-droppingly awful: 'gender reveal' parties.[27] If you haven't already heard of these, they go something like this: at about twenty weeks into a pregnancy, it is usually possible to tell the sex of the child you are expecting from an ultrasound scan, thus, apparently, triggering the need for an expensive party. There are two versions, and both are a marketing dream. In version 1, you decide to remain in ignorance and instruct your ultrasound technician to put the exciting news in a sealed envelope and send it to your gender reveal party organiser of choice. In version 2, you find out for yourself but decide to break the news at the party. You then summon family and friends to the event via invites bearing a question such as 'A bouncing little "he" or a pretty little "she"?', 'Guns or Glitter?' or 'Rifles or Ruffles?' At the party itself you might be confronted with a white iced cake which can be cut open to reveal blue or pink filling (it may also be decorated with the words 'Buck or Doe? Cut to know').

Or there could be a sealed box which, when opened, will release a flotilla of pink or blue helium-filled balloons; a wrapped outfit from your nearest nursery store which will be opened to reveal the pink or blue creation into which you will stuff your newborn; even a piñata which you and your guests can hammer away at until it releases a flood of pink or blue candy. There are guessing games which appear to involve toy ducks (Waddle it be?) or bumblebees (What will it bee?), or some sort of raffle where, on arrival, you put your guess in a jar and win a prize once the reveal is made. Or (the front-runner for the most tasteless) you are given an ice cube containing a plastic baby, and in a 'my waters have broken' race, you try to find the quickest way of melting your ice cube to reveal whether the baby is pink or blue. In case you think I am making this up, here is a direct quote from one website advising on how to host one of these parties: 'Go simple with pink and blue cocktails, candles, plates, cups, napkins – you name it. (I even put pink and blue guest towels in the bathroom!) In lieu of the Super Bowl, have a Baby Gender Reveal Bowl.'[28]

So, twenty weeks before little humans even arrive into it, the world can already be tucking them firmly into a pink or a blue box. And it is clear from the YouTube videos (yes, I became obsessed) that, in some cases, different values are attached to the pinkness or blueness of the news. Some of the videos show existing siblings watching the excitement of 'the reveal' and it's hard not to wonder what the three little sisters made of the screams of 'At last!' that accompanied the cascading blue confetti. Just a harmless bit of fun, maybe, and a marketing triumph, for sure, but it is also a measure of the importance that is attached to these 'girl'/'boy' labels.

Even efforts to level the playing field get swamped in the pink tide – Mattel have produced a STEM Barbie doll to stimulate girls' interest in becoming scientists. And what is it that our Engineer Barbie can build? A pink washing machine, a pink rotating wardrobe, a pink jewellery carousel.[29]

You might wonder why on earth any of this matters.[30] What it all comes down to is the debate as to whether pinkification is signalling a natural biological divide (fixed, hard-wired, not to be interfered with) or reflecting a socially constructed coding mechanism (possibly associated with past social needs but with the potential to be reconstructed in the light of changing social requirements). If it is really the sign of a biological imperative then perhaps it should be respected and supported. If we're looking at a social set-up, then we need to know if the associated binary coding is still serving the two groups well (if it ever did). Apart from gender signalling ensuring that men don't accidentally use lavender and chamomile shower gel, are our journeying girl brains helped out by being directed away from construction toys and adventure books, and those of their boy counterparts from cooking sets and dolls' houses?

Perhaps we should check first if the power of the pink tide has some kind of biological basis. As mentioned in Chapter 3, a female preference for pink has already been put in evolutionary terms. In 2007, a team of vision scientists suggested that their finding of this preference was linked to a need long ago for the female of the species to be an effective 'berry gatherer'.[31] Responsiveness to pink would 'facilitate the identification of ripe, yellow fruit or edible red leaves embedded in green foliage'. An extension of this was the suggestion that pinkification is also the basis of empathy – aiding our female caregivers to pick up those subtle changes in skin tone that match emotional states. Bearing in mind that the study, carried out on adults, used a simple forced choice task involving coloured rectangles,* this is quite a stretch, but it clearly struck a chord with the media, who variously hailed the finding as proof that women were 'hard-wired to prefer pink' or 'modern girls are born to plump for pink'.[32]

However, three years later the same team carried out a similar study in 4–5-month-old infants, using eye movements as a measure of their preference for the same coloured rectangles.[33] They found no evidence of sex differences at all, with all babies preferring the reddish end of the spectrum. This finding was not accompanied by the media flurry that greeted the first one. The study with adults has been cited nearly 300 times as support for the notion of 'biological predispositions'. The study with infants, where no sex differences were found, has been cited fewer than fifty times.

Parents will still exclaim that there must be something fundamental about this preference for pink when they find that, despite their best efforts at 'gender-neutral parenting' for their daughters, all is swept away by the pink-princess tide mentioned above. Children as young as three will allocate genders to toy animals based on their colour; pink and purple ones are girl animals and blue and brown ones are boy animals.[34] Surely there must be a biological driver behind the emergence of a preference this early and this determined?

---

* Forced choice means you compare items and you *have* to pick one – options such as 'don't mind', 'don't care', 'don't know' are not available. In this case you were shown two rectangles and you had to indicate which you preferred (so you might not really like either but disliked one less than the other).

But a telling study from American psychologists Vanessa LoBue and Judy DeLoache tracked more closely just how early this preference did emerge.[35] Nearly 200 children, aged seven months to five years, were offered pairs of objects, one of which was always pink. The result was clear: up to the age of about two, neither boys nor girls showed any kind of pink preference. After that point, though, there was quite a dramatic change, with girls showing an above-chance enthusiasm for pink things, whereas boys were actively rejecting them. This became most marked from about three years old onwards. This tallies with the finding that once children learn gender labels, their behaviour alters to fit in with the portfolio of clues about genders and their differences that they are gradually gathering.[36]

As we know, brains are 'deep learners', anxious to work out the rules and avoid 'prediction errors'. So if their owners and their newly acquired gender identities venture out into a world full of powerful pink messages helpfully signposting what you can and can't do, can and can't wear, then it would take a really compelling rerouting project to divert this particular tide. So we could indeed be looking at a brain-based process, but one that has been triggered by the world in which it finds itself.

What about the evidence of a pink–blue divide being a culturally determined coding mechanism? Why (and when) pink became linked to girls and blue to boys has been a matter of quite earnest academic debate. One side has claimed that this used to be the other way round, and that, until the 1940s, blue was actually seen as the appropriate colour for girls, possibly because of its links with the Virgin Mary.[37] This idea has been critiqued by psychologist Marco Del Giudice, who, after a detailed archive search via Google Books Ngram Viewer, claimed to find little evidence of for blue-for-girls/pink-for-boys claim. He dubbed this the Pink–Blue Reversal and, naturally, an acronym (PBR) has followed; he's even awarded it the status of a 'scientific urban legend'.[38]

But the evidence for some kind of cultural universality for pink as a female colour is really not that powerful. Examples from Del Giudice's own review suggests that any kind of gender-related colour coding was established little more than 100 years ago and seems to vary with fashion, or depending on whether you were reading the *New York Times* in 1893 ('Finery For Infants: Oh, pink for a boy and blue for a girl') or the *Los Angeles Times* in the same year ('The very latest nursery fad is a silky hammock for the new baby ... First on the net is laid a silk quilted blanket, pink for a girl, blue for a boy'). To add to the confusion, the *El Paso Herald* published this letter in 1914: 'Dear Miss Fairfax: Will you kindly tell me the colour used for baby boys? Anxious Mother.' Which elicited this response: 'Pink is for boys. Blue for girls. It used to be the opposite but this arrangement seems more suitable.' Hardly a consistent message (and, unfortunately, there were no psychologists around at the time to check out any matching *Blue* Frilly Dress phenomena).

So the jury is still out in this recasting of the nature–nurture divide in terms of the biological versus the social origins of pinkification. Those who challenge the notion that there is some kind of essential link between girls and pink can find themselves seriously under fire. An article by Jon Henley in the *Guardian* in 2009 tells the story of the two sisters who started the Pink Stinks campaign, highlighting the consumer culture that was supporting harmful stereotypes. In response, one suggestion in the comments under the article was that the sisters should wear T-shirts saying 'I am a left-wing communist loony trying to brainwash girls'.[39]

In terms of understanding the significance of pinkification for our journeying brains, the key issue is not, of course, pink itself but what it stands for. Pink has become a cultural signpost or signifier, a code for one particular brand: Being a Girl. The issue is that this code can also be a 'gender segregation limiter', channelling its target audience (girls) towards an extraordinarily limited and limiting package of expectations, and additionally excluding the non-target audience (boys). Tricia Lowther, a Let Toys be Toys campaigner, points out that the kind of toys that are now coded pink-for-girls are almost universally associated with dressing up (so emphasising the importance of appearance) or with domestic activities such as cooking or hoovering, or looking after fluffy pets or baby dolls. No problem with that, but it also means that these little princesses are *not* playing with creative construction toys or having adventures as superheroes.[40]

'Agents of socialisation', such as that notorious Barbie, can convey career-limiting messages to girls. Aurora Sherman and Eileen Zurbriggen showed that girls who had played with 'Fashion Barbie' dolls were less likely to choose male-dominated careers such as firefighter, police officer, pilot as possibilities for themselves than girls who had played with a more neutral toy (and both sets of girls showed pretty low career aspirations anyway).[41]

Paradoxically (and in fairness to the other side of the argument), sometimes pink appears to serve as a kind of social signature that 'gives permission' for girls to engage with what would otherwise be seen as a boy domain. But, as per my STEM Barbie example, pinkification is all too often linked with a patronising undertow, that you can't get females to engage with the thrills of engineering or science unless you can link them to 'looks and lipstick', ideally viewed through – literally – rose-tinted glasses.

Toy Story

This very clear demarcation between boys and girls, colour-coded from the outset, of course also applies to toys. The kind of toys that children play with can have significant effects on the kind of skills they may develop or role-play they will indulge in, so any process which narrows the choices for either boys or girls should be viewed with alarm.

The whole issue of the increased gendering of toys and the contribution this is making to the sustaining of stereotypes has been the focus of much concern in recent years, even to the extent of the White House holding a special meeting to discuss it in 2016.[42] Might toy choice be a major chicane for our journeying brains? Or have they already been set on this route before birth? Do toy choices *reflect* what is going on in the brain? Or do they *determine* what is going on in the brain?

Researchers in this area can be pretty firm about the status quo in this aspect of children's behaviour: 'Girls and boys differ in their preferences for toys such as dolls and trucks. These sex differences are present in infants, are seen in non-human primates, and relate, in part, to prenatal androgen exposure.'[43] This statement neatly encapsulates the sets of beliefs about toy choice in children, so let's explore the story of toys, who plays with what and why (and whether or not it matters) in those terms.

The issue of toy preference has acquired the same kind of significance as the pink–blue debate. From a fairly young age, possibly as young as twelve months, it appears that boys and girls show preferences for different kinds of toys. Given the choice, boys are more likely to head for the truck or gun box, whereas girls can be found with dolls and/or cooking pots. This has been adopted as evidence for several different arguments. The essentialist camp, supported by the hormone lobby, would claim that this is a sign of differently organised brains following their differently channelled pathways; for example, an early preference for 'spatial' or construction-type toys is an expression of a natural ability. The social-learning camp would claim that gendered toy preference is the outcome of children's behaviour being modelled or reinforced in gender-appropriate ways; this could arise from parent or family gift-giving behaviour or it could be the outcome of a powerful marketing lobby determining and manipulating their target market. A cognitive-constructionist camp would point to an emerging cognitive schema, where fledgling gender identities latch onto objects and activities that 'belong' to their own sex, scanning their environment for the rules of engagement that specify who plays with what. This would suggest a link between the emergence of gender labelling and the emergence of gendered toy choice.[44]

These are arguments about the *causes* of toy preference, what toy preferences mean to those trying to understand sex/gender differences, be they parents or cognitive neuroscientists (or both). But there are other arguments that are about the *consequences* of toy preference. If you spend your formative years playing with dolls and tea sets, will that steer you away from the useful skills that playing with construction kits or playing target-based games might bring you? Or might these different activities just be reinforcing your natural abilities, offering you appropriate training opportunities and enhanced talents for the occupational niche that will be yours? Looking particularly at the twenty-first century, if the toys you play with carry the message that appearance, and quite often sexualised appearance at that, is the defining factor of the group you belong to, does that have different consequences from playing with toys that offer the possibility of heroic action and adventure?[45] With respect to our own particular quest in this arena, might any of these consequences of early toy choice be found not only at the behavioural level but also at the brain level?

As ever, the causes and consequences issues are entangled. If gendered toy preference is an expression of a biologically determined reality, then the interpretation tends to be that it is inevitable and shouldn't be interfered with, and that those who challenge it should be sent away with the mantra 'Let boys be boys and girls be girls' ringing in their ears. Specifically for researchers, it would mean that sex differences in toy preference could be a very useful index of sex differences in underlying biology, a genuine brain–behaviour link. On the other hand, if gendered toy preference is actually a measure of different environmental input it would be possible to measure the different impacts of that input and, perhaps more importantly, the consequences of changing it.

However, before we launch into the pros and cons of the various theories attached to toy preference, we need to look at the actual characteristics of these differences. Is it a robust difference, reliably found at different times, in different cultures (or even just in different research studies)? Who actually decides what is a 'boy toy' and what is a 'girl toy'? Is it the children who play with them or the adults that supply them? In other words, whose preferences are we actually looking at?

'Of course I would buy my son a doll'

Among adults, there appears to be pretty widespread agreement as to what constitutes male-typed, female-typed and neutral toys. In 2005, Judith Blakemore and Renee Centers, psychologists from Indiana, got nearly 300 US undergraduates (191 females, 101 males) to sort 126 toys into 'suitable for boys', 'suitable for girls' or 'suitable for both' categories.[46] Based on these ratings, they generated five categories: strongly masculine, moderately masculine, strongly feminine, moderately feminine and neutral. Interestingly, there was fairly universal agreement between males and females about the toys' genders. There were ratings disagreements about only nine of the toys, with the largest difference concerning a wheelbarrow (rated as strongly masculine by men and moderately masculine by women); similarly, there was a bit of arm wrestling over horses and toy hamsters (rated moderately feminine by men and neutral by women), but there were no incidences of cross-gendering. So it would appear that 'toy typing' is pretty clear-cut in adult minds.

And do children agree with these ratings? Do all boys choose boy toys, all girls choose girl toys? To consider this, let's look at a lab-based study on this issue. As in many other instances we have looked at, what questions are asked, how they are asked and how the answers are interpreted can give us pause for thought in assessing claims that toy preference is one of the most robust sex differences that psychologists have found.

Brenda Todd, a psychologist from City University in London, researches children's play. Her group was interested in the emergence of preferences for 'gender-typed' toys, so they began by surveying ninety-two men and seventy-three women, aged between twenty and seventy years, in order, rather like the study above, to identify how adults might gender toys.[47] The participants were asked which toy first came to mind when thinking about a young girl or a young boy. For a boy, the most common response was 'car', followed by 'truck' and 'ball'. For a girl it was 'doll', followed by 'cooking equipment'. Teddy bears were identified as a female toy, but the researchers then argued that baby boys got teddy bears as well, so they elected to include a pink teddy and a blue teddy in their offerings. (You may ponder why the researchers, having rightly identified the need to get some outside confirmation of how to label the toys they were testing, decided to override the answers they got. And, additionally, to throw the whole pink–blue scenario into the mix.) Nevertheless, in the final selection a doll, a pink teddy and a cooking pot were given 'girl' labels, and the 'boy' labels were awarded to a blue teddy, a car, a digger and a ball.

Once these adult-labelled toys are field-tested on children, do all little boys obligingly head for the car/digger/ball/blue teddy bear? And all little girls for the doll/cooking pot/pink teddy bear? The toys were given to three groups of children: one aged 9–17 months (identified as the age when children first start to engage in independent play), one aged 18–23 months (when children show signs of acquiring gender knowledge), and one aged 24–32 months (when gender identities become more firmly established). The test involved an 'independent play' scenario, where the chosen toys were arranged in a semicircle round each child and an experimenter encouraged their participants to play with any of the toys that they wanted to. An elaborate coding procedure gave a measure of toy choice.

The boys were more obliging to the researchers in picking the 'boy toys', showing a steady age-related increase in the amount of time they played with the car and the digger. If (as you should be) you're wondering what happened to the blue teddy bear and the ball, the researchers decided (post hoc) to drop the former as there was 'no significant sex difference in play'. They also decided to drop the pink teddy as well, because the older children didn't play with either bear. And they then noticed that there were an uneven number of toys in their two categories, so they also dropped the ball (even though it actually showed a sex difference, with boys playing with it more than girls). So now it was the car and digger versus the doll and cooking pot. As you will recall, these were the top two for each group in the survey mentioned above. So the data being reported were now just from the choices between the most stereotypical toys (with no neutral or even less strongly gendered toys for comparison). Actually, in the 'debriefing' section of their report on this study, the researchers claimed this as a strength of their study, the outcome of a 'decision made in order to avoid the dilution of dichotomous sex differences in toy choice by introducing a third option'.[48]

So there is an element of self-fulfilling prophecy in the reported findings that, at all ages, the boys played longer with the toys that had been labelled 'boy toys', and the girls with the 'girl toys'. Interestingly, there was a little twist in the overall picture. For boys, a steady increase in play with boy toys paralleled a decrease in play with girl toys, but the story was different for girls. Although the younger girls appeared to be more interested in girl toys than boys were in boy toys, this interest wasn't sustained in the middle group, where there was actually a drop in the amount of time they spent with girl toys. And the girls showed an increase in the amount of time they played with boy toys as they got older. The authors of the study helpfully interpret this thus: 'Although girls initially *much prefer* female-typed toys, this preference settles to a *merely strong* preference [my italics].'[49] So, even though the researchers cheerfully admitted to stacking the odds with respect to the gender labelling of the toys they used, their little participants did not show the kind of neat dichotomy that might be expected. Given the emphasis put on toy choice as a powerful index of the essential nature of gender differences, together with the contemporary insistence from the gendered-toy marketing lobby that they are merely reflecting the 'natural' choices of boys and girls,[50] this kind of nuance in the whole toy story saga should really be given more air time.

Perhaps the matter might be settled by a recent research article that reports a combination of a systematic review and a meta-analysis of a range of studies in this area, together with an analysis of the effects of key variables, such as the age of the children in the various studies, whether or not the parent was present, even how gender-egalitarian the various countries were where the studies took place. The article looked at sixteen different studies, encompassing twenty-seven groups of children (787 boys and 813 girls) overall.[51] If anything could confirm the reliability, universality and stability of toy preference, might this be it?

The overall conclusion was that boys played with male-typed toys more than girls, and girls with female-typed toys more than boys. There was no effect from the presence of an adult (thus controlling for a 'nudge' factor), the study context (home or nursery) or the geographical location (so it would appear to hold true in different countries). But we were not given any details about what these toys were or who decided their 'gender'. The authors of this review included their own study, the one we've just looked at, where the gender typing of toys could be characterised as rather less objective than might be hoped. To be fair, the authors did raise this concern themselves, noting, for example, that jigsaw puzzles could be classified as 'girly' in one study and neutral in others. Nor were we given any information about whether the children had siblings, and what kind of toys were to be found in their home environment. So we don't know who or what sorted toys into their different categories, or what kind of experience children had of toys (however labelled) prior to being volunteered for one of these studies. Bear this in mind when considering one of the review's overall conclusions that 'the consistency in finding sex differences in children's preferences for toys *typed to their own gender* [my italics] indicates the strength of this phenomenon and the likelihood that [it] has a biological origin'.[52]

We should also think about the messages our little gender detectives are picking up about what toys they are 'allowed' to play with, given the assumption in the kind of studies we have looked at above that children are given a free toy choice. But is this free rein symmetrical? Girls heading for the toy trucks? No problem! Boys selecting a tutu from the dressing-up box? Hold on a second.

Even if there is an overtly egalitarian message, children are pretty astute at picking up the truth. A small-scale study by Nancy Freeman, a teacher education expert from South Carolina, illustrated this neatly.[53] Parents of 3–5-year-old children were quizzed on their attitudes to child rearing and were asked to indicate their agreement or disagreement with statements such as 'A parent who would pay for ballet lessons for a son is asking for trouble' or 'Girls should be encouraged to play with building blocks and toy trucks'. Their children were then asked to sort a pile of toys into boy toys and girl toys and also to indicate which toys they thought their father or mother would like them to play with. There was agreement about which toys were which, divided along predictably gendered lines, with further agreement of parental approval for playing with matched-gendered toys: tea sets and tutus for the girls; skateboards and baseball mitts for the boys (yes, some of these children were only three years old). Where the disconnect emerged was that these little children had very clear understanding of the level of approval they would get for playing with a 'cross-gendered' toy. For example, only nine per cent of five-year-old boys thought their father would approve of them choosing a doll or a tea set to play with, whereas sixty-four per cent of the parents had claimed they would buy their son a doll, and ninety-two per cent didn't think ballet lessons for boys were a bad idea. With a rule-scavenging brain on the lookout for gender clues, these children have either misread the message or, as Freeman proclaims in the title of her paper, are good at picking up 'hidden truths'.

What happens if you deliberately invent the labels of toys as 'for boys' or 'for girls'? This was tested on another group of 3–5-year-olds, fifteen boys and twenty-seven girls.[54] Children were presented with a shoe shaper, a nutcracker, a melon baller and a garlic press, either in pink or blue, with the objects randomly labelled as 'for girls' or 'for boys'. Children were asked how much they liked the toys and who they thought would like and play with them. Boys were much less affected by either the colour or the labels, rating them all as just about equally interesting. Girls, however, were much more gender-label compliant at one level, quite strongly rejecting the blue boy toys and approving of the pink girl toys. But they also showed a significant shift in approval rating for so-called boy toys if they were painted pink, for example, earnestly indicating that other girls might just like the 'boyish' garlic press if it could be produced in pink. The authors describe this as a 'giving girls permission' effect, where the effect of boy labelling can be counteracted with a girlish colour wash. What a dream result for the marketing industry!

So, with respect to toys at least, girl choices do seem to be affected more by the social signals, in this case verbal and colour gender labels. Why might the same not be true for boys – why would they not be equally enthused by a 'girly' melon baller if they could have it in blue? Could it be that, while girls are generally not *discouraged* from playing with boy toys, and, indeed, may occasionally be given permission to pick up the odd hammer (as long as it has a soft pink handle, of course), the reverse is not the case, with evidence of active intervention, particularly from fathers, if boys appear to be choosing to play with girl toys?

An increasing concern in the twenty-first century is the power of marketing in determining toy choice. Given that we know that children are anxious to fit in with their social circle and that they are always checking the rules of that circle, then they will respond strongly to messages about 'gender-appropriate' toys (and, of course, shoes and lunch boxes and pyjamas and bicycles and T-shirts and superheroes and school bags and wallpaper and Halloween costumes and sticking plasters and books and duvet covers and chemistry sets and toothbrushes and tennis rackets – feel free to add your own pointlessly gendered product of choice!).

The extreme gendering of toys as a recent phenomenon has received much attention. Those of us who had our children in the 1980s and 1990s feel that the marketing of toys to *their* children is much more gendered now than it was then. But, according to Elizabeth Sweet, who has made a detailed study of the history of toy marketing, this may be because we were then experiencing the effects of the second wave of feminism.[55] She points out that there was clear evidence of gendered toy marketing in the 1950s with a focus on fitting little humans into their stereotypical roles – toy carpet cleaners and kitchens for the girls, construction sets and tool kits for the boys. Between the 1970s and the 1990s, gender stereotypes were much more actively challenged, and this was reflected in more egalitarian toys (which could, of course, be good news for any attempts to reverse the gendered toy marketing (GTM) trend). But that seems to have been swept away in recent decades, partly due, Sweet feels, to the deregulation of children's television, so that children's programmes could be commercialised and used as marketing opportunities, driving the 'need' for Rainbow Brite or She-Ra or the next Power Ranger.

Grassroots campaigns such as Let Toys Be Toys have reflected increasing concern about the potential power of GTM, particularly where it might be encouraging a self-construct emphasising the prime importance of physical appearance for girls. Research has suggested a link between the perils of this kind of perfectionism and mental health problems such as eating disorders.[56] In addition, if the messages conveyed by such stereotyped toys serve to limit the choices of either gender, then it is a source of stereotyping we could do without.

So it is clear that boys and girls *do* play with different toys. But an additional question should be – why? Why do boys prefer trucks and girls dolls? Is it because they're meekly complying with the social rules their families, social media and marketing moguls are pressing upon them? We know that boys and girls are offered different toys by their parents and that a boy's toy cupboard is likely to be different from a girl's from as young as five months. So, if you're out looking for the rules of engagement, then toy choice is pretty heavily signalled. Our junior gender detectives are acutely tuned in to what is expected of them, and claiming that toy choice is actually inevitable, as with pink preference, ignores the power of the social signals that our highly sensitive deep learners are being bombarded with from extraordinarily early on in their lives.

But perhaps these toys are serving some kind of innate need, some kind of training opportunity to ensure that you are well prepared for your biological destiny. If the toys encourage 'cradling', might you play with dolls because some primordial driver (social prior constructor) knows that that will make you a better mother? If your toy of choice is a 'manipulation' toy, is this a response to your 'engineering' gene?[57]

Back to the berry gatherers and the horizon-scanning hunters perhaps? Have the social rules just evolved to incorporate toys to ensure that males and females acquire the distinct and 'appropriate' skills necessary for future social roles? To examine this proposition, we would need to see if there is some kind of innate driver behind toy preference. We would need to examine toy choice in very young infants who would supposedly not have been exposed to socialising influences, or even in non-humans, again on the assumption that we wouldn't need to take socialisation factors into account.

Not those bloody monkeys again

Newborns can't reach or grasp for anything. They are hostage to the toys that their caregivers give them. These caregivers will probably have their own ideas as to what is appropriate for their tiny charges, even if it is just to ensure that it is the toy that was given by whoever is about to visit.

We know that the apparent preference shown by newborn boys for mobiles and girls for faces has been generally refuted and never replicated. Gerianne Alexander from UCLA measured eye gaze time and frequency in 4–5-month-olds looking at dolls and trucks, with the frequency measure suggesting a girl preference for dolls.[58] But as we saw above, there is already evidence of gender differences in babies' toyed environment from as young as five months, so it is hard to extract a definitive answer to the question about whether toy preference is present from birth. Throw some older siblings into the mix, together with more or less gender-aware grandparents or childminders, and it is hard to see just how you might get proof of this assertion. The idea, of course, is that newborn infants supposedly offer a chance to look at pre-socialisation behaviour, though those gender reveal parties would suggest these babies are not entering a world without expectations.

But there is (again, supposedly) another way of finding out what toys might be chosen by 'unsocialised' individuals. In my experience, whenever the 'innateness' of children's toy preference is debated, at some point someone will say: 'But what about the monkeys?' This is because a compelling 'monkey myth', accompanied, in some cases, by a convincing little video clip, has entered the public consciousness as proof that toy preferences are not socially constructed but are really biologically based. I once appeared on a Sky News breakfast programme following up a claim that a shortage of carers could be 'cured' by getting boys to play with dolls.[59] They asked me to do a sound check just when the presenter announced in my earpiece that they would be showing this monkey clip prior to my appearance. So somewhere in the archives of Sky News is a recording of my exasperated, and apparently clearly audible, tones exclaiming: 'Not those bloody monkeys again!'

Various versions of this video show male monkeys eagerly grabbing wheeled toys, almost appearing to 'brmm-brmm' them along the ground like little boys with toy trucks, whereas their female counterparts can be seen cradling doll-like toys. As, it is claimed, monkeys can't possibly have been exposed to gender socialisation processes, this 'clear-cut' gender divide is proof that toy preference is a reflection of some kind of biological bias, a 'natural' expression of gender-based predispositions either to 'manipulate' or 'cradle', with a whole raft of downstream consequences for lifestyle choices and future careers.

There are two oft-quoted studies in this effort to disentangle 'nature' from 'nurture'. One of these is by Professor Melissa Hines, now director of the Gender Development Research Centre at the University of Cambridge.[60] Together with Gerianne Alexander, she studied toy preference in vervet monkeys. A large group of monkeys (male and female) were offered six different toys, one at a time (a police car, a ball, a doll, a cooking pan, a picture book and a stuffed dog), and the amount of contact time the monkeys had with each toy was measured. The findings were then reported in terms of toy gender categories, with the police car and the ball being deemed 'masculine', the doll and the cooking pan 'feminine' and the other two toys neutral. This 'genderisation' was obviously for the benefit of the researchers, monkeys, one assumes, being unfamiliar with the concept of cooking utensils – or, come to that, police cars.

What was found was that the male monkeys spent more time with one of the neutral toys (the dog) and roughly equal amounts of time with the 'masculine' ball and police car and 'feminine' cooking pan. The female monkeys spent most time with the cooking pan and the dog, followed by the doll, with the least time with the ball and the police car. So the monkeys' 'gender' wasn't really neatly aligned with those of the toys they made contact with. But the overall summary of the findings, while statistically correct, rather obscured this fact, referring to simple overall comparisons which showed that females spent more time with the feminine toys and males with masculine toys. There was no mention of the overall winner, the gender-neutral furry dog, or the males' attraction to the cooking pan.[61] The paper also contained images of a female monkey with the doll (even though that wasn't their overall favourite) and a male monkey with the police car (again, not their favourite). When the toys were regrouped along non-gendered lines, according to whether the toys were animal-like (dog, doll) or object-like (pot, pan, book, car), no sex differences in monkey toy preference were found.

The second example that is frequently pulled out in defence of the 'nature' camp is a later study, this time with rhesus monkeys, which involved a simpler comparison where the monkeys were presented with a choice between plush (or soft) toys and wheeled toys.[62] In this case there was a more explicit hypothesis about what toy preference might be demonstrating, targeting the opportunity for either 'active manipulation' or for 'cradling'. The female monkeys didn't appear to distinguish much between plush or wheeled toys, whereas male monkeys did show a marked preference, clearly scorning the plush toys for the chance to interact with the wheeled ones.

It should be noted that while the females did play with the wheeled toys less than the males (touching them on average 6.96 times compared with the males' 9.77 times), there was quite an overlap in the scores (a moderate effect size of 0.39). It should also most particularly be noted that nearly half of the original group of male monkeys and nearly two-thirds of the females actually couldn't be bothered with the toys at all, interacting with them so rarely that they were dropped from the study.

In summarising their results, the authors state that 'the magnitude of preference for wheeled over plush toys differed significantly between males and females'.[63] While, again, this is statistically true, it rather masks the fact that both males and females showed a pretty similar level of interest in the wheeled toys (and that, although the males did play least of all with the plush toys, there was an enormous variability in this effect, so some males were quite enthusiastic about the Winnie-the-Poohs and the Raggedy Anns).

The authors of both these studies strongly emphasise that male monkeys 'show more interest in boys' toys than do female monkeys'. But, as we have seen, in the first study the differences reflected the fact that the female vervets weren't that keen on one of the boy toys (the police car), whereas in the second study the male rhesus monkeys didn't prefer the girl toy, but the female monkeys were pretty happy with either kind (although, in the spirit of full disclosure, it should be noted that one of the wheeled toys was a supermarket trolley).

You might understandably be rolling your eyes by now and thinking 'enough of the monkeys'. But these monkeys do not go away. A news item on whether encouraging boys to play with dolls might increase the number of carers in the UK? Roll out the 'monkeys with toys' clip. A BBC *Horizon* programme on whether your brain is male or female? A quick visit with an armful of toys to a monkey sanctuary is a must-have. In a debate between Elizabeth Spelke and Stephen Pinker about women's natural aptitude for science (or lack of it), the monkey findings were one of the pieces of evidence quoted by Pinker as proof of the biological basis of sex differences in scientific aptitudes.

So our search for clear-cut toy preference in pre-socialised individuals, be they human or simian, has not yet revealed a sound basis for the suggestion that this is a good proxy measure for underlying innate sex/gender differences. So, rather than look at the 'toy choice' side of this biology-equals-destiny (aka dolls versus trucks) equation, let's have a closer look at the biology side.

Hormonal hurricanes

The hunt for evidence of the innate aspects of toy preference has taken us, so far, to research into human infants and monkeys. A third strand of this search has been to look at the effects of prenatal hormones, particularly prenatal androgen exposure. As we saw in Chapter 2, the claims for the masculinising effects of these hormones has gone beyond the mere determination of genitals to the organisational shaping of brain structure and function and thence behaviour.[64] It is obviously ethically difficult to explore the causal role of hormones in humans by manipulating hormone levels and watching the effects, so researchers have turned to 'natural' sources of such information, where foetuses have been exposed to high levels of opposite-sex hormones, such as girls with congenital adrenal hyperplasia (CAH). These girls have been identified as an 'ideal' opportunity to investigate the power of biological forces over social pressures, or, of course vice versa. Do these girls show that exposure to 'masculinising' hormones will trump society's drive for them to be 'feminine'? Do CAH girls play differently, and with different toys, form their unaffected sisters? Evidence is emerging that this aspect of their behaviour certainly appears to be less strongly gendered.[65]

A recent study by Melissa Hines and her team in Cambridge has thrown some interesting light on the potentially differing contributions of biologically-based developmental processes and socialisation pressures.[66] She looked at toy choice in the context of self-socialisation processes, manipulating the clues that might tell children whether a toy was for a girl or for a boy, either by labelling them as such, or by allowing the children to watch the choices that other females and males were making.

The study involved CAH girls and boys aged four to eleven years, together with matched control groups, both boys and girls. Gender labels were attached to neutral toys – a green balloon, a silver balloon, an orange xylophone and a yellow xylophone. The children were told that balloons and xylophones of one colour were for boys and balloons and xylophones of the other colour were for girls. They then had a chance to play with them. The amount of time the children spent with each of the toys was timed and afterwards they told the researchers which of the two balloons and which of the two xylophones they liked best.

The children were also engaged in a 'modelling' protocol. They watched four adult females and four adult males choosing one object from sixteen pairs of gender-neutral objects (such as a toy cow or a toy horse, a pen or a pencil). In each case the female 'role models' always chose the same of each pair, with each male choosing the other. Children were then asked which object from each of the sixteen pairs they preferred.

The control children showed the expected effects of the labelling and modelling, with the girls playing with and preferring those objects labelled as for girls, or picking those objects which had been chosen by the female adults. And the same applied to the boys. But the CAH girls showed significantly reduced play time with and preferences for those toys which had been identified as 'for girls' either via labelling or via modelling.

Hines and her team interpreted these findings as reflecting hormone effects on self-socialisation processes, specifically in girls. The reduced preferences shown by the CAH girls were deemed to reflect a reduced susceptibility to the kind of socialisation pressures that would be flagged up by toy labelling or perceptions of the actions of 'gender-matching' adults.

This complements the study we looked at earlier, where the girl-specific cross-gendering effect of pink was interpreted as demonstrating girls' greater compliance with social rules, reading the pinking of a toy as 'giving them permission' to adopt it, even if only temporarily, as a gender-appropriate choice. Perhaps a fundamental sex difference might be found in a differential sensitivity to social rules, a greater drive to comply with such rules? Or perhaps this reflects a greater socialisation pressure for girls to conform? Or is one entangled with the other? Hold that thought – we'll come back to it in Chapter 12.

This model offers a major rethink of the earlier simplistic, rather unidirectional processes of brain organisation and acknowledges a central role to external factors (much in the same way that epigenetics has transformed our understanding of the relationship between a genetic blueprint and the phenotypic outcome). This gives us a much more flexible theoretical perspective with which to interpret findings to date, and offers an understanding not only of how atypical hormone activity might be reflected in gender-related behaviour, but also of the emergence of such behaviour in the first place.

The consequences of toy choice

What if toy choice is not a manifestation of a predetermined process, part of a journey to an appropriate endpoint, but is actually a determinant itself of that endpoint? Could the toys you play with, perhaps thrust upon you by the agents of a gendered world, actually guide you down a particular path – or, more worryingly, could they divert you from one?

Boys show evidence of superior visuospatial processing skills as early as four or five years of age,[67] and this ability seems to be the most robust of all the (very small) sex differences we have been discussing,[68] albeit one that is showing some signs of diminishing, and can be made to vanish entirely if you test it differently.[69] But, as we will see, there is a focus on this particular ability (or the lack of it) as the reason for the underrepresentation of women in science subjects. So if we are hoping that our little girl might grow up to be a scientist we should check that this brain route remains clear.

We know that particular parts of the brain are involved in spatial processing – but does experience of spatial tasks (which can involve construction toys and videogames) change these parts of the brain? The answer is a firm 'yes', as we have seen in the Tetris and juggling tasks we looked at in Chapter 5 – recent research has shown that what were apparent sex differences in spatial cognition were actually due to videogame-playing experience.[70] When the data were reanalysed with game-playing experience as a main effect, the differences were much more powerful (and, interestingly, didn't interact with sex differences, so game-playing girls were just as superior as game-playing boys).

Psychologists Christine Shenouda and Judith Danovitch showed that Lego blocks are a player in this debate too, with an association between a Lego construction task and stereo-typical attitudes towards girls and what they might play with.[71] As mentioned earlier in this chapter, girls as young as four were significantly slower at completing the task than other girls if they had previously been exposed to a 'gender activation' task (colouring in a picture of a girl holding a doll). In another experiment, having been read a story about a child, gender unspecified, who won a block construction competition, the girls were asked to repeat the story and the researchers noted what pronoun the children used when they referred to this competition winner. A masculine pronoun was used in nearly three of out five cases (59 per cent of the time), more than twice as often as a neutral one (27 per cent) and no less than four times as often as a feminine one (14 per cent). If girls this young are being driven away from useful experiences with construction toys, then the existence of this kind of gendered cueing deserves attention. When training on games like Tetris can be shown to dramatically alter the brain and the associated behaviour, missing out on these kinds of experiences is a real route changer for our journeying brains.

Roads not taken

There are strongly gendered messages out there, perhaps more potent than ever before. Gender signalling is in place even before our little humans arrive and their very earliest experiences will be of colour-coded signposts as to which route is open to them and which is not, which training opportunities will be available and which will not.

We have explored the very earliest points at which brains encounter the world. We have seen how unexpectedly sophisticated baby brains are, particularly with respect to the kind of adult-like networks that underpin social behaviour – eye gaze radar, for example, tuned from very early on to the nuances of who is or might be a significant other. In parallel, we have witnessed the extraordinarily advanced understanding of the rules of social engagement shown by very tiny humans – down with the Hinderers, and long live the Helpers! We have seen how the old nature versus nurture, innate versus learned, debates really do not capture the multiply entangled factors that our journeying brains will encounter. And one consistent strand in this tangle is that these brains will encounter very clear gendered messages about what is 'for' girls and 'for' boys, messages of the 'girls will be girls' and 'boys will be boys' type. These messages can be conveyed by external or internal stereotypes, by gendered beliefs about male and female aptitudes and 'appropriate' roles, firmly entrenched in a sense of self that is being constructed from day one (if not before). Our focus on pink and toys serves as an insight into how early this process begins. There is an intriguing glimpse that girls might be more susceptible to such gendering – more readily pouring themselves into society's she-mould. And that boys, despite their tutu-buying fathers' protestations, are pretty clear that they would be wise to steer clear of tiaras. So the gendered signposts and diversions in the worlds to which our developing brains are exposed are present and powerful from the very beginning.

But as we grow older we don't grow out of or away from the power of stereotypes – they can continue to mould our brains and our behaviours throughout our lives.

PART FOUR

Chapter 10:
Sex and Science

In the developed Western world, one of the gender gaps most examined and exclaimed over is the underrepresentation of women in the so-called STEM subjects: science, technology, engineering and maths. This can be illustrated by statistics from many different levels of science and across many different countries. The UNESCO Institute of Statistics 2018 report shows that, globally, only 28.8 per cent of science researchers are female. The figures for the UK (38.6 per cent) and for North America and western Europe (32.3 per cent each) show that, even in the more developed countries, women make up only around a third of the science research workforce. With respect to industry across the world, only 12.2 per cent of board members in STEM fields are female. Across the full range of the STEM workforce in the UK, a 2016 report found that there were just over 450,000 women; if there were gender parity the number would be 1.2 million.[1]

A recent review of attitudes to science across Europe reported that, at the current rate of increase in female-held science professorships, the UK will have to wait until 2063 for gender parity among academic professors, and Italy will wait until 2138.[2] At university level, in 2016, fifteen per cent of computer science and seventeen per cent of engineering and technology first-year undergraduates were female (compared to just over eighty per cent who gain entry to subjects allied to medicine, indicating that ability in science subjects is not an issue). In forty-four per cent of all state schools, no girls at all do A-level physics (although sixty-five per cent of girls have physics in their top four grades at GCSE).[3] At the other end of the scale (and of possible relevance), a recent CBI report showed that only five per cent of primary school teachers (of whom eighty-five per cent are female) hold a science or science-related degree of some kind.[4]

These gender gaps will not be news to most people – but what we still have not managed to answer is why they exist. Why are there fewer women in STEM subjects at university level and beyond? When do these gaps start to appear in our lives? And what do these gaps mean about women's and men's abilities, interests and, above all, brains? The example of the lack of women in STEM jobs provides a potent illustration of just about all the issues we have been looking at. Essentialist views of what women's brains are capable of (or not) are entangled with robustly gendered and stereotyped attitudes about science and scientists, and the effects of this can distract and divert our journeying brains. The issue of women's underrepresentation in STEM subjects is not just worrying on a social level (it is estimated that there is a shortfall of about 40,000 STEM graduates in the UK each year), but it also showcases the roles of stereotypes about science and scientists, about science and the brain, and about science and sex differences in the emergence of such gaps and, more importantly, in their apparent resistance to attempts to reduce them. What is science? Who can do it and who can't?

Sexing science – science is not for women

What do we think about if we are asked to describe science? The Science Council have defined it as 'the pursuit and application of knowledge and understanding of the natural and social world following a systematic methodology based on evidence'.[5] The last part is particularly important, stressing that scientific activity is about data, about finding ways of generating objective measurements of what is going on in the world around us in our attempt to understand it. It is a system (hang on to that word) that *should* remove us from the confusion of multiple and often contradictory anecdotes from people with prejudices, preconceived ideas or personal or political agendas of some kind.

The writer and scientist Isaac Asimov came up with a more user-friendly definition: 'Science does not purvey absolute truth, science is a mechanism. It's a way of trying to improve your knowledge of nature, it's a system for testing your thoughts against the universe and seeing whether they match.'[6] Generally, science is seen as a systematic way of asking questions, of generating and testing theories. It can explain the status quo (What causes tides? Why is the sky blue?) or it can be about discovery (of gravity, radioactivity, DNA's double helix). It can be seen as a force for good (antibiotics, cancer treatments) but also as a potential force for harm, meddling with nature (GM crops, pesticides, cloning) or creating the means for catastrophic destruction (nuclear weapons, chemical warfare).[7]

The notion that scientific knowledge is in some way different from general knowledge, in that it is acquired by the application of set rules and principles, goes way back to Aristotle. What we now know as modern science had its beginnings in the seventeenth century, but even before then there were always recognisable scientific activities in institutions such as monasteries and universities. Many such institutions were for men only, and it was rare for women to be given any kind of formal education, so although Science itself was often personified as a woman, it was an activity almost exclusively involving men.[8]

Historically, the engagement of women in science has appeared and disappeared in parallel with its changing fortunes, from its initial manifestation as a form of fashionable hobby to its establishment as a highly respected, widely recognised, rather elitist (and often extremely lucrative) profession. Science began moving from a rather unregulated pursuit of knowledge, accessible to anyone who had the means and the education to pursue it, to being an institutionalised profession, pursued in exclusive societies from which women were quite explicitly excluded. The

Royal Society was founded in 1660 as a learned society for 'natural philosophers' and physicians, but the first petition for female membership was not until 1901, with the first female members not actually elected until 1945.[9]

But when women did start gaining access to education or had the means to pursue their own academic interests, we often find them as specialists in scientific subjects. Astronomy was a particular favourite and a book solely on the topic of women astronomers was published in 1786.[10] Though there was still a whiff of sexism; topics such as geology and astronomy were seen as 'safer' for women as the classics and history might encourage political activism. But, overall, the involvement of women in science was not at that stage seen as unusual or problematic.[†]

As we've seen a few times in this book, the rise of 'essentialist' movements in the nineteenth century deemed that men and women had different biologically based qualities, and that those of women were most definitely inferior to men and certainly rendered them incapable of high levels of scientific thought. So a woman showing any kind of interest and ability in, for example, astronomy or mathematics was now more likely to

---

* The physicist Hertha Ayrton was nominated in 1902 but it was decided that, being married, she could not be regarded as a 'person' in the eyes of the law and was therefore not eligible.

† In the seventeenth and eighteenth centuries, interest in science and scientific pursuits was quite common among those women who had the money and time to pursue it. There is no evidence that they were viewed as in any way inferior – as well as proficient astronomers, they were acclaimed as being excellent mathematicians. Schiebinger describes how the *English Ladies Diary* (published 1704–1841), initially with a fairly broad remit to teach 'Writing, Arithmetik, Geometry, Trigonometry, the Doctrine of the Sphere, Astronomy, Algebra, with their dependants, viz, Surveying, Gauging, Dialling, Navigation, and all other Mathematical Sciences', morphed into a journal solely for 'enigmas and arithmetical questions' in response to the enthusiasms of its readers. In 1718 its editor wrote that women have 'as clear Judgments, as sprightly quick wit, penetrating Genius, and as discerning and sagacious Faculties as ours, and to my Knowledge do, and can, carry through the most difficult Problems'.

be described as a two-headed gorilla than praised for her 'sprightly wit and penetrating genius'.[11]

Women were caught in a pincer movement. Not only was it now deemed that their bodies as a whole and their brains in particular were unsuited to any form of taxing mental exercise, but they were deliberately excluded from those institutions where the newly emerging profession of scientist was being formed.

Beyond the physical barring of women from these scientific institutions, another way of excluding them from science is by generating worldviews of its defining characteristics, and of the requirements for its successful practice, which then turn out to be inconsistent with women's abilities, aptitudes and preferences. One version of this comes in the argument that women are underrepresented in science subjects because their interests lie elsewhere. Women are more interested in people than things, and therefore don't choose STEM subjects, which allegedly fall firmly into the 'things' category.[12]

If you recall, we looked at how this People versus Things variable is measured in Chapter 3. Although it is clearly a flawed metric, it remains a popular myth in the women-and-science arena, and is at the core of many arguments about causes (and cures) for gender gaps in STEM subjects. When, as we shall see, this is linked to a biological argument, that women's lack of interest in Thing-like occupations is linked to the brain organisation associated with prenatal hormones, this can lead to suggestions along the lines of letting nature take its course and ceasing attempts to address these gender gaps.

This also brings us back to Simon Baron-Cohen's concept of systemising and empathising. Given the definition of systemising, it will come as no surprise to see how mappable it is onto measures of the characteristics of science (especially subjects such as engineering, physics, computer science and maths) and onto the personality profiles of scientists. The Thing versus People dimension was not initially devised to apply to science in general, or to STEM subjects in particular. Similarly, the E–S dimension was not (to put it in simplistic terms) about Science versus Arts. However, the close mapping of systemising behaviour onto the characteristics of 'hard' science (unsurprising, given its definitional criteria) generated this link.

Research from Baron-Cohen's lab found that a 'systemising' style was a significantly effective predictor of being a physical sciences student, but that sex/gender itself wasn't.[13] This is somewhat surprising, given the explicit connection between sex and E–S made by Baron-Cohen. Perhaps the authors of the paper were somewhat surprised as well, as their summary of their findings suggested that sex/gender *was*, in fact, a relevant variable: 'Thus, individuals with low systemizing scores (predominantly females) may be less likely to pursue scientific academic disciplines, presumably a result of difficulties in dealing with domains in which systemizing is required.'[14] So we still have that issue of a take-home message that does more to sustain a stereotype than to reflect a rather more subtle reality.

An additional aspect of the empathising–systemising dimension and its role in essentialist explanations of gender gaps in science is that it is firmly mapped onto different brain types. Those whose empathising skills are stronger than their systemising ones are type E; those with systemising skills stronger than their empathising ones are type S; and those with an equal distribution of both are 'balanced' or type B.[15] Baron-Cohen nailed his colours to the sex/gender mast at the beginning of his book: 'The female brain is predominantly hard-wired for empathy. The male brain is predominantly hard-wired for understanding and building systems.'[16] This gives us a clear steer to a brain stereotype, and a gendered one at that.

When you look at the links being made between male brains, systemising and science, with the additional assertion that biological characteristics are fixed, it is easy to see how a misinformed stereotype of the natural, even essential, connection between sex and science can emerge. We must acknowledge the additional caveat that you don't have to be a woman/man to have a female/male brain, but our questing, rule-scavenging guidance systems may not hover too long on the semantic niceties of a 'male brain' not meaning 'the brain from a man'.

Take-home messages about sex/gender differences, particularly when they conform to pre-existing stereotypes, often come across louder and clearer than subtler qualifications.

Science is about brilliance

Another aspect of this ingrained science stereotype lies in the belief that a 'raw, innate talent' is necessary to excel in any scientific discipline. This was vividly captured by Sarah-Jane Leslie from Princeton University in a study that measured an 'ability belief' by surveying more than 1,800 academics across thirty disciplines.[17] Participants were asked to rate their agreement with statements such as 'Being a top scholar of [x discipline] requires a special aptitude that just can't be taught' (measuring a belief in some form of innate ability), or 'With the right amount of effort and dedication, anyone can become a top scholar in [x discipline]' (measuring a belief that hard work can bring success). The resulting ability belief scores in the different academic fields were then compared with the percentage of female PhD students in each discipline (as a practical measure of a gender gap). You might not be surprised to read that the greater the belief in the need for innate talent in a discipline, the fewer female PhD students there were in that discipline.

Leslie and her team also sneaked in a statement to identify sexist elements: 'Even though it is not politically correct to say it, men are often more suited than women to do high-level work in [x discipline].' Members of those disciplines (who were both male and female) that endorsed the notion of success being based on some kind of raw, innate talent were more likely to agree with this kind of statement. In science subjects, the discipline with the highest field-specific ability belief scores were engineering, computer science, physics and maths, in other words the core STEM subjects, those very areas where there is so much hand wringing about the underrepresentation of women. So, we have an endorsement of the status quo (not really women's work) and, by inference, a biology-based explanation.

Leslie has characterised this raw, innate ability by linking it to a notion of 'the Beam' in science research circles.[18] This is a special gift possessed by only a few individuals who seem to carry with them a laser-like, invisible beam of talent, which they can shine on problems that others have been struggling with over long periods of time, and near-instantaneously arrive at a solution. She illustrated this by comparing the 'feral genius' of Fox Mulder from *The X-Files* with the hard-working, rule-bound Dana Scully. A good excuse for some television watching would be to spot parallels in the many police or forensic procedurals such as *CSI* or *Criminal Minds*, noting also the respective genders of the workhorses and the feral geniuses.

Related to this is the popular metaphor in science of the 'eureka' or light-bulb moment, when a solution is alleged to have presented itself in a flash of inspiration.[19] Although two well-known examples of this, the Archimedes bath story and the Newton falling apple incident, are probably apocryphal, there are more reliable such tales, including Fleming's discovery of penicillin (spotting that the mould that had contaminated his antibiotic trials was itself working as an antibiotic) and Descartes' discovery of the concept of Cartesian co-ordinates (tracking the position of a fly crawling across a ceiling by referring to its distance from two of the walls).

How does a discovery that's attributed to a flash of inspiration or a light-bulb moment affect assessments of the quality of that discovery? Does it also contribute to the perception of the inventor as a genius, as opposed to a dogged workhorse? These ideas were tested in a series of studies by Kristen Elmore and Myra Luna-Lucero, which looked at the 'inspiration' versus 'effort' metaphor on assessments of Alan Turing's work with computers.[20] One group of participants read a passage that described Turing's work in light-bulb terms – 'an idea that struck him like a light bulb turning on' – whereas another group read about an idea that 'took root', like a 'growing seed that had finally borne fruit'. When asked to rate the exceptionality of Turing's work, the light-bulb group rated it much more favourably than the seed group.

A second study introduced a gender dimension. The invention here was in the field of wireless communication technology and told the story of Hedy Lamarr, best known as a Hollywood film star (*Samson and Delilah*) but also an accomplished inventor. She and the composer George Antheil devised a 'frequency hopping' technique that manipulated radio frequencies to prevent classified messages being read when intercepted (the basis of today's encryption techniques for mobile devices). This story was briefly presented either in light-bulb terms of 'a bright idea for a signal that would jump across multiple frequencies', or in more effortful terms of 'the seed of an idea for a signal that would jump across multiple frequencies'. The first version was illustrated with a picture, showing either Lamarr or Antheil, and a light-bulb; the second version had the same choice of inventor, this time with a picture of a small sprouting seed. Participants from each group were asked to rate the genius and exceptionality of the inventor and his or her idea.

It emerged that these ratings depended on whether the readers had been looking at the female or the male inventor. The seed metaphor significantly increased the assessment of Hedy Lamarr as a genius, whereas it significantly decreased that of her male partner. On the other hand, the light-bulb metaphor left Lamarr's readers unimpressed, but increased George Anthiel's genius ratings. The researchers suggest that this reflects the congruence between the expectations of how men can be successful, making use of that inborn extra 'something' that conjures a solution out of thin air, as compared to women's road to success, which more likely involves dogged effort and hard work.

The key aspect here is the view of 'effort' in great ideas. Generally speaking, people appear to believe that the work of geniuses is associated more with inspiration than effort, but this intersects with whether the genius is male or female. To be hailed as a genius, a man's idea has to come across as effortless, as achieved in one inspirational moment. Any suggestion that hard work or effort was involved devalues this achievement. For females, the expectation is that their achievements are almost invariably associated with nurturance and persistence, and a jolly good pat on the back is deserved when this pays off. Here, any hint of a light-bulb moment could be dismissed as a flash in the pan, a stroke of luck.

What does this all mean for women in science? If there is a worldview that the road to the top is lined with inspirational moments, and that, by the way, females are significantly less likely to have that 'certain something' associated with such moments, how much confidence might this instil in women that they are just as likely to succeed in science as men? Similarly, if effort and determination (those 'grindstone' adjectives that, as we shall see below, are much more likely to be found in letters of recommendation for females) are viewed as rather incidental qualities in the generation of successful ideas, then, as a female, you might wonder just what you could ever bring to the table in this particular institution.

Sarah-Jane Leslie's team have also looked at this by manipulating 'messages of brilliance' via hypothetical internship adverts and measuring their effect on women's interest in the posts, their assessment of how anxious they thought they might be in post, and whether or not they thought they might belong in the position's context.[21] The job descriptions emphasised either brilliance ('intellectual firecracker', 'sharp, penetrating mind') or dedication ('great focus and determination', 'someone who never gives up'). A key finding was that the messages about brilliance had negative effects on women but not on men. Women showed less interest in the 'brilliance' internship than the 'dedicated' one, and reported that the former kind of internship would make them more anxious. Men's interest or levels of anxiety didn't differ between the two. Related manipulations demonstrated how highly women rated their need for a feeling of belonging and of being like others, and that their concerns about potential mismatches arose because they had compared themselves unfavourably with others. So women themselves, consciously or unconsciously, buy into the notion that there are certain jobs, professions, careers where some kind of innate brilliance is required, and that as women they are unlikely to have that gift.

Born to do science?

Another wrinkle in the sex and science stereotypes story we are following here is that we can see these sorts of effects very early on and we can track their consequences on the diverted paths of our journeying brains. We can find sex differences in the perceptions and expectations that teachers have of their little pupils' aptitudes and abilities and, rather sadly, in the perceptions and expectations those little pupils have of themselves.

We have seen that very young children show evidence of quite sophisticated scientific skills such as awareness of numbers and quantities and laws of motion, with no consistent evidence of sex differences in maths- and science-type abilities displayed by infants. However, as mentioned in Chapter 8, recent findings suggest that there is some evidence of early (and very small) sex differences in a specific science-related skill, mental rotation.[22] Mental rotation is viewed as a key skill for success in a range of science-based activities such as architecture, engineering and design, so any kind of advantage here could give you a useful edge.[23]

We've also explored the evidence for some sex differences in toy choice among toddlers (although these are characteristically small and overlapping), with boys early on directed to objects that might enhance spatial cognition, such as construction toys, or which might index systemising-type interests, such as puzzles or mechanical toys. Although there are, of course, ongoing disputes about where such behaviours come from, with both biological and socialisation factors nominated, whatever the cause, the outcome is that there are greater science-related 'training opportunities' for boys in the early years.

It's possible, then, with respect to a minor advantage in a particular spatial skill and a higher level of spatial experience, that boys may well have a small headstart in the world of science. However, a close look at the wide range of statistics available shows that gender gaps do not exist at kindergarten level, but only start to appear in 6–7-year-olds and then get larger.[24] As we will see, it is clear that this is not wholly due to the emergence of some inherent skill but is associated with strong external forces driven by stereotypical views on who can do science (and who can't). And it can come not only from those responsible for nurturing the emergence of whatever talents there are, but also from the possessors of those talents themselves.

You might think that it's only after years of exposure to negative stereotypes about women's intellectual capacities that your ever-helpful predictive brain might pick up on the idea that, on the whole, women don't do science. Or the idea that those who do are not going to go far, and anyway you'll be very lonely and isolated if you put yourself in sciencey situations. Sadly, though, fledgling versions of these kinds of beliefs seem to be established very early on in life. In another study from Leslie's group, they examined gender stereotypes about intellectual abilities in children between five and seven years of age.[25] Using storytelling and picture-matching techniques they discovered that, at five, children tended to award the most positive 'really, really smart' ratings to models of the same gender as them (an 'own-gender brilliance' score) but, by the age of seven, girls were significantly less likely to equate brilliance with females, however they were depicted. Did these beliefs affect children's behaviour? Separately, 6–7-year-old children were introduced to two unknown videogames. Having been given the rules, they were also told that the games were either for 'really, really smart children' or for children who try 'really, really hard'. They were then asked if they liked the game and would be interested in playing it. Girls were significantly less interested than boys in the game that was presented as for smart kids, and this was related to their own-gender brilliance score. The less they believed that girls in general could be smart, the less likely it was that they themselves would express an interest in doing something that was for 'smart' people. If your established prior is that your firmly fixed female gender schema does not include a 'really, really smart' tag, then to avoid uncomfortable prediction errors you need to steer clear of anything that is labelled as only for 'really, really smart' people.

Maths is generally included as one of these things that is for 'really, really smart' people, and is not tagged as 'for girls' in our brains. The stereotype that maths is a male domain has been well demonstrated in adults, at the explicit level but also as an implicit belief.[26] If, for example, in a paired association test, the word 'maths' is paired more rapidly with the word 'male' this has been taken as a measure of a more powerful mental link between these terms than (say) a combination such as 'language' and 'male'. In this way, even if a participant explicitly denies any stereotypical beliefs, it is possible to demonstrate that such beliefs are there, even if their owner is not aware of them.

Psychologist Melanie Steffens and colleagues used this approach with nine-year-old children.[27] The presence of general gender stereotypes about boys and girls had already been demonstrated in 6–8-year-olds, and the aim of the study was to see if there was evidence of gendered stereotypes about more specific topics, such as maths or science. They also collected data about the children's performance in maths and science subjects and, additionally, asked the children about whether or not they thought they might choose to carry on with maths at a higher level. The results showed that girls had much stronger maths–male associations, much lower associations of themselves with maths or maths-type words, and much stronger intentions to drop maths. Did this just reflect the fact that they were struggling with maths? No – in fact, there weren't any sex differences in the grades the children were achieving. So, sadly, nine-year-old girls think maths is not for them, and that they will probably give it up, even though they are performing as well as their male counterparts.

Interestingly, the boys did not show any maths gender stereo-typing, so they had apparently not made the link that the girls had. Rather like the suggestion we saw in looking at toy prefer-ences and the power of pink, this might be another example of girls being more aware of the social 'rules', in this case a stereo-type about who does maths.

Another factor feeding into this could be the attitudes of parents, who have been shown to believe that maths is more important for their sons than their daughters, and are more likely to encourage boys to do higher-level science classes than their girls.[28] And, as we saw in Chapter 9, looking at very young children's awareness of their parents' likely approval ratings of their toy choice shows that children are well tuned to what is expected of them (or not), despite what those same parents will claim.[29] So, overtly or covertly, the owners of our journeying brains will be getting different recommended routes to Destination Science.

Teachers obviously have a role to play in children's acquisition of science knowledge, but it appears they also have a powerful influence over who might think of themselves as potentially successful in science. A recently reported longitudinal study on children in Israel looked at the effects of a very early 'teacher bias', calculated as the difference between marks awarded on an external blind-marked matriculation exam and those given on an internal teacher-marked version of the same type of test.[30]

The key finding here is the effect of this teacher bias in the assessment of maths performance. In the first stage of testing, girls outperformed the boys in the external exam. With respect to the teachers, there was clear evidence of a systematic bias in favour of boys, with teachers over-assessing boys' ability and under-assessing the girls'. These children were then followed up two and four years later. There were clear sex/gender differences, in high school scores, in matriculation results and, most marked, in who chose to do optional advanced-level courses. In maths, this was 21.1 per cent of the boys as compared to 14.1 per cent of the girls; in physics, it was 21.6 per cent of the boys compared to 8.1 per cent of the girls; and in computer science it was 13.0 per cent of the boys as compared to 4.5 per cent of the girls.

The researchers then modelled these data with a large range of other information to see what might be causing these differences. Could it be the size of the class, whether or not it was mixed ability? Could it be the teachers' qualifications? Could it be the level of education of the parents? Could it be to do with how many siblings the children had? None of these affected the outcome measures as profoundly as the initial teacher bias score (and we should bear in mind that the girls were outperforming the boys at the outset of this particular educational journey). It is clear that gendered expectations, even if unfounded, proved to be a powerful driving force in who arrives at the 'scientist' endpoint, with downstream consequences for higher-level employment, earnings and, of course, the overall impression (and stereotype) of who can do science.

With respect to science, then, there appear to be some strong canalising forces which can divert young females quite early on down a path that will skirt round the sciences, particularly maths. If you and your teachers think you can't, then there is a strong possibility that you won't.

Science's chilly climate

Another factor could be that science doesn't offer a very welcoming environment for women. Even if we've got past the deliberate gatekeeping of the past, the overwhelming message is that science, by its very nature – requiring raw innate brilliance and flashes of sheer genius on the one hand, but commanding a systematic rule-bound approach on the other, with the subject matter being things rather than people – is no place for a woman.

If you are a social being, you try to match yourself to the group you have learned you belong to; to choose an environment where you will find like-minded members of that group; to match your skill set to an environment in the hope that you will fit in. If you are confronted with a 'chilly climate', where people don't feel you belong, and you have the impression that there aren't many 'people like you', then it is perfectly understandable if, on the whole, you steer clear. If you are the only girl at a maths or physics open day at a university you might have a rethink about your UCAS choices (or you might, of course, be thrilled at the other opportunities this might offer).

Women appear to take more note of the environment in which they might be working. The American psychologist

Sapna Cheryan and colleagues tested recruitment outcomes by showing potential computer science candidates either a 'typical' computer science classroom – full of *Star Trek* posters, science fiction books and 'stacked soda cans' (presumably very neatly stacked) – or a neutral classroom with nature posters and water bottles.[31] Women were much more likely to express interest in computer science if they had been in the neutral room. These researchers also manipulated the content of virtual introductory computer science classrooms, one full of stereotypical objects associated with computers, and one without. Only eighteen per cent of the female undergraduates chose the former classroom as compared to more than sixty per cent of the males. Other studies have shown that female recruitment to science summer schools can be affected by manipulating the male-to-female ratio in taster videos, with girls less inclined to sign up if the majority of students shown were male, whereas male students didn't seem to mind either way.[32] These data support the concept that women are more sensitive to the social context of the choices they might make, the signals that this might be somewhere where they might 'belong' (or not).

This brings us to the Gender Equality Paradox, which is becoming something of a hot topic nowadays. A paper published in 2018 investigated STEM enrolment between 2012 and 2015 in sixty-seven countries using an international database.[33] It revealed that, universally, there were fewer women than men obtaining STEM degrees, ranging from 12.4 per cent in Macao to 40.7 per cent in Algeria (with the UK and US at 29.4 per cent and 24.6 per cent respectively). These findings were then related to a measure of gender equality, the World Economic Forum's Global Gender Gap Index, based on gender inequalities in areas such as earnings, health, seats in parliament, financial independence and so on. This was where the apparent paradox emerged: in those countries with the *most* gender equality, the gender gap in STEM enrolment was highest. Finland (where 20.0 per cent of STEM graduates are female), Norway (20.3 per cent) and Sweden (23.4 per cent) were the prime examples of this puzzle.

Measures of school performance in science and maths revealed vanishingly small sex/gender differences (with an average overall effect size of −0.1). For science, the biggest difference was in Jordan, with girls outperforming boys (an effect size of −0.46); and for maths, boys in Austria showed the biggest difference (an effect size of +0.28); but in the overwhelming majority of the countries assessed there was very little difference between boys and girls. So the lack of women in STEM higher education does not come from a lack of ability. The data for reading showed a different story: in all countries measured, girls performed better. In this instance some of the effect sizes were quite large (−0.76 for Jordan, −0.61 for Albania) and in all instances the sex/gender differences were greater than those for science and maths.

The authors of the paper focussed on the availability of a different academic strength as a potential answer for the Gender Equality Paradox. They generated a 'best subject' index for all participants, ranking performance scores on science, maths or reading to identify the strongest subject for each participant. There were marked sex/gender differences here, with fifty-one per cent of girls having reading as their strongest subject as opposed to twenty per cent of boys; science was the strongest subject for thirty-eight per cent of the boys and for twenty-four per cent of the girls. So, although girls were generally as good as boys at science, they were markedly better at what might be seen as a more humanities-based skill.

The next link in the chain of this particular argument was that, in less developed countries, factors such as economic necessity, and the acknowledgement that a STEM education was likely to be of better value in terms of future employment and earnings, would have priority in the career path chosen by girls as well as boys. In more gender-equal countries, however, girls had the freedom to choose those subjects which they thought would suit them best, that is, the ones they were good at. Overall life satisfaction could be given priority over economic necessity. Press coverage of the paper offered 'painting and writing' as the kind of choices that might be made. Do I detect a whiff of the old 'complementarity trap'?

Something of an unexamined footnote is that there were also data on measures of self-beliefs in science ability and enjoyment of science. You might not be surprised that, overall, boys had higher levels of self-belief in their science ability; you might be more surprised that this is particularly true in more gender-equal countries – the very countries where girls were choosing not to do science. How accurate were these boys' assessments of their ability? Comparing these assessments with their perform-ance scores, it emerged that in thirty-four of the sixty-seven countries covered there was evidence of boys overestimating their science ability, whereas there was only evidence of this tendency in girls in five of these countries. And, again, it was the more gender-equal countries where this overconfidence manifested itself in boys.

Picture this. You have a choice of pursuing a subject where your self-confidence has been undermined from a very early age, where the stereotypical message (and the reality) is that it is something that members of your in-group don't do, due to a lack of the 'essential' skills required (even if your performance scores should suggest otherwise). You are sensitised to messages about the 'chilly climate' that might await you. What would you choose to do? Those who blame the girls for not wanting to do science might just think about science itself.

Sexing the scientists

But what about the people in science, the scientists themselves? Even if the culture feels rather alienating and exclusionary, if you have the right set of skills, personality and temperament to bring to the table, then surely there will be a niche for you? Science in this day and age must be an educated, informed, enlightened institution which has a clear-eyed, ungendered view of women as scientists, mustn't it?

But of course, another aspect of the stereotyping of science is the stereotyping of scientists. You might be surprised to learn that, allegedly, the term 'scientist' was first publicly coined to describe a woman, the Scottish polymath Mary Somerville.[34] Exponents of this discipline having previously described themselves as 'men of science', they realised that, once they came across the surprising phenomenon that women could also produce scientific papers, they would have to find a different term to refer to them.

This early coinage doesn't appear to have impacted modern-day impressions of what and who a scientist is. Back in 1957, researchers were interested in systematically measuring the image of a scientist among high school students in the United States.[35] They sampled over 35,000 essays on the subject of science and scientists written in answer to some open-ended questions (with the questions themselves giving a startling insight into gendered thoughts about career choices at the time). The stated aims of the study included the following (with, as you may guess, italicised emphases from me):

1) When American secondary-school students are asked to discuss scientists in general, without specific reference to their own career choices or, *among girls, to the career choices of their future husbands*, what comes to their minds and how are their ideas expressed in images?

2) When American secondary-school students are asked to think of themselves as becoming scientists (boys and girls) or *as married to a scientist (girls)*, what comes to their minds and how are their ideas expressed in images?

Students were asked to complete statements including 'When I think about a scientist, I think of …' and 'If I were going to be a scientist, I should like to be the kind of scientist who …' Alarmingly, there was a separate version of this question for the 'lady participants' in the study: 'If I were going to marry a scientist, I should like to marry the kind of scientist who …'

So what kind of composite image was generated from these thousands of essays? The researchers built up the following characterisation from the responses:

> The scientist is a man who wears a white coat and works in a laboratory. He is elderly or middle aged and wears glasses ... He may wear a beard, may be unshaven and unkempt ... He is surrounded by equipment: test tubes, bunsen burners, flasks and bottles, a jungle gym of blown glass tubes and weird machines with dials. He spends his days doing experiments ... He is a very intelligent man – a genius or almost a genius ... One day he may straighten up and shout: I've found it, I've found it![36]

But, of course, this was 1957 and things have moved on from these early days, in terms of both the questions asked and the answers given – haven't they?

One way of tracking the answer to this question involves looking at a very simple test involving drawings. You might feel that these do not really count as data, but they have proved surprisingly useful in getting access to personal mental models and revealing personal beliefs, and are, additionally, a practical way of measuring these in children, allowing insights into how early stereotypical views about scientists might develop.

This was the aim of psychologist David Chambers in the 1980s when he devised the 'Draw-a-Scientist Test'.[37] Children were asked to 'draw a picture of a scientist' and these drawings were then analysed to see the extent to which they contained what were defined as the standard images of scientists. These standard characteristics were: lab coat (usually but not necessarily white); glasses; facial hair (including beards, moustaches or abnormally long sideburns); symbols of research (scientific instruments and laboratory equipment of any kind); symbols of knowledge (principally books and filing cabinets); technology (or the 'products' of science); and finally relevant captions (formulae, taxonomic classification, the 'eureka!' syndrome etc.). The study took place over eleven years and images from 4,807 children aged between five and eleven years in 186 classes were analysed. The youngest children produced refreshingly stereotype-free drawings. 'Defining' features started to emerge in drawings from 6–7-year-olds, most commonly lab coats and equipment, but also beards and glasses. With 9–11-year-olds, some if not all of the features were present in all drawings. But, tellingly, of the more than 4,000 images produced, only twenty-eight of them were of women, all of which were drawn by girls (so the other 2,327 girls drew men scientists).

This test has been used many times across the world and the findings with respect to the stereotypical gender of scientists are universally similar: scientists are male, bearded and bald.*[38]

And things don't seem to be changing much with time (and despite the increasing numbers of women who can be found in all forms of science, even if they are woefully underrepresented). A 2002 study showed that the portrayal of scientists as male has, as the researcher put it, largely endured (as, apparently, has the presence of facial hair as a key defining characteristic.)[39] Any downward shift in the percentage of drawings of males is mainly accounted for by an increase in the number of 'indeterminate' portrayals, which perhaps might offer a glimmer of hope! The task is, obviously, very open-ended and it has been suggested that it was unfairly eliciting stereotypes.

Perhaps, today, we have imperceptibly arrived at a stereotype-free world of science and too much effort is being put into overcoming barriers that have actually disappeared? This was tested out in a 2017 study with a new version of the test called the Indirect Draw-a-Scientist Test.[40] This time it contained the following instructions: 'Imagine how scientific research is conducted. Present what you see in a drawing. Add a short description below.' The authors were quite excited by what they

---

* The test has been carried out in Australia, Bolivia, Brazil, Canada, Chile, China, Colombia, Finland, France, Germany, Greece, Hong Kong, India, Ireland, Italy, Japan, Mexico, New Zealand, Nigeria, Norway, Poland, Romania, Russia, Slovakia, South Korea, Spain, Sweden, Taiwan, Thailand, Turkey, the United Kingdom, the United States and Uruguay.

saw as a dramatic change in the frequency with which scientists were represented as female under the indirect instruction version – however, I was disappointed to see this was actually only an increase from 7.8 per cent to a still measly 15.8 per cent.

The current generation, however, must surely be much more clued up about the different variations and types of scientists that exist, via media representations of forensic scientists, computer scientists, pathologists, wildlife biologists – so perhaps the test should acknowledge this? Using the same drawing protocol, but being more specific about what kind of scientist you were interested in, did show some increases in the percentage of times that a female was depicted, but still in much lower numbers than males. A 2004 Draw-an-Engineer study produced sixty-one per cent male drawings and thirty-nine per cent female ones;[41] for a study on depictions of environmental scientists in 2003, twenty-two per cent were of females.[42] And a 2017 draw-a-computer-scientist study produced seventy-one per cent male images and twenty-seven per cent female ones.[43] All admittedly more than the 0.06 per cent female images in the very first Draw-a-Scientist Test thirty years ago, which may just have been a tad influenced by the somewhat gendered design of the study itself, of course, but still an indication of the staying power of this particular stereotype, that scientists are first and foremost male.

How do you do science – and do women have what it takes?

Another way of measuring the gendering of science is to see what particular personality characteristics have been associated with successful scientists and measuring the overlap between these and the personality characteristics of men or women. 'Agentic' or 'action' traits such as 'persistence, confidence, competence, competitiveness, ambition and drive' have frequently been associated with success in science, as opposed to 'communal' traits, characterised as, for example, being 'selfless, supportive, aware of the feelings of others, family oriented, having a need for social acceptance and a desire to avoid controversy'.[44]

Psychologist Linda Carli and colleagues measured the extent to which undergraduate respondents rated men, women and successful scientists as possessing either 'agentic' or 'communal' traits.[45] As predicted, there was a close overlap between the perceived traits of successful scientists and those of men as ambitious, analytical and agentic beings, and very little overlap with their courteous, communal, passive, tactful and (naturally) talkative sisters. This picture was generated by both men and women, regardless of the kind of institution they came from (single sex or mixed) or the subjects they themselves were studying (sciences, humanities or social sciences). So the rather grim take-home message is that women are not perceived as having the right personal qualities to be successful scientists, and that this perception is held not only by men but by women themselves – even if they are studying science! So whether we are looking at stick men with beards and glasses, or carefully crafted rating scales, there is a clear message in our gendered waters that males have what it takes to be successful scientists and females don't.

A crucial question that arises here is: does the existence of such stereotypes actually affect how science is conducted and by whom? Does it matter if there is a mismatch between some theoretical notion of what makes a 'good scientist' and what makes a 'good woman'? This kind of mismatch, where there is a lack of fit between the profile of a role (however inaccurate) and the profile of anyone aspiring to fill that role (or already in it), is called 'role incongruity' by social psychologists.[46] It was initially proposed to explain prejudice towards female leaders, where inconsistencies between stereotypical female characteristics and the stereotypical characteristics of leaders can lead to negative evaluation of the behaviour of women in leadership roles. If they demonstrated dominant, directive, competitive behaviours appropriate to being a leader, then they violated expectations as to how a woman should behave; if they demonstrated the nurturance, warmth, supportiveness stereotypically associated with being a woman, then they were seen as incompetent leaders.[47]

It's been suggested this kind of double whammy could also be at work in science, where the lack of fit between what is seen as a typical female and what is seen as a typical successful scientist could certainly lead to prejudice and discrimination (overt or covert).[48] Clearly, if you confronted search committees or appointments panels with this as an issue, there would be stout denials, references to objective performance metrics and carefully crafted job descriptions, allusions to gender equality initiatives and any number of HR checks and balances. And yet, there *is* good evidence of some kind of imbalance in how women scientists are treated.

One study in Scandinavia reported that women had to be 2.5 times more productive than men to get the same score on a points-based system for awarding postdoctoral fellowships.[49] Looking at the 'competence' scores awarded by reviewers to applicants for Medical Research Council grants in Scandinavian countries, it was noted that for key measures of impact (number and quality of publications, how often they had been cited), only women applicants with a score of 100 impact points or more received equal competence ratings to any of the men, but the men they were equated with had scores of twenty impact points or less. As the authors note, Scandinavia has a certain reputation for equal opportunities so if this kind of thing is going on there, you have to wonder about the rest of the world. Perhaps this might be contributing to the Gender Equality Paradox we talked about earlier?

What about the kind of references that might be given to support job applications? Having written and read many of these in my time as an academic, I know how important they are in giving some kind of added value to the search committees trawling through dozens if not hundreds of pretty similar CVs. You do your best to paint a picture of an exceptional must-have student who has already shown remarkable talent and persistence, will go far, is a great team player and creative thinker, and so on. Linguists Frances Trix and Carolyn Psenka examined over 300 such letters for faculty applications to an American medical school.[50] They noticed that those for female applicants were significantly shorter than those for men, and barely covered the basics (one example was just five lines long and merely assured the reader that 'Sarah' was 'knowledgeable, pleasant, and easy to get along with').* They dubbed these 'letters of minimal assurance'. But an interesting aspect, in view of the stereotypical light-bulb versus seed view of success in science that we saw above, was the much higher inclusion of what the authors called 'grindstone' adjectives in letters of recommendations for women. These included words such as 'meticulous', 'conscientious', 'thorough' and 'careful'. Men's letters more often had what Trix and Psenka called 'standout' adjectives, such as 'superb', 'exceptional' and 'unparalleled'. The researchers did not feel there was any evidence of negative intent in what the letter writers were producing, rather that it reflected a form of unconscious bias, of different ways of viewing males and females, which could colour the decisions to be made by the appointing team.

Even if women do make it past these barriers, they then seem to find it harder to reach the highest levels in the scientific profession, or to be awarded the highest levels of recognition. A 2018 paper on the Nobel Prize archives (currently available 1901–1964) for nominations in science (physics, chemistry and medicine or physiology) showed that of the 10,818 nominations, just 98 were for women.[51] Of those, only five (Marie Curie, Irène Joliot-Curie, Gerty Cori, Maria Goeppert Mayer and Dorothy Hodgkin) were actually awarded a Nobel Prize. Some were nominated many times; Lise Meitner was nominated twenty-seven times in physics and nineteen times in chemistry, but never awarded the prize.[†]

---

* Ten per cent of the letters about female applicants were less than ten lines long, whereas eight per cent of the letters about males were over fifty lines long.

† There's no evidence of any men having been turned down that often. You might hope that opening the next fifty years of the nomination archives will show some improvement, but a quick count of the number of male science laureates since 1964 (350) and the number of female (12) does not show much promise. Although, of course, 2018 did see two more female laureates.

These kinds of data are not necessarily direct evidence of discrimination as there could be several additional factors at work. However, lab-based studies could provide such evidence. A widely cited paper from Corinne Moss-Racusin and her team at Yale provides a powerful illustration of this.[52] Over one hundred members of biology, chemistry and physics faculties at high-ranking universities were given application materials for the appointment of a student to a laboratory manager position. All the details were identical except that half the faculty were given applications with a male name (John) and half with a female name (Jennifer). You can probably guess the outcome.

Significantly more of the faculty (male and female) rated John as more competent and more hireable (at a higher salary). They were also more prepared to offer career mentoring to John. Using the Modern Sexism Scale, which includes factors such as knowledge and explanations of sex segregation in the workforce, the researchers were also able to elicit a measure of any pre-existing sexist bias in their participants. This measure showed that the higher the levels of this pre-existing bias, the less competence and hireability was perceived in Jennifer's application, and the less mentoring they would be prepared to offer her. Again, this was true of both male and female faculty. Finally, and rather paradoxically, Jennifer was described as more likeable by the team (remember, all details apart from the names were identical in the applications). So it wasn't to do with some kind of generic hostility towards women – the fictional Jennifer was obviously a pleasant person, she just didn't have much future as a scientist.

Perhaps this kind of bias might be overcome if information about candidates wasn't actually identical but contained evidence of different abilities? In one experiment, student 'employers' could hire student 'employees' to carry out a maths task.[53] All the students had previously completed one version of the task so they knew what it was like and their own level of performance. 'Employers' could hire 'employees' solely on the basis of their appearance (via an online photo), or on the basis of their appearance together with some information about how good these potential 'employees' might be at the task. The 'employers' (both male and female) picked twice as many men as women if the only information was appearance; and they generally stuck by their decisions even when they were provided with information that showed that the women they hadn't hired had performed better on the maths task. So, if you are a female, even being better than the male applicants at what the job requires does not overcome the knee-jerk reaction that this is a job for the boys.

The gender gap data we have looked at makes it clear that women don't do science. There is certainly historical evidence that they *did* do science, but as Londa Schiebinger has chronicled, they were gradually excluded, certainly by the gatekeeping activities of nascent scientific societies and by a pervasive view that it wasn't a suitable sphere of activity for females. Perhaps this might just be judged as a backward glance to the bad old days. But contemporary reviews of gender imbalance in appointments and achievements at the highest level suggest that some form of discrimination, conscious or unconscious, is still at work. The face of science is still that of a male domain, peopled by individuals whose characteristics bear striking similarities to those of a stereotypical male, agentic and systemising with innate access to light-bulb-like flashes of genius. There is dispiritingly early evidence that not only do the parents and teachers of potential scientists buy into this 'men only' picture of science and scientists, but so do these potential scientists themselves.

There is another strand to this argument, which echoes the story we have been telling all along: perhaps the maleness of science merely reflects a natural outcome of the hand that biology has dealt. However inconvenient this 'truth' might be, the reality is that women don't do science or at least cannot be found in the higher echelons of science. Could this be because, when it comes down to it, they lack the necessary 'natural' aptitude?

Chapter 11:
Science and the Brain

The gender gap data makes it clear that women *don't* do science – but this doesn't mean they *can't* do science. In order to understand the essentialist approach to this gender gap, we can start by looking at the 'greater male variability' hypothesis (GMV).

This is another of those Whac-a-Mole themes that seem to characterise sex difference research. It refers to the claim that if you look at the upper and lower ends of any distribution of measures of intellectual ability, you will find more men: more male geniuses, more male idiots. This idea was first proposed in psychology circles by Havelock Ellis in 1894: noting larger numbers of men than women in homes for the mentally deficient, and much larger numbers of men in the spheres of eminence and high achievement, he concluded that there was a greater innate 'variational tendency' in males.[1] (You might notice that this rather overlooks the possibility that the greater eminence at one end might have reflected greater opportunities, and the higher rates of institutionalisation at the other end could have reflected different levels of available social support networks.)

Perhaps unsurprisingly, discussion about the implications of this greater variability focussed on the right-hand end of the distribution, on the high achievers, rather than the left-hand end. This variational tendency had obvious implications for the expectations for men and women: men are more likely to be geniuses, and women are more 'average'.

Reference to the GMV hypothesis is often made in explanations of gender gaps in achievement: even if, *on average*, women and men perform equally on some kind of task, be it maths, logic or chess, the high achievers will be found several rarefied standard deviations to the right of the distribution and more of them will be male.

An assumption behind the claim to greater variability in males is that it is a fixed, cultural universal that should be stable over time and evident in all groups, in all countries. In fact, none of these criteria are met. A meta-analysis in 2010 of international studies of maths skills showed that, in the US, gender gaps at the high end had all but disappeared, in most other countries there was no difference, and in some (Iceland, Thailand, the UK) there were more females than males among the highest scorers.[2]

Even today, there are attempts to demonstrate the evolutionary validity of the GMV hypothesis, with claims along the lines that women are picky about who they mate with and only go for the top half of some kind of mate-ability ranking, resulting in a rarefied top end of the distribution.[3] This leaves the bottom half to merrily reproduce with anything going, resulting in highly variable outcomes, some of them the sum of just about everything negative you might think of. Actually, the mathematical paper making these claims was withdrawn, amid claims of political conspiracy, but a helpful mathematical blogger pointed out that the underlying maths assumptions were of highly dubious quality.[4] But there is no doubt that this myth will surface again, in gender gap explanations or in backlashes against diversity initiatives or in any other fora where it appears to be necessary to fall back on a centuries-old hypothesis, even if discredited.

We've already come across the breathtakingly misogynistic pronouncements of earlier centuries, but fast-forward to our own time and, as we have seen, there remains a general 'essentialist' undercurrent in discussions about the gender gap in STEM. Mainly, this remains underground, firmly attached to stereotypes about women and science, mixed in with other dogmas about the nature of science and scientists, perhaps unconsciously driving employment decisions and career choices.

Every now and then, a more overt declaration of this kind of 'blame the brain' belief surfaces in a very public way. Two oft-quoted examples of this are the infamous Larry Summers speech in 2005 and the Google memo of 2017. What characterises both is not only an expression of the belief that, when looking at high-level achievement in science, women just don't have what it takes, but also that this is a problem based on their biology.

Firstly, Larry Summers, then president of Harvard, chose to talk about 'the issue of women's representation in tenured positions in science and engineering at top universities' at a conference on 'Diversifying the Science and Engineering Workforce'.[5] His arguments were firmly based in the field of the GMV hypothesis, suggesting that, at the high end of science, you might be looking at people who were performing four standard deviations above the mean, and the data indicate that this is populated by about five men to every one woman. So, like it or not, if your territory was the high end, be it business or science, you would find more men than women. One of his explanations for this gender gap was the 'different availability of aptitude at the high end'.

There was a considerable outcry following Summers' declaration. As well as more general expressions of outrage in the media about what was seen as an outdated and discriminatory stance, the academic community got together to address the issues raised.[6] There were a few classic mistakes Summers made. For example, he based his estimates of a 5:1 male-to-female ratio at the high end on data from tests which, at the very conference he had been attending, had been identified as 'not highly predictive with respect to people's ability' – which, astonishingly, he in turn noted in his own speech. Indeed, the authors of the work he was referring to, Kimberlee Shauman and Yu Xie, themselves took issue with his interpretation of what they had shown.[7] They had been looking at career progression in women and had generated the data on gender gaps to which

Summers referred. They noted that the gender gap in mathematics was small and had been declining since the 1960s. In a later paper, Xie explicitly stated that 'the declining trend ... casts doubt on the interpretation that the gender gap in math achievement reflects innate, perhaps biological, differences between the sexes' and added that 'President Summers failed to cite the following finding: gender differences in neither average nor high achievement in mathematics explain gender differences in the likelihood of majoring in science/engineering fields'.[8] (In fact, Shauman and Xie thought their data showed that the chief barrier to women's progression in science was parental responsibilities and coined the term 'the leaky pipeline' to describe the problem.)

So here we have someone who cheerily acknowledges that he might be looking at a 'dodgy dossier' of data, misrepresents what it is showing and then misinterprets it anyway. He was taken to task by the researchers who produced the data and, later, by a wide range of top psychologists, who also criticised him for his circular arguments.[9] With a pounding as firm as that you might think that this particular mole had been permanently whacked.

Not so fast. In the summer of 2017 a (very long) memo from a Google employee, James Damore, made it into the public domain.[10] It had apparently been penned in a fit of frustration following attendance at a diversity training course which the author had clearly not enjoyed. He effectively told Google that they were wasting their time (and his, one assumes) on equal opportunity initiatives to increase the number of women in their workforce. Channelling his 'inner Larry Summers', Damore asserted that 'the distribution of preferences and abilities of men and women differ in part due to biological causes and ... these differences may explain why we don't see equal representation of women in tech and leadership'. It didn't take long for this memo to be leaked, for the employee to be identified, for a huge media storm to erupt, and then for Damore to lose his job.

Damore's target was wider than Summers' and included preferences as well as aptitude; he was much more explicit about the biological bases, calling them 'universal across cultures' and 'highly heritable', and citing prenatal testosterone as a primary causal factor. The key 'preference' dimension he seems to be targeting is our old friend People versus Things. He is less specific about which 'ability deficit' is apparently a problem for women (and for Google), but he states that men are suited to coding because of their systemising skills, so it seems he has mapped the systemising–empathising dimension onto a male–female dichotomy, with the systemising males having what it takes to be a successful coder. The empathising aspect he links to the people-liking tendencies he's reserved for women. Rather extraordinarily, later on he calls for Google to de-emphasise empathy, feeling it may cause a tendency to 'focus on anecdotes, favour individuals similar to us, harbour other irrational and dangerous biases'.

He also focussed on a large-scale study, covering fifty-five nations, on sex differences in personality traits, which reported that on average women display higher levels of neuroticism and agreeableness.[11] The data from this study also showed that sex differences were larger in the more developed countries, which the authors interpreted as due to men and women being able to naturally express their true selves when less constrained. Damore warned against allowing men to be more 'feminine', suggesting this could result in them quitting high-level science and leadership positions for 'traditionally feminine roles' (left unspecified). He concluded his message with a range of suggestions as to how his viewpoint should be incorporated into future diversity programmes.

A lot has changed since the Larry Summers speech (if not the minds of people such as Damore) and the speed of the response to this memo was dramatic, with online articles, blogs, tweets, Facebook posts pouring out almost instantaneously.[12] Given the length and content of the memo, there was a lot on which to comment. Setting aside general observations on diversity programmes and the rights and wrongs of sacking Damore, at least some of the debate was about the science he quoted.

There were people on his side. Those in the evolutionary psychology camp felt his case was well made, possibly related to his enthusiastic endorsement of their thinking in his memo.[13] You could see where he might be coming from with respect to competitiveness and drive for status, consistent with evolutionary psychology's man-the-hunter arguments, but it is hard to see the evolutionary story behind extraversion, agreeableness and neuroticism, all characteristics which Damore names as responsible for women's lack of presence in the higher reaches at Google. Debra Soh, a sexual neuroscientist and science writer, clearly feels she has the right to speak for all neuroscientists, and to dismiss huge swathes of critical neuroscience in voicing her support for Damore:

Within the field of neuroscience, sex differences between women and men – when it comes to brain structure and function and associated differences in personality and occupational preferences – are understood to be true, because the evidence for them (thousands of studies) is strong. This is not information that's considered controversial or up for debate; if you tried to argue otherwise, or for purely social influences, you'd be laughed at.[14]

However, on the other side there was much criticism of Damore's use of apparent evidence. In an echo of the problems with Summers' speech, David Schmitt, the lead author on the personality paper Damore had been so enthusiastic about, did not think his findings actually supported Damore's case. He pointed out that the size of any differences was generally small, and where it was measured, how it was measured and other contextual factors needed to be taken into account. In a rather more trenchant dismissal of this use of his research he also observed: 'Using someone's biological sex to essentialise an entire group of people's personality is like surgically operating with an axe. Not precise enough to do much good, probably will cause a lot of harm.'[15]

Other commentators noted that, as with Summers' observations, no attention had been paid to the plasticity of the human brain, the potential role of experience in determining performance on any range of measures, including those which would certainly be relevant to success in science.[16] The point being that even if there was some substance to claiming a biological basis for the kind of aptitudes in which women were apparently lacking, it was not fixed and insuperable in the way that Damore was suggesting. He was channelling the 'limits imposed by biology' mantra as opposed to the 'potential offered by biology'.

Damore's identification of coding as a guy thing was easily dealt with by pointing out the preponderance of women in the computing field in countries such as India, and linking the disappearance of girls from computing to a cultural phenomenon, the advent of the home computer in the 1980s and its marketing as a gaming system for men.[17] There was pretty thorough 'fisking' of his essentialist opinions, with systematic filleting of the misrepresentations and fallacies in these arguments. Two authors who had researched and published widely on this issue for many years summed up the general feeling thus: 'We have been researching issues of gender and STEM (science, technology, engineering and math) for more than 25 years. We can say flatly that there is no evidence that women's biology makes them incapable of performing at the highest levels in any STEM fields.'[18]

So, Groundhog Day for a firmly essentialist opinion, based on dubious data, misrepresenting the science quoted, and appearing to ignore the widely published research stressing the importance of context and experience on the emergence of aptitudes and preferences in both males and females. Summers' and Damore's widely publicised assertions have been criticised with clear, and very firmly stated, conclusions as to the misguided nature of the assumptions underpinning these public statements about women in science. These two infamous pronouncements and the backlash against them encapsulate just about all the issues in the ever-ongoing debate about women, biology and science. Unfortunately this seems to include the stickability of such thinking and its propensity to re-emerge in almost unchanged form.

So, even today, with major technological advances that should allow us to really get a handle on individual differences in the brain, there are still eighteenth-century thinkers coming up with eighteenth-century answers. Although we have trenchant (and repeated) denials that women's biology renders them unfit for science, the underlying 'blame the brain' maxim that has been with us since the days of Gustave Le Bon is proving remarkably hard to shift. So let's look at the quality of the evidence that is marshalled in its support.

Stereotyping the science brain

The notion that the 'male brain' is a necessary source of the kind of systemising skills that put you on track for a Nobel Prize has entered the public consciousness. Stimulated by Simon Baron-Cohen's systemising–empathising model, the last five years or so have seen a search for the underlying neural correlates of these types of processing.[19] A principal motivation has been the insights this might provide into autism spectrum disorders, which Baron-Cohen has described in terms of an 'extreme male brain'.[20] As he has also stated that the male brain is hard-wired for systemising and the female brain for empathy, sex differences naturally figure prominently in the analysis and interpretation of research in this area.

So is there such a thing as a 'science brain'? A 'maths brain'? And, by extension from the stereotypes of science and scientists, is it a male brain?

I was once sent a cartoon by a colleague about sexing a cat. Two men have found a cat and they want to work out if it is female or male. The next frame shows them watching the cat trying to parallel-park. You might like to count up the number of times an article in the press on sex differences in the brain is illustrated by a man holding a map and gazing confidently in what is clearly the right direction, sometimes with a female companion bearing a puzzled frown and a crumpled, upside-down chart, pointing anxiously the opposite way.

A major focus in discussions on women in science has been on spatial cognition, a skill which is commonly associated with success in STEM subjects.* Spatial cognition is a general capacity ranging from the ability to navigate round our environment, to create and read maps and plans, to the ability to mentally manipulate objects, symbols and abstract representations, to identify patterns and work in many dimensions (and parallel-park). It has been claimed that the sex differences in this ability are one of the most 'robust' of all sex differences.[22] From early studies of the effects of brain injury, through hormone manipulation studies, to identifying the neural real estate underpinning spatial skills and mapping the functional brain networks activated by spatial tasks, a key focus on the study of spatial cognition has been why women are so bad at it.

The idea that spatial cognition is a fixed brain-based skill has become another Whac-a-Mole meme in the whole sex/gender differences debate, particularly with respect to the extent to which prenatal hormones organise the male and female brain. Performance on visuospatial processing tasks has been taken as an index of the extent of masculinisation of brains that have been exposed to high levels of testosterone.[23] Evolutionary psychologists have weighed in with suggestions that men's superior spatial skills are linked to the hunting, spear-throwing and wayfinding skills they needed in the past.[24] So examining the extent to which a 'spatial brain' is biologically determined (with themes of 'raw innate ability' or 'expectations of brilliance' much more common in the STEM world) or a product of

---

* Indeed, in 2009, Wai and colleagues reported on a longitudinal study of 400,000 US high school students, whose academic progress was tracked over eleven years.[21] They found clear evidence of a link between their early spatial ability and success in university-level STEM subjects or STEM-related careers. It's rather surprising that neither Lawrence Summers or James Damone picked up on this, as it might have made a much firmer foundation for their claims.

gendered training (think Lego and videogames) could offer a real insight into what the sex differences in this set of skills really are and where they might come from.

High levels of spatial ability can be a practical skill, making you good at finding your way in strange places or being able to read maps. It can also make you good at tasks that require understanding of the relationships between different parts of objects, such as construction or architecture. It can also be a theoretical skill, so you might be good at understanding certain branches of mathematics. Almost universally, wherever you look, at different times and in different cultures, male superiority in this particular skill has been claimed. Even when gender gaps are diminishing elsewhere, this male–female difference allegedly stands firm. It is hailed as the most robust of all sex differences, perhaps the last bastion of male superiority.

Actually when you see the term 'spatial cognition' or 'visuo-spatial processing', more often than not what scientists are really talking about is performance on a mental rotation task (MRT), testing the ability to mentally rotate a 3D figure to see if it matches a second version, which we've encountered a few times already in this book.[25] This is certainly the test most commonly used as a general measure of spatial processing, the one that normally shows the largest sex difference (although, as ever, we are still talking overlapping scores here), that has been demonstrated in very young children, that seems to have shown the most stability over time (although there is evidence of this diminishing) and to show the most consistency cross-culturally.

If you recall, in Chapter 8, there was a suggestion that there were early sex differences in mental rotation ability, with 3–4-month-old boys looking longer at pairs of images when one had been rotated.[26] It has been suggested that this might reflect the consequences of prenatal exposure to testosterone.[27] Equally, there is good evidence that experiential factors such as toy choice, sports participation and computer games can affect mental rotation performance. Intriguingly, a recent study showed that, in infant boys, there is a positive correlation between testosterone levels and mental rotation ability, an effect not shown in girls.[28] On the other hand, there was a negative correlation between parental gender-stereotypical attitudes and mental rotation performance in girls – the more traditional the parental attitudes to gender were, the worse the girls did on a mental rotation task. The researchers suggested that biological factors were at play in the boys and socialisation factors in the girls. So, evidence of an early difference in a spatial skill, but with inexplicably mixed attributions of the causes. But here we are still looking at correlations between variables, and these are only indirectly linked to brain processes. Perhaps examination of what is going on in the brain when someone is carrying out an MRT may shed more light.

Does an MRT reveal a neatly ordered set of brain areas where activity is triggered, with the degree of activation closely associated with the level of performance? Or perhaps two sets of brain areas, one male and one female, matching the different ability levels? Well, hopefully you will have grasped by now that that is almost never the case, but with the study of spatial cognition perhaps we can at least extract some general principles from brain imaging studies that should then form the backdrop to all the other questions that need to be asked.

Very early studies of the effect of brain damage on behaviour located spatial processing in the parietal cortex, that part of the cortex between the visual areas in the occipital lobes and the executive areas of the frontal lobes.[29] Although damage to either hemisphere in the parietal area causes problems with spatial cognition, it is most evident with damage to the right side of the brain. If you recall, one of the early neuromyths was that men's 'uncluttered' right hemisphere gave them the visuospatial edge over women, whose right hemisphere also had shared responsibility for dealing with language demands. Although this notion has largely been dismissed, certainly within the scientific community, it can still be found in certain outdated textbooks or in works of the 'neurotrash' genre.[30]

With respect to mental rotation, it is indeed the right parietal lobe where increased activity is most consistently found, but it is usually accompanied by left hemisphere activation as well. So should we home in on the parietal cortex as the brain structure that underpins this robust sex difference? A study in 2009 found that men's superior MRT performance was associated with a greater surface area in the left parietal cortex whereas women's poorer performance was associated with more grey matter depth, again in the left parietal cortex.[31] So a bigger parietal cortex in men helps them do better, but a thicker one in women seems to get in the way. But, bearing in mind the 'size matters' debates we looked at in Chapter 1, we should be cautious about attributing too much explanatory significance to this. And, as ever, we need to bear in mind that we might be looking at the consequences for our plastic malleable brains of different visuospatial experiences.

Does this mean that the answer lies in children, with fewer years of brain-changing experience on the clock? There are, understandably, fewer brain imaging studies of MRT performance in children, but one fMRI study in 2007 compared children aged nine to twelve and adults doing the same version of an MRT.[32] In children the researchers found activation patterns in the right parietal cortex similar to those in adults, although the adults were more likely to show left hemisphere activation as well. But what was interesting was that there were no sex differences in the children, either in MRT performance or in their brain activity, whereas there *were* differences in adults' brain activity, with women showing more frontal and motor-related activity. It could be that the task was more child-friendly (the stimuli were animal images such as seahorses and dolphins) or it could be that the children were prepubertal, but given the emphasis among biological determinists on the *early* emergence of any sex difference and what we now know about the role of experience in shaping the brain, this result makes a powerful case against the essential innateness of spatial abilities.

We often assume that everyone tries to solve a problem the same way, some more efficiently than others. But some of the brain findings tell a different story. When men and women are matched for how well they do on an MRT, the male participants, on average, showed greater activation in the parietal cortex but the females demonstrated more frontal activation.[33] What's been inferred from this is that men solve the problem in a holistic fashion, but women take a more linear approach, possibly by counting the components that make up the image to be rotated. (This sounds like a dangerously systemising method to me but perhaps, for the time being, we could let that pass.) This latter tactic is more time-consuming, so anyone employing it would take longer to reach a solution. We are still looking at a sex difference, though, so, as well as looking at strategic differences, there are other factors that should be taken into account before we can agree that how men and women deal with space is indeed the robust difference it is claimed to be.

As ever, sometimes it is not about the question itself (are men better than women at spatial tasks?) but how you ask it. When different versions of the classic MRT are used, supposedly robust sex differences diminish or even disappear. This has been shown on paper-folding versions of the test, and on versions using real 3D objects or photographs of them.[34] In Chapter 6, we looked at the effect of stereotype threat on brain activity and mental rotation.[35] One study showed that if you describe the task differently, as a perspective-taking task as opposed to one requiring mental rotation, this affects both brain activity and task performance. This might suggest that this fundamental sex difference is not so fundamental after all.

Training the brain – and a reminder that toy choice matters

Along the same lines, differences in MRT performance might not be as stable as has been supposed. It can improve with relevant training, which has been shown to reduce sex differences or even remove them altogether, so it is certainly a malleable skill.[36] This suggests that the supposedly robust difference may not be associated with a fixed biologically based sex difference but is actually the effect of different levels of spatial experience. We have already seen that boys are more likely to have construction-type toys or to play target sports with strong spatial elements, so perhaps they are showing the benefit of these early 'training' opportunities?

Behavioural clues come from watching people play computer games. One study, in 2008, showed that just four hours of playing a Tetris-like game brought about significant improvements in MRT performance, in women more than men.[37] Another study in the previous year by Jing Feng and colleagues at the University of Toronto looked at sex differences in MRT performance as a function of previous experience in videogame playing.[38] This showed that experienced players did much better at MRT tasks than non-players, and that the sex differences in the player group were very small. It looked like being good at MRT tests might be more of a function of how much time you spent on your Xbox than your XX or XY genotype. The researchers confirmed this by putting another group of students through ten hours of training on an action videogame, with pre- and post-training MRT tests. Both males and females showed significant improvement, the females more than the males, dramatically reducing the pre-training gender gap.

Are these behavioural clues matched by changes in brain activity? As we've seen previously, Tetris has also been brought into the brain scanner, where it's been shown that significant brain changes in both structure and function can be brought about by the training. In a cohort of twenty-six girls, quite widespread increases in thickness were found across the cortex, most particularly in part of the left temporal lobes and the left frontal lobes. Before-and-after differences in blood flow showed that the Tetris-trained girls showed some reduction in activity in the right hemisphere, consistent with the kind of changes shown when becoming more expert at a new skill.[39]

These kinds of studies tally with other studies of brain plasticity we've looked at, and we've seen how other spatial skills (such as the navigational expertise required of taxi drivers and the hand–eye co-ordination necessary for juggling), although clearly related to specific activation patterns in the brain, are changeable as a function of experience, at both the brain and the behaviour level.

Stereotyping yourself

We have seen elsewhere how belief in stereotypes (such as the 'excellence/light-bulb' factor) can affect individuals' self-perception of their ability and the lifestyle choices they might make as a result. It appears that it can also, in classic stereotype threat fashion, affect the very performance that supposedly characterises the 'inferior' category in the first place. Psychologist Angelica Moè looked at MRT performance as a function of the type of explanations given in advance of the test.[40] After getting baseline measurements from her participants (95 female and 106 male), she divided them into four groups. Every group was told the following: 'This test measures spatial abilities. These are very important in everyday life, i.e., for finding a route on a map, orienting in a new environment, describing a road to a friend. Research has shown that men perform better than women in this test and obtain higher scores.' Then, different explanations were issued to the different groups (except for the control group). The genetic explanation read: 'Research has shown that male superiority is caused by biological and genetic factors.' The stereotype explanation read: 'This superiority is caused by a gender stereotype, i.e., by a common belief in male superiority in spatial tasks, and has nothing to do with lack of ability.' The time limit explanation read: 'Research has shown that women are generally more cautious than men and require more time to answer. Hence, their poor performance is owed to time limit and has nothing to do with lack of ability.' Then MRT performance was measured again.

Participants who were given the stereotype and the time limit explanations showed significant improvement. Moè suggested that these are 'externalising explanations', that poor performance is nothing to do with lack of ability, but with some kind of stereotype (that you can ignore) or some kind of strategic choice rather than basic incompetence. She attributed the improved performance to a 'relief' factor, that how you did on the task could be put down to external factors and didn't suggest you were born to be inferior. The 'genetic' group, who got the 'your performance is a measure of your innate ability' message, showed a decline in performance. It should be noted that there were sex differences (males better than females) both before and after the instructions phase, so it wasn't as though the externalising explanations counteracted these, but they did show that this kind of performance could vary as a function of the beliefs you had about what the test was measuring, again undermining any 'robustness' statements.

Interestingly, parallels have been drawn between the effects of general stereotype threat and 'maths anxiety', a problem that is particularly relevant to achievement in science.[41] It is also a problem to which women appear to be particularly prone. There has been a suggestion that maths anxiety is actually related to poor spatial processing ability, so it is just a realistic assessment of your chances of doing badly. But if we look at how spatial processing ability is measured it is often via a self-report questionnaire, the Object Spatial Imagery Questionnaire.[42] This includes self-rating on statements such as 'I am good at playing spatial games involving constructing from blocks and papers' and 'I can easily imagine and mentally rotate 3-dimensional geometric figures'. So it is not really a measure of ability, but rather a measure of your *belief* in your ability. So maths anxiety could well be showing that women who *believe* their spatial processing ability is poor are understandably anxious about undertaking tasks where spatial skills are required. We're back in the land of stereotypes and self-fulfilling prophecies.

At the brain level, studies of maths anxiety, maths performance and stereotype threat have shown how closely intertwined these processes are (and their effects on behaviour). In an EEG study investigating maths anxiety, increasing stereotype threat via instructions ('We will be comparing your score to other students for the purpose of studying gender differences in math') activated affective centres in the brain and increased attention to negative feedback.[43] Students in this group also gave up quicker and did not make use of offered online tutorials. And, as you have probably guessed, they did worse than their unthreatened peers.

An fMRI study showed more directly that activating stereotype threat was associated with differential recruitment of brain resources.[44] Women who were given neutral instructions before completing a maths task showed activation of the areas usually associated with maths, including the parietal and prefrontal areas, whereas women who were primed with the gender stereotype of poor maths performance in females activated areas more usually associated with social and emotional processing. Their performance deteriorated over the time of the testing, again as opposed to the non-threatened group.

Hormones and spatial skills

We have already noted some correlations between hormone levels and spatial skill in human infants. Is there any evidence of some kind of causal link, that altered hormone levels may cause altered visuospatial performance?

Direct manipulation of hormone levels is obviously rare in human studies and it is hard to model the entanglement of spatial ability with socialisation, experience, training opportunities and exposure to stereotypes that we have been looking at in this chapter. The findings from research with transsexuals undergoing hormonal treatment have been mixed, and generally report non-significant differences in performance on spatial tasks between treated and untreated transsexual participants and controls. One of the better-designed studies did show reduced brain activation in parietal areas in male-to-female transsexuals, although their performance did not differ from male controls.[45]

Studies of developmental disorders such as CAH have provided evidence of higher spatial ability in girls who have been exposed to higher levels of testosterone, confirmed by a meta-analysis of such studies, but it has been suggested that this may be an indirect effect of increased interest in toy choice and 'masculine' activities. A study headed by Sheri Berenbaum from Pennsylvania State University tested this model on a group of CAH girls and boys compared with their unaffected siblings.[46] Using a range of spatial ability tests, including MRT, the study showed that CAH girls scored significantly higher than their unaffected sisters but lower than unaffected males. These girls also had significantly more male-type hobbies and further analyses showed that it was this variable that predicted MRT ability. So superior MRT performance does seem here to be a downstream consequence of spatial experience, potentially linked to some kind of early preference for the kind of hobbies that offer this. Recall that, in typically developing infant girls, MRT performance was negatively affected by the stereotypical views of their parent.[47] Also recall that CAH girls seemed to be less affected by the kind of gendered permission being given by colour-coded toys.[48] So these studies may offer some new insights into the entangled factors, both biological and social, that are contributing to spatial ability.

Updating the 'map reading' stereotype

The claim that spatial cognition is the one area where evidence of sex differences is reliable and well established (and could thus serve as an appropriate forum for considering all aspects of such differences, especially women's underrepresentation in science) has not withstood further scrutiny. It's a popular and long-standing stereotype, but it looks as though there's a much greater degree of similarity than was originally thought. Where there are differences, they may be a function of how this set of skills is assessed, of who has had what relevant experiences, and, entangled with these and other factors, of the role of self-belief and stereotype threat. Hints of innate differences and causal biological differences are inextricably entangled with gendered expectations and gendered experiences. Using 'space behaviour' as our lens into the male or the female brain appears to be misguided. In short, the 'map-reading' stereotype needs updating.

Far from being the gold standard proof that female–male aptitude and ability differences are rooted and fixed in their different biologies, spatial cognition provides a detailed and ongoing case study of the power of the world to shape such individual skills, further entangled with the social context in which those skills might be used. You might have the cortical and cognitive wherewithal to succeed in science, but a chilly and unwelcoming climate could turn you away.

As we've seen, social stereotypes have a self-sustaining characteristic whereby, once they become part of an individual's or a society's social guidance system, they will determine that the individual or their society behaves according to the messages embedded in the stereotype. This reinforces the 'truthiness' of the stereotype and further strengthens its stickability. Stereotypes are not just inert reflections of a society's belief system; their very existence can influence the behaviour of members of that society: either the behaviour of society in general towards those groups who are characterised by the stereotypes, or the behaviour of members of those groups themselves. With our predictive brains out there looking for rules, stereotypes can eagerly be adopted as a readily available guidance system with respect to, in this instance, who does science. A readily established prior will be that certain types of humans *don't* do science, sustained by a belief that this is because they *can't*. So, let them avoid prediction errors and *not* do science. If they are confronted with doing science, feedback processes generate distracting warning systems which have negative effects on performance. This can triumphantly reinforce the accuracy of the 'don't do/can't do science' prior and enhance the future power of the prediction error signalling system and the inflexibility of this prior.

The underrepresentation of girls in STEM subjects is a worldwide problem.[49] This loss of human capital is having negative effects on science and the science community; clear evidence that this is not due to lack of ability flags up a waste of human potential, with fully capable individuals turning (and being turned) away from fulfilling career paths. Historically presented as a simple consequence of biology, it is now clear that this shortfall arises from a complex entanglement of brains and experiences, self-belief and stereotypes, culture and politics, unconscious and conscious bias.

And our understanding of this process has implications for a more general understanding of how brains get to be gendered, how the guidance rules in our gendered world can shape our brains.

Chapter 12:
Good Girls Don't

We're raising our girls to be perfect and our boys to be brave.
Reshma Saujani, founder of Girls Who Code

From the moment of birth (and even before) our brains are confronted with different expectations from families, teachers, employers, the media and, eventually, ourselves. Even with the emergence of amazing brain imaging technology, men and women are still being sold the concept of the male brain and the female brain, whose innate differences will determine what they can and can't do, what they will and won't achieve.

We are now much more aware of the core role that becoming a social being plays in our development, how our predictive brain is constantly on the lookout for social rules of engagement, how key to our well-being are our self-identity and self-esteem. Importantly, it's also clear how this can be threatened by encounters with negative stereotypes or social rejection.

This is where we might find explanations for the gender gaps that have been the centre of so much attention for so many years. Has all this gendered journeying changed the vehicle navigating the terrain? So, even if we make Herculean attempts to even up the route or remove the more ill-informed signposts, might we have a brain that is no longer fit for purpose, whose way of dealing with the outside world is too entrenched, whose priors are too firmly established to be changed?

Let's revisit what we know about the social brain and see if the sex/gender differences on the 'recommended' routes to being a social being could be impacting on journeying brains.

Alarm systems in the brain

Much is made in brain-based populist literature about the difference between the highly developed, information-processing part of our brains and the more primitive, irrational and emotionally charged part. One can characterise the cognition system, particularly the prefrontal cortex, in Sherlock Holmes-type terms: rigidly rational, implacably logical, a focussed executive system in charge of planning and problem solving. The more impulsive and occasionally overexcitable affective control system, mainly composed of the limbic system, has been associated with a range of metaphors, such as 'the beast within'. The sports psychiatrist Steve Peters, in his book *The Chimp Paradox*, has dubbed this part of the brain the 'Inner Chimp', characterising it as a more primitive, emotion-driven brain system which is generally held in check by the evolutionarily younger, rational, frontal lobe systems (but whose power it might be useful to harness if you want to be an elite athlete with a will to win at any cost).[1]

A common model of the relationship between these two systems is that the older, more volatile emotional system has to be monitored, kept in check and, ideally, generally overruled by the Holmesian cortex, with its cool, detached evidence-based approach to life's problems. We now know that cognitive processes such as learning, memory and action planning – and even more basic perceptual processes – do not occur in an affect-free vacuum. The Holmes part of our brain is actually more in touch with our emotional underpinnings, frequently consulting with them, comparing notes and even making or changing decisions based on the pretty primitive 'feel-good' or 'feel-bad' input from the brain's lower layers. As we saw in Chapter 6, this is particularly true of the networks in the social brain. Although the higher-level aspects of being social, such as self- and other-reference and self- and other-identity, are focussed in various areas of the prefrontal cortex, we know these are closely linked to our limbic system, exchanging positive and negative information and constantly updating our social coding catalogues.[2] These interlinked systems form part of our theory of mind network, and are key for our mind-reading abilities and our intentionality detection skills.

But there is a third part of this chain, in effect the bridge between Sherlock Holmes and our inner chimp. It is based on a structure that has figured frequently as we have unpacked the various components of our social brains: the anterior cingulate cortex. If you recall, this sits right behind the front part of the brain, and is structurally and functionally tightly linked to the prefrontal cortex, and also to the emotional control centres such as the amygdala, insula and striatum.[3] It has been suggested that its unusual spindle-shaped nerve cells may be linked to the kind of high-speed communication needed in sustaining activity in the social brain.[4]

So what special function does this well-situated and specially endowed area of neural real estate have? It's become clear that the anterior part of the cingulate cortex is involved in an extraordinarily wide range of tasks. On the one hand it has a key role in cognitive control, reliably activated when someone makes an error (such as in a Go/NoGo task); on the other hand it appears to be key to evaluative mechanisms, responding differently to the different positive or negative 'colouring' associated with task feedback, and showing marked emotional change if it gets damaged.[5]

A very influential review, published in 2000 by neuroscience researchers George Bush (yes, he does get that a lot), Phan Luu and Michael Posner, reported that detailed meta-analyses of a range of studies in this area suggested that you could broadly map anterior cingulate functions into two areas.[6] Any type of activation associated with cognitive tasks was linked to the dorsal part of the anterior cingulate (dACC), whereas activation associated with emotion was more likely to be found in the ventral areas (vACC). The model that was proposed on the basis of this review stressed the role of the dACC as an error detection system; its links with emotion centres give it an evaluation role, so the consequences of errors are registered and behaviour adjusted accordingly. The ever-busy dACC is also monitoring difficulties associated with conflicting responses (such as we have seen in Stroop or Go/NoGo tasks), when a decision needs to be made about which response might or might not lead to a mistake being made.

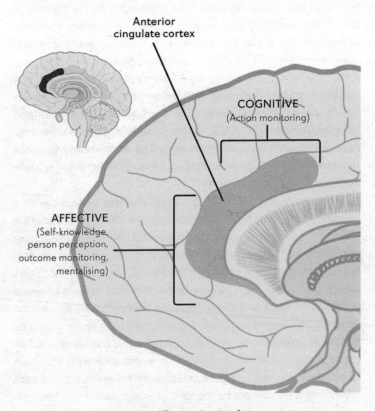

Figure 3: The anterior cingulate

Bush and colleagues felt that these 'error-evaluate' and 'conflict-monitoring' roles for the dACC did an admirable job of tying together research findings, but did note some puzzles. One of these was that there were some research findings that showed clear evidence of *anticipatory* activity in the dACC (for example, after task instructions had been given but before any stimulus was presented) which didn't fit in with their model of the dACC as monitoring ongoing events. Now, of course, we could link this aspect of cingulate activity to the role of establishing a prior – could it be that the dACC is coding an event as the sort of situation where a mistake just *might* be made?

A slightly different take on the dACC's role was taken by Matthew Lieberman and Naomi Eisenberger, of the Cyberball task. If you recall, they proposed a gauge or 'sociometer' system as part of the social brain that is constantly monitoring our self-esteem levels.[7] This alarm system will be activated in circumstances where these levels could be depleted below what is needed to sustain our social well-being, primarily circumstances which signal social rejection or exclusion. Lieberman and Eisenberger put the dACC centre stage in their sociometer network, based on their observation that the response to social rejection was the same as that to physical pain, both reliably associated with ACC activity.*[8]

Lieberman calls the ACC the brain's 'alarm system', with a cognitive detection system, which keeps track of problems that will require responses, evaluates errors and checks out conflicting messages, and an emotional sounding mechanism, which will flag up any problems and drive the brain's owner to switch on/switch off/change tack or do whatever it takes to keep its social activities on track and the self-esteem tank full. The focus is on avoiding any events that threaten self-esteem in some way. The system is linked to the prefrontal cortex and has some level of control over how much distress is associated with the pain of plummeting self-esteem; more activity here is associated with lower levels of distress in the face of social pain.[9]

Sounding the alarm will set off a chain of events. Suppose, for example, that someone had suggested you ask for a promotion at work. Initially flattered, you start updating your CV and

---

*Bush had explicitly excluded pain studies from his review of ACC activity, as not relevant to his cognition versus emotion divide, but Lieberman and Eisenberger felt that the pain response actually exactly encapsulated the involvement of both processes, signalling both the occurrence of a harmful event and the distress associated with it. They noted that in physical pain studies, those who reported higher levels of distress associated with their pain had more ACC activity, whereas those with more activity in the prefrontal cortex reported less distress. So there was a top-down control mechanism which could modulate this aspect of ACC activity.

checking out the promotion criteria. (Already, it is clear you are not a member of the 'just go for it' school.) Then your ultra-cautious inner critic starts sounding alarm bells – whoa, just stop a minute and think what this might entail. Just how many of those promotion boxes do you tick? Will you be working with different people? Are they your sort of people? What happens if you make a mistake (much more likely if you're working at a higher level)? Is this really 'your' sort of job? Think how comfort-able you are with what you're doing – the pay is lousy (and you're pretty sure you get paid less than some of your colleagues, but let's not rock the boat) but you find the work easy and you've done it so long you very rarely get anything wrong. You're known as a safe pair of hands – don't put yourself in a position where people think they made a mistake in promoting you. Yes, you should bin that application – phew, lucky escape.

In neuroscience terms, the primary response will be a 'stop' or inhibitory one. A mistake or potential mismatch having been flagged, the ongoing behaviour needs to be 'switched off' and alternative responses sought. Additional systems can then be brought into play; perhaps a reappraisal, involving the prefrontal cortex, will dampen down the affective component and prevent too much drain on the self-esteem stores. So rather than a chimp-like 'Feel the fear and do it anyway' response, there is a wimp-like 'Feel the fear and get out of there as fast as you can' command. So, just what determines whether we get a chimp or a wimp?

The sociometer and the 'inner limiter'

In the chapter on the social brain, we came across the notion of an internal gauge or sociometer, measuring our self-esteem levels and alerting our dACC-based alarm system if the readings are in the dangerously low zone.

In his book about the social brain, Matthew Lieberman describes problems with the two alarm systems in his house. These comprised a doorbell which didn't ring (a faulty sounding system) and a smoke alarm where a malfunctioning sensor meant the alarm went off in the absence of any smoke (a faulty detection system).[10] In this same spirit, I'd like to tell you a bit about a faulty alarm system in my life in order to illustrate a possibly malfunctioning component we need to be aware of in our brain's sociometer. My house is quite an old building, gradually converted by our predecessors over decades, with the many and varied parts of the electrical wiring system reflecting this history. Shortly after we moved in, the installation of an additional light in the porch necessitated yet another (extraordinarily expensive) electrical circuit. What followed was weeks of frequent, unexplained and apparently random total power cuts. Several (similarly expensive) visits from puzzled electricians later, it emerged that the problem was in the newly installed trip switch – proudly fulfilling all the nuanced control requirements of a modern porch light, but constantly panicked by the vagaries of the Edison-era electrics to which it now found itself linked as well. Touching that slightly stiff switch in the upstairs bedroom? Someone opening the airing cupboard door? Maybe thinking about using the iron? Any slight hint of unusual activity and our super-sensitive trip switch took the line of least resistance (sorry) and switched everything off. So unlike Matthew Lieberman's smoke detection system, the sensor wasn't really faulty but just oversensitive.

Just like this trip switch, I think that there can be quite marked individual differences in the threshold above which our sociometer alarm system might be triggered. As we will see, some people can shrug off a job rejection with a touch of efficient post hoc rationalisation; others will be plunged into a Slough of Despond, their sociometer plummeting swiftly into the red zone. The outcome of an activated sociometer may not just be a short-term stop-at-the-red-traffic-light incident; it may, over your lifetime, steer you away from potentially positive events, or prevent you from making any life-affirming decisions.

What sets the threshold for this system? Is there some kind of internal mechanism that we are born with, or might the rules of our outside world be incorporated into the mechanism? And if the rules are gendered, do we get a gendered sociometer?

The sociometer is presented as a reactive system and not a predictive one. But might it also have a barometer-like function, where the readings of the gauge can give a forecast of what is in store? As mentioned above, Bush and colleagues commented on the anticipatory nature of ACC activity in some tasks, and our twenty-first-century model of the predictive nature of our brains suggests that monitoring systems are on the lookout for what *might* happen as much as what *is* happening. A recent review of dACC activity from the University of Oxford neuroimaging group specifically identified an updating role in its control of behaviour.[11] By flagging up past errors or rewards, it guided decisions about whether carrying on with the same patterns of behaviour was advisable, or whether it was time to try a different tack.

In many circumstances, particularly with respect to social activities, we may be calling on past or even present events to get a handle on the rules of engagement. But very often our social musings are to do with predicting the future: how someone *might* receive your job application, what *might* happen if you changed jobs, asked for a promotion, put your hand up in class, how you *might not* fit in or succeed, or *might not* enjoy the event you've been invited to. As an example, the kind of reactions to stereotype threat that we met earlier can be seen as an anticipatory response; your group- and your self-identity antennae are twitching with the realisation that you are in a situation where a blow to esteem is at stake and you might perform poorly, make mistakes, let the side down.

This part of the system can also malfunction, in that the anticipation may not match what actually does happen. A drop in pressure doesn't always signal rain, so it may be safe to go out without an umbrella. But if the gauge is set to err on the side of caution, then the inhibitory warnings will be heeded much more than necessary. And, of course, failure to discover whether or not reality did match the anticipation will reinforce this avoidant behaviour, as no prediction error will be registered. If you don't go out, you won't get wet. Additionally, if the estimated cost of the anticipated consequences (a blow to self-esteem) is set at a much higher value than the benefit of ignoring them (I could prove everyone wrong by going for that job) then our conflict-monitoring, behaviour-limiting dACC will win the day.

For some people, this anxious anticipation is so overpowering that they are reluctant or even unable to engage with the potential vicissitudes of everyday life,[12] which suggests that the predictive part of their sociometer is both overactive and focussed on negative outcomes. We might consider one version of the sociometer which would be less than useful in our social brain network. This version is oversensitive and may unnecessarily slam on the behavioural brakes; it is driven by a predictive cost–benefit analysis which is permanently set on 'the game is not worth the candle'. To use a motoring metaphor, it is as though the speed-control limiter is set too low, and our brains will stay well below even the most minimal of speed limits, cautiously manoeuvring us along an ultra-safe inside lane.

So we have a brain-based system, an inner limiter, which normally acts as an adaptive and influential control centre in the social brain, but whose settings have been altered to make it an overactive brake on ongoing behaviour. Based on the sociometer model, the heart of this system is the dACC. The consequences of our cautious limiter will therefore be evident in problems with self-esteem, anxiety and overinhibited behaviour. As we will see, it is possible to characterise sex/gender differences in social brain processes in terms of overactivity in dACC systems, which can help to explain where those gender gaps in power and achievement might be coming from.

Self-esteem

When we looked at the social brain in Chapter 6, a sense of self or self-identity was seen as a core outcome of our social brain's activities. And these activities were focussed on doing whatever it would take to ensure that this self-identity was positive, searching for the best ways of guaranteeing high levels of self-worth and self-esteem. Our brain-based sociometer will keep a running check on our self-esteem levels, avoiding the perils of social rejection and the consequent activation of the same pain mechanisms that would be active if we were to break an arm or a leg. It has been suggested that maintaining or improving levels of self-esteem may be nearly as vital to our well-being as adequate food or shelter. Pathologically low levels of self-esteem are associated with a range of mental health problems such as depression or eating disorders.[13] It is probably because of this core role of self-esteem in so many areas of behaviour that it is, arguably, one of the most widely studied constructs in the modern social sciences', as was claimed in 2016 in a huge cross-cultural study of self-esteem, with a running total of over 35,000 studies on self-esteem or measures of self-identity.[14]

A near-universal finding from these thousands of studies has been that there is always a gender difference in self-esteem, with men consistently scoring higher.[15] And this is not just in WEIRD countries (Western, Educated, Industrialised, Rich and Democratic, of course). This huge study tested nearly one million people online, in forty-eight countries, and found significant gender gaps in every single one. In all the countries, women had lower self-esteem scores, though, as you might expect, the effect size was not the same in each country. Where were the biggest differences? In the top ten were Argentina, Mexico, Chile, Costa Rica and Guatemala (showing that cultures in some South and Central American countries have a major self-esteem problem in their female populations), followed by the UK, the US, Canada, Australia and New Zealand (suggesting it is by no means just a South American problem). The smallest differences were found in Asian countries such as Thailand, India, Indonesia, China, Malaysia, the Philippines, Hong Kong, Singapore and South Korea.

The researchers had also collected a range of socio-political variables such as GDP (gross domestic product) per capita, Human Development Index data (life expectancy, levels of literacy and educational enrolment), and Gender Gap Index data (gender differences in economic participation and opportunity, educational attainment, political empowerment, health and survival), which were all taken into consideration and assessed to see how they might have contributed to the variations in self-esteem gaps they had found. The universality of lower female self-esteem could be explained by biological factors which obviously weren't measured here, but the range of differences suggested there might be some additional exacerbating or protective factors at work. Perhaps paradoxically, the overall picture that emerged was that the wealthier, more developed and more egalitarian a country, the bigger the gender gap.

As the authors pointed out, this paralleled the findings from another big study, this time looking at personality differences, which showed that sex differences were larger in prosperous, healthy and egalitarian cultures.[16] Here the interpretation was that innate differences might be able to 'naturally diverge', the 'true biological nature' of sex differences no longer masked by socio-political factors. In fact it was this study that captured the fancy of James Damore, the author of the Google memo that I mentioned earlier, although David Schmitt, the lead researcher who carried out the study, felt that Damore had misunderstood and misinterpreted it. But Schmitt certainly emphasised a biological basis for the findings reported. Could the self-esteem differences (or rather, similarities, as every country showed a deficit in female self-esteem) be related to similar factors, yet again blaming biology for some kind of alleged deficiency? There had been little research into biological sources of gender differences in self-esteem, although as we shall see, it *is* possible to explore such differences in terms of the neural sociometer proposed by Lieberman and Eisenberger, particularly as it can encompass the kind of social factors explored here.

Another possible explanation was couched in terms of 'social comparison' processes. In some cultures, how you match up to members of groups other than your own is a key feature of self-identity, a sort of process of checking out the competition. This is more common in Western developed cultures; in non-Western cultures it's more common to compare yourself only with members of your own group. The smallest self-esteem differences found in the study above were in Asian countries such as Thailand so the researchers suggested that women here were culturally 'protected' from the negative consequences of cross-gender comparisons.

So there appears to be a worldwide sex/gender difference in levels of self-esteem. Could this be the basis of gender gaps in achievement or even in engagement with the potential sources of achievement? So far, explanations of gender gaps have been couched in terms of brain-based cognitive skills, genetically determined, hormonally organised, fixed and context-independent. Revisiting the alleged differences in such skills in the twenty-first century has revealed that they are either too small to explain the kind of gender gaps we are looking at, or are diminishing, or maybe never actually existed in the first place.[17] Perhaps we should turn our attention to brain-based social processes instead? Might the variations in self-identity flagged by gender differences in self-esteem offer another source of explanations?

What might be the brain mechanisms behind this low self-esteem? We know that actual social rejection, which lowers self-esteem, activates pain mechanisms involving emotion-processing systems and a prefrontal cortex–ACC partnership.[18] Alexander Shackman and his team from Richard Davidson's lab at the University of Wisconsin–Madison have focussed on the ACC as a hub where information about the negative consequences of activity can be linked to 'action control centres', which will inhibit actions to avert the pain they are causing.[19] So in the ACC we have a social coding system which can be linked to a social action (or inaction) system.

A focus on the negative is characteristic of the normal functioning of the sociometer – we are driven more to avoid our self-esteem levels reading empty than to register when they are full. But an abnormal focus on the negative is characteristic of clinical depression, where numerous studies have reported greater reactivity to negative feedback, greater processing of negative facial expressions such as sadness or fear, better memory for negative images or events.[20] With respect to clinical conditions associated with low self-esteem such as social anxiety disorder and depression, where there is a much higher incidence in females, there has been a focus on self-criticism or a negative view of the self as a key feature in such disorders.[21] So as well as an external focus on the negative, this focus is turned inwards as well.

Self-criticism is a form of negative self-evaluation, directed to various aspects of the self, such as appearance, behaviour, thoughts and personality attributes. There is good evidence that excessive self-criticism is a vulnerability factor in developing depression, is correlated with level of severity, and is predictive of future episodes and even suicidal behaviour.[22] One's 'sense of self' or self-worth may not always be positive, and there can be days (or unfortunately for some, long stretches of time) when our sociometer is reading low or empty. Our self-esteem is based on our evaluation of so many different attributes, including our physical appearance and intellectual ability, our past achievements and hope for future ones, membership of the 'right' in-groups, or, nowadays, comparison with social media's celebrities and success stories. So there are many, many ways in which we can find ourselves falling short of the standards we have set ourselves or of the standards we believe are expected of us. In some of us, this can lead to a constant barrage of self-criticism and negative self-judgement. This powerful inner critic is the dominant voice; if things go wrong it is clearly our fault and a measure of our general inferiority. We are in a state of heightened error monitoring, usually associated with negative affect and response inhibition. In other words, we shake our heads in shame, shut down and switch off. Psychological studies have consistently found that women are more self-critical and much more likely to underrate their work performance and have a greater fear of disapproval than men.[23]

As we know from our look at social brain function, error monitoring is a key feature of the ACC–prefrontal cortex axis.[24] So it should be possible to track the brain bases of self-criticism and discover the extent to which this kind of negative self-evaluation is reflected at the brain level. If you remember from Chapter 6, this is what research carried out in the Aston Brain Centre showed when we were studying the brain bases of self-criticism and self-reassurance.[25] We found that the 'critical voice' was associated with activation in the error-monitoring and response inhibition system, the prefrontal cortex and ACC. The 'reassuring voice' did not activate the error-monitoring system but was associated with activation in brain areas consistent with empathic behaviour. And higher levels of activation in the error-monitoring, behavioural inhibition system were seen in those participants who rated themselves as typically being self-critical in their day-to-day life.

So, it seems a highly active 'inner critic' could be switching on those parts of our self-referencing system which are on the lookout for mistakes, constantly flagging up errors and putting the brakes on those aspects of our behaviour that might lead us into painful social melees, instead diverting us down quiet, safe little side streets.

Rejection sensitivity and self-silencing

As we know, the pain of social rejection activates the same areas as physical pain, a measure of how important our sense of belonging is to our well-being. Given the aversive nature of such experiences, we need a sensor mechanism that will keep us on the lookout for the rejection possibilities – a rejection prior. The desire to avoid rejection is therefore generally adaptive, but in some cases the mechanism appears to be overactive. This 'rejection sensitivity' is defined as 'the tendency to anxiously expect, readily perceive, and intensely react to rejection'.[26]

A rejection sensitivity questionnaire can give a measure of high or low rejection sensitivity (RS), with respondents indicating how concerned or anxious they might be about potential rejection in a range of different situations, such as 'You approach a close friend to talk after doing or saying something that seriously upset him/her', or 'You ask your supervisor for help with a problem you have been having at work'.[27] For those high in RS, rejection itself (whether real or perceived) can lead to a range of different behaviours. A common response can be aggression, measured in lab-based studies by what is intriguingly called the 'hot sauce paradigm'.[28] Basically, individuals who have just experienced an experimentally induced rejection by their previously unknown partner in a lab-based scenario are given the opportunity of allocating an amount of 'hot sauce' to said partner, together with the 'accidentally revealed' information that this partner really dislikes hot sauce. The amount they give them is taken as a measure of their aggression towards them. The use of this and other measures indicates that, for some high RS individuals, rejection will be followed by aggression.

For others, however, the response is more likely to lead to withdrawal and negative affect, possibly even tipping into clinical depression. Social rejection is strongly associated with the onset of depression. The kind of adverse 'internalising' responses associated with depression are much more characteristic of women, who are twice as likely as men to suffer from clinical depressive disorders. Such behaviour has been described as a tendency to 'self-silence', to inhibit any thoughts and feelings or preferred actions that might be perceived as likely to lead to conflict and rejection. This 'self-silencing' concept was developed in the 1990s by psychologist Dana Crowley Jack and was described as associated with a fall in self-esteem and feelings of a 'loss of self'.[29] It particularly related to significant relationships and described the process whereby women felt they had to sacrifice their own needs or not state their own feelings if they perceived that these might cause conflict.

This self-silencing has been noted strongly not just in women, but also in minority groups. Bonita London's SPICE (Social Processes of Identity, Coping and Engagement) research lab at Stony Brook University in New York state has studied the mechanisms associated with social identity threat in minority groups and/or institutions where there is an imbalance of presence and power, particularly educational or business-based institutions.[30] The researchers specifically investigated the link between RS and self-silencing. With respect to women, they proposed a model of gender-based RS to account for individual differences in how women perceive and cope with gender-based evaluative threats in competitive, historically male institutions.

In order to measure the various manifestations and consequences of this kind of RS they developed a Gender RS questionnaire. Participants were presented with various scenarios such as 'Imagine that you are starting a new job in a corporate office. On the first day, the manager arranges an office meeting to introduce you as a new employee', or 'Imagine that you have worked at your job for nearly a year. A position is open for a manager and you approach your boss to ask for the promotion'. They then had to indicate on a six-point scale how anxious or concerned they would be about being treated differently or experiencing a negative outcome because of their gender. They found that both men and women reported themselves as familiar with the type of situations depicted in the questionnaire, but that women were significantly more likely to anxiously anticipate gender-based rejection, which the researchers described as a form of hyper-vigilance. By looking at other types of scenarios, such as race-based situations, they demonstrated that the women didn't show gender-based anxiety here, so it wasn't that women were just more likely to expect rejection of any kind.

When looking at coping strategies to deal with rejection, they also found that women working in academia were much more likely to use self-silencing, assessed by an adaptation of a 'Silencing of the Self' questionnaire to capture self-silencing in academic contexts, than to speak out (risking confrontation) or to seek help. One outcome of this process was higher levels of academic disengagement in women, a withdrawal from participation in academic activities and reduced take-up of additional support systems such as open office hours or additional tutorials. Mindful of the accusation that lab studies often don't reflect real life, the researchers followed a group of males and females for three weeks using a daily diary format just as they entered a top law school. They found that the women showed significantly higher RS than their male peers and were more likely to attribute negative events to their gender. The overall finding from this series of studies was of much higher levels of RS in women, leading to forms of self-silencing and avoidance of evaluative opportunities, which, in the long run, might compromise any chances of success the situation may have offered them. These findings parallel the reluctance of women to engage in more challenging aspects of STEM that we saw in the last chapter.

The ultimate type of self-silencing might be the wish to become anonymous. A study carried out in 2011 showed that women's maths performance was better in an experimentally induced stereotype threat situation when they were allowed to complete a test under a fictitious name.[31] The researchers were investigating whether stereotype threat reflected anxiety about self-reputation more than the reputation of the group with which you were identified. They found that women, who overall had reported higher levels of concern about the effects on their self-reputation of doing worse in a maths test than men, did significantly better on a 'threat-enhanced' maths test under an assumed name (male or female) than women who attempted it under their own name. No such effect was found for men. So if there was a way of 'disconnecting the self' (or the 'L'eggo my ego' effect, as the authors wryly dubbed it) from a potential self- or group-threat situation, women benefited far more than men. Yet another indication of the greater impact of external evaluation on women and their need to avoid loss of self-esteem, and more work for the sociometer.

So what brain mechanisms might be at work? As Lieberman and Eisenberger have shown, the interaction between the prefrontal cortex and the dACC is associated with not only the experience of pain but also its degree. Participants reporting more pain (social or physical) showed more activation in the ACC than those with a more active prefrontal cortex.[32] Perhaps this type of system might underpin the kind of pain *anticipation* associated with rejection sensitivity? Researchers at Columbia University investigated this by using paintings representing themes of rejection or acceptance.[33] Mirroring actual pain responses, images of rejection were associated with higher levels of prefrontal and ACC activity than images of acceptance. But different patterns of activity within these regions differentiated the low- from the high-RS group; low-RS individuals had higher levels of activity in the prefrontal areas, showing a similar response to individuals instructed to down-regulate or reappraise negative responses to aversive images. This suggests that the high-RS group were unable to make use of the same sort of process and could not rethink their fears.

A similar study from Eisenberger and Lieberman's lab looked at ACC activity in reaction to disapproving facial expressions, identifying these as socially coded cues of potential harm, rather than the actual harm which could be signalled by angry or fearful faces.[34] The study showed that high-RS participants had more dACC activity in reaction to disapproving faces but not to faces showing anger or disgust, so their response was only to the socially negative expressions. Similar to the Columbia study, there was a negative correlation between RS and prefrontal activity, again suggesting that high-RS individuals are less able to activate appraisal or down-regulation resources.

Rejection sensitivity clearly has a profound effect on those who experience it, and appears to activate a self-protective switch-off mechanism resulting in disengagement and self-silencing. The results from brain imaging studies suggest that this system is based around the dACC, consistent with its role in monitoring self-esteem.[35] This system can apparently be modulated by input from the prefrontal system, reducing the levels of distress associated with the pain of rejection. But there appear to be individual differences in the availability of this modulating influence, with the result that the inhibitory power of the dACC is unchecked, like an oversensitive speed limiter. So the greater RS in women could reflect atypical dACC activity. This is consistent with the ongoing work in Eisenberger's lab.[36] Here they have shown that girls who had had previous episodes of depression, when experiencing a scanner-based social rejection scenario, showed increases in dACC activity, as well as increases in depressed mood.

Rejection sensitivity clearly has a profound effect on those who experience it. And the consequence for women seems to be to activate an introverted and inhibitory 'switch-off' system, resulting in withdrawal, failure to engage and self-silencing. Extreme versions of this response are characteristic of clinical depression.[37]

Self-esteem and stereotype threat

Besides the consequences of rejection or even just the fear of it, another source of attack on self-esteem stores, with consequences for performance and behaviour, can be seen in the process of stereotype threat.[38] Stereotype threat effects have been demonstrated in males as well as females,[39] so it would be necessary to establish that lower levels of self-esteem in women are not just related to greater susceptibility to stereotype threat and/or different responses to it. This was investigated in a series of studies carried out by psychologist Marina Pavlova.[40] The aim was to induce stereotype threat in a previously neutral task and to measure any resultant effects, including gender differences. Participants carried out a simple story card arrangement task and were either given explicit positive messages, like 'Males are usually better at this task' (with the implicit shadow message being 'So females are usually worse'), or explicit negative messages, like 'Males are usually worse at this task' (with the implicit shadow message being 'So females are usually better').

The result of this was clear differences between males and females. In the 'female negative' condition, females showed significantly worse performance than controls and males significantly better. In the 'female positive' condition, females showed some performance enhancement, but there was little change in the males in this group. In the 'male positive' condition, there was increased performance from the males, but quite dramatic underperformance in females, who were receiving the implicit message that they would probably do badly on the task. Where there was a 'male negative' message, a rather paradoxical effect was shown. Males showed some deterioration in performance but so did females, despite the fact that they should have been responding positively to the message that, as males generally did worse, females should do better.

Overall, then, males responded as expected to explicit messages that, as males, they would do better or worse on the task, but were less responsive to the implicit messages. Females, on the other hand, were more adversely affected by implicit messages, showing lower levels of performance with an implicit negative instruction (this is a task done better by males) but also with a supposedly positive message that could have been inferred from the message that males usually did worse. The researchers suggested that females might have interpreted this as meaning that if men did badly on this task then they (as women) were likely to do even worse. There was no debriefing of participants to check this out, but the same effect was not seen in men, so women were clearly responding to a different take-home message. In fact, of the four conditions, women's performance suggested they were taking negative messages from three of them, with only the explicit 'women are better at this' message resulting in a small improvement. These findings are consistent with Bonita London's findings of increased rejection sensitivity in women that we looked at above, indicating some form of hyper-vigilance in women to potential negative evaluation in gender-based situations. Unless the message is pretty clearly spelled out, males seem blissfully unaware (or at least much less susceptible) to the possibility of failure, whereas women seem ever on the lookout for it, even to the extent of reinterpreting a possibly positive message.

As we've seen before, one of the consequences of stereotype threat is that the brain engages task-irrelevant networks, those associated with emotional coding and self-reference; in other words the familiar limbic areas and prefrontal–ACC partnership.[41] We saw this in the last chapter when looking at the brain bases of maths anxiety. When participants were told that the maths task they were about to embark on was 'diagnostic of your math intelligence' it resulted in very different patterns of response to those shown when the task was described as a measure of preferred problem-solving strategies. And there was much greater responsiveness to negative feedback, and faster disengagement from potential sources of support. This is all very consistent with the behavioural consequences of RS and also with the patterns of brain activity associated with this process.

So brain activity during stereotype threat and related situations is consistent with the action of some kind of Facebook-like profile updating, with a particular focus on negative feedback associated with errors. Evidence that women are more susceptible to negative stereotype threat, either real or inferred, together with their greater susceptibility to RS, suggests that they have a much more active or at least more sensitive inhibitory or 'inner limiter' system. As we know, these activities are focussed around the ACC, part of a powerful socially focussed behavioural control system, which also codes positive and negative values in our outside world. What other aspects of behaviour might be associated with an overactive error-evaluate system, with an overcautious, risk-averse approach to life?

Sugar and spice and all things nice

The development of self-identity is linked to an understanding of those aspects of your behaviour that will gain positive acknowledgement, that are 'appropriate' for the group to which you belong. Maintaining these patterns of behaviour should ensure that you will continue to be accepted by your significant in-group, and avoid the real pain of social rejection.

A very early measure of 'good behaviour' in children is the ability to self-regulate, to direct their behaviour and attention to the task in hand.[42] This may involve paying careful attention to rules and inhibiting inappropriate behaviours, such as running around, shouting out and so forth. It is often related to school readiness (so you are expected to have it quite young) and also to early achievement in school. Teacher and parent reports suggest that this ability emerges earlier in girls and that girls remain better behaved in classroom situations than boys.

But, as we know, self-report is not always reliable. A group of American researchers have devised a more direct measure of self-regulation behaviour based on the 'head, shoulders, knees and toes' game.[43] The participating children are taught that if the researcher shouts 'Touch your head!' you have to touch your toes and vice versa, or if you are told to touch your knees you have to touch your shoulders. This results in the accronym HTKS – for Head touch Toes, for Knees touch Shoulders. The idea is that the children have to pay attention, remember the rules, and inhibit the first response in order to do the opposite (rather like the British game 'Simon says'). A study carried out in Michigan looked at performance on this self-regulation task in five-year-olds in the autumn and spring terms of their kindergarten year.[44] Girls outperformed boys at both stages of testing, confirming teacher ratings that were obtained at the same time. The HTKS task was also used in a cross-cultural survey comparing self-regulation in the US with Taiwan, South Korea and China.[45] The choice of Asian countries was partly motivated to compare cultures where there has been a history of very specific gendered behavioural expectations, with girls expected to be more passive and submissive. This study showed that, although the teacher ratings reported girls as more self-regulated, they didn't actually do better than the boys on the HTKS task, the direct measure of behaviour. So there is a fairly universal impression among teachers that girls are more self-regulated, which isn't always backed up by reality. But we know that teacher expectations can, of themselves, serve as powerful biases in the production of behaviours, leaving a clear message that girls are well behaved and good at self-regulation.

An aspect of self-regulation is that you will invariably have to inhibit some patterns of behaviour, possibly those associated with spontaneity and impulsiveness, and focus on those that will earn you personally the most Brownie points, boosting your self-identity, and give a positive image of the set you belong to, enhancing your group identity. It will certainly help you avoid negative or unpleasant events that might elicit disapproval.

A long-standing concept in personality psychology, first devised by Jeffrey Gray back in the 1970s, is that of a behavioural inhibition system (BIS), which is sensitive to negative events in the outside world and will inhibit those patterns of behaviour associated with punishment or non-reward. BIS-like behaviour is assessed by self-report questionnaires, including items such as 'I feel pretty worried or upset when I think or know somebody is angry at me' and 'I worry about making mistakes'. This contrasts with the behavioural activation system (BAS), a reward-seeking system ('I crave excitement and new sensations'), often associated with impulsive behaviour. The role attributed to the BIS is to process threat and to halt ongoing behaviour which might lead to negative consequences; so, in contemporary predictive brain terms, to establish a 'warning' prior. Females show higher levels of BIS-type behaviour, which is also associated with higher rates of disorders such as anxiety and depression.[46]

As you may have already guessed, these BIS functions are consistent with the conflict-monitoring, action interruption and self-regulation functions that have been identified as characteristic of the dACC. Studies have shown that higher BIS scores are indeed associated with the amplitude of error-related and Go/NoGo brain responses whose sources lie here.[47] So there is accumulating evidence that the kinds of inhibitory self-regulation process more commonly found in girls are linked to increased activity in a self-esteem monitoring system, a system based in the ACC. But where might this self-regulation come from? Are girls born well behaved, anxious to please, risk-averse or unfeisty? Do they arrive in the world with a preset inner limiter which will steer them cautiously down safer pathways? Or is there something in their world which might be nudging them down these routes?

As a junior academic I had the unenviable role for several years of being the admissions tutor for our undergraduate psychology course. This meant that I had to plough through thousands of UCAS applications and give them the thumbs-up of an offer or the thumbs-down of a rejection. It also meant that I had to scrutinise in detail the personal statements and references that accompanied these applications, the latter often giving me more insight into the referee than the applicant (one personal favourite was 'This young man needs to be left up the creek without a paddle'). Mainly, of course, referees assured me that the applicants were paragons of whatever virtues it was felt we admissions tutors were looking for. But there could also be a heavy element of 'damning with faint praise', where you got the impression that the referee couldn't tell you outright not to bother but couldn't find anything academically valuable that might influence your decision. For me, 'always nicely turned out', 'helpful with the younger children' or 'work always well presented' fell firmly into this category. And (I put my hands up here – this is only a personal impression and with the benefit of much hindsight) I believe I only ever saw it in references for girls. Though this would be consistent with the study we looked at in Chapter 10, where the 'letters of minimal assurance' ('Sarah is easy to get along with') in medical school applications were much more common for female applicants.

Are girls praised for different things than boys? Given the role of social feedback in the formation of self-identity, it is important to understand if such feedback is unevenly distributed. Within education, there certainly seems to be some sort of asymmetrical praise system in place. Whereas boys are praised more for getting things right, girls get praised more for good behaviour.[48] Similarly girls are criticised more for making mistakes, whereas boys get criticised for bad behaviour. This means overall that more positive attention is paid to girls' good behaviour than to their academic ability (with the reverse effect for boys).

The Stanford psychologist Carol Dweck has proposed a 'mindset' model for understanding human motivation.[49] Broadly speaking, a 'fixed mindset' indicates a deterministic belief that your skills portfolio comprises the hand that nature dealt you. This will pretty much determine your progress through life's challenges and there is little you can do to change things. Alternatively a 'growth mindset' relates to a belief that your skills can always be developed, that you will embrace challenges, welcome criticism and always be willing to learn. The development of fixed or growth mindsets is linked to the kinds of praise meted out at key stages of development. Although the theory has proved controversial, with difficulties in assessing the suggested intervention strategies in educational settings, the background research has provided some insights into the different ways in which praise is handed out to girls as opposed to boys.

Dweck suggests that a constant emphasis on non-intellectual aspects of work, such as 'neatness' or 'speaking clearly', can have the effect of devaluing praise (if there is any) on the outcome of the work itself. Just saying something is tidy doesn't give you much insight into how well you've grasped the basic principles of your maths problems or history homework. And there was a big imbalance in the extent to which boys and girls received this kind of feedback, with lots more positive feedback given to girls about neatness and so on, whereas for boys, much less attention was paid to these non-intellectual aspects of their work. With regard to actual content of the work, in an echo of the educational observations noted above, girls were more likely to have attention drawn to their mistakes, whereas boys were more likely to receive praise when they got things right. So girls and boys were getting different messages; for girls, doing well wasn't a measure of ability, but of having good handwriting and making effective use of highlighter pens and rulers. The take-home message could be that their 'always nicely turned out' homework could counteract their basic lack of ability, evident from how many times teachers had to draw attention to their mistakes. On the other hand, boys were getting a 'you've got talent' message whenever possible, with an infrequent and fairly muted sigh on the side about any scruffiness.

Another issue that educational psychologists have noticed is that 'person praise' ('you must be really smart') has a different effect on the consequences of failure than 'performance praise' ('you must have worked really hard').[50] Person praise seems to be very motivating while someone is getting things right, but if they start to get things wrong, they are more likely to be demotivated and to give up on the task in hand (an 'I seem to have lost my mojo' sort of response). Those given performance praise, on the other hand, deal with failure better and are likely to persist. The explanation given is that person praise emphasises aspects of self-identity more than performance praise, so that if your feel-good factor has come from person praise, then failure means your sense of self-worth is taking more of a knock, or it elicits a feeling that this task is obviously not the kind of thing that you are good at. Performance praise, on other hand, relates more specifically to the task in hand, so a bit more knuckling down might just get the job done.

Gender differences in the effects of these different kinds of praise have been demonstrated in 9–11-year-olds.[51] In one study, following some successes and some failures with different kinds of puzzles, the children were offered one of the puzzles they had failed to solve as a free gift at the end of their sessions. Girls who had been given the person praise when they succeeded were much more likely to reject the puzzle they had failed at than girls who had had the process praise. Boys, on the other hand, were more likely to want to take home the puzzle they had failed at, especially if their puzzle performance had been associated with person praise.

So, even where boys and girls are similarly praised, praise that has elements of self-reference can have negative down-stream consequences for girls when they encounter failure. Interestingly, the researchers repeated this study with 4–5-year-old pre-schoolers and found no gender differences. So this differential sensitivity to praise is not evident in the early years.[52]

If we look at the gender gaps in social behaviour and measures of social esteem we begin to get a very different picture for males and females. And this is clearly linked to different patterns of behaviour, arising from different sensitivities in a brain-based inner limiter mechanism. This mechanism drives a self-setting, self-organising process, basing its thresholds and triggers on the rules of social engagement it absorbs. Its thresholds will be set and reset according to the schedule of rewards and punishments, of approval and disapproval, that it encounters in the outside world. It will be super-sensitive to the different social messages it picks up, to the gendered world it encounters, and it will adjust its settings accordingly.

What might this mean for people who appear to start life with no distinguishable differences other than some physical paraphernalia associated with reproduction, and with apparently similar sets of cognitive skills? If markedly different messages, or settings data, are input, this can result in a markedly different portfolio of responses. If your world sets very different limits on your performance then your inner limiter may drive you down a very different pathway.

Chapter 13:
Inside Her Pretty Little Head – A twenty-first-century update

It's clear we've come a long way from Gustave Le Bon's idea that women are 'closer to children and savages than to an adult, civilized man' and that breakthroughs in technology, such as fMRI, in this century and the last have given us a more complex and fine-tuned idea of how our brains work. The arrival of fMRI offered the opportunity for much better access to what was going on in the brain and should also have impacted on the quest for answers to the age-old question about whether women's brains are different from men's. In Chapter 4, we saw from tracking a hype cycle that misinterpretation of the exciting new images meant that fMRI didn't quite succeed in overturning the stereotypes or in challenging the status quo. A tide of neurohype and neurotrash washed the promise of brain imaging into the Trough of Disillusionment, and the new technology of neuroimaging, combined with a supporting cast of psychologists and neuroendocrinologists, contributed more to the sustaining of stereotypes than their deletion. Perhaps now, some years later, we have reached the Slope of Enlightenment at last?

Neuroimaging has gone through a process of putting its house in order, and there are new models of how the brain works and interacts with its world, as we've seen throughout this book. The last decade or so has seen focussed attempts on the part of the cognitive neuroscience community to address the 'neuro-foolishness' that brought a certain amount of disrepute to their activities.[1] Attempts to educate and inform both themselves and their listeners have been aired quite extensively, and there have been dramatic improvements in the quality and the quantity of the techniques available and how their outputs are interpreted.

So, how has the study of sex, gender and the brain fared in this clean-up? There should be a much better chance of finding new answers to the, by now, pretty old questions about the brains of men and women – shouldn't there?

Have we taken out the neurotrash?

We know that the output of early brain imaging research into sex differences was enthusiastically and often mistakenly adapted by our purveyors of neurotrash. This was often despite, but sometimes because of, what researchers were actually saying they had found. The early neurohype was fed by understandable but misplaced enthusiasm, and later fuelled by the emerging use of press releases to 'big up' the findings from universities or research centres. Following the wave of criticism that brought us to the Trough of Disillusionment, researchers are now more aware that care is needed both in the spin they put on their own findings in their published papers and in the spin they allow their marketing department. If the effect sizes in their data are only small, for example, then terms such as 'fundamental', 'significant' or 'profound' should really not be used.

To see if we have come as far as we think we have, let's look at a 2014 published study that studied sex differences in patterns of connectivity in the brain, which as we know has become a new area of focus for neuroscientists, as opposed to rehashing the same old 'size matters' debates.[2] The technique the researchers used meant that they could make 34,716 comparisons; of these, only 178 showed differences between males and females, with (as the researchers did indicate) quite small effect sizes (0.32). That is to say, only 0.51 per cent of the differences they tested revealed differences between males and females.

Astonishingly, though, the authors still described these as 'prominent' differences. They did note in the paper that 'on the whole male and female brains are more alike than different' but the title of the paper and the keywords listed both include the words 'sex differences' – so there is a pretty strong likelihood that this paper would land up in the 'proven' pile of sex differences evidence, despite the actual evidence being largely to the contrary. Drawing on an amazing data set, 1,275 participants in all, the only exclusion criteria were medical conditions or problems with the scanning or behavioural data that had been collected, and the only additional information used for the resulting 722 participants was their sex and age (312 males, 410 females, aged eight to twenty-two). There was no additional information about years in education, occupation, or socio-economic status. So certainly not the Slope of Enlightenment we would have hoped for.

Even without researchers themselves adding to the problem, sometimes the media go further and make up their own spin. A story of sex differences without a mention of the brain? That can easily be remedied!

Take a recent survey published in 2014 that tracked changes in sex differences in certain cognitive skills over several decades and across several different parts of Europe.[3] There was evidence of overall increases in skill over time, as you might expect from wider access to education between the 1920s and the 1950s. In some cases you could see gender differences decreasing or disappearing; in others (such as episodic memory), you could see greater increases in women over time, resulting in greater gender differences in this particular skill. The authors of the study put this down to societal shifts, concluding that 'our results suggest that these changes take place as a result of women gaining more than men from societal improvements over time, thereby increasing their general cognitive ability more than men'. There was also evidence of a sustained, but diminishing, gender gap in numeracy in favour of males.

312 The Gendered Brain

But guess what? It was the existence of *this* gap (and not its diminution) that the *Daily Mail* focussed on. Their headline read: 'Female brains really ARE different to male minds with women possessing better recall and men excelling at maths.'[4] Assuming that their readers might not make it back to the original study, they helpfully interpreted this particular finding for them as follows: 'It is thought the differing strengths can be explained by differences in the biology of the brain as well as in the way the sexes are treated by society.' Yet a scan of the original text reveals that neither the word 'brain' nor the word 'biology' appears. This takes us beyond misinterpretation to near fiction-alisation, all in the name of upholding the status quo.

The 'Chinese whispers' problem

Even reliable and valid research findings can fall foul of the 'Chinese whispers' problem. The pipeline from science to science journalist isn't always straightforward – sometimes diverging via press officers and journal editors, and experts that the journalists have been able to corral for comment. Add in the online science 'trawling' systems, which scoop up hot-topic headlines and put their own spin on them, and with so many hands to pass through, stories can become completely mangled, with the final version sometimes bearing very little relation to its origins.

Catchy headlines can hide the truth from the casual or unwary eye. 'Brain regulates social behaviour differences in males and females,' trumpeted an article in *Neuroscience News* in 2016.[5] The article was helpfully illustrated with the classic cross-section of two brain-containing human heads, one pink, one blue (with added male and female symbols just in case you didn't get the colour coding). The original study had shown that different neurochemicals influenced aggression and dominance behaviour differently in females and males. The article made reference to the significance this could have for understanding and treating the 'prominent sex differences' in depression and anxiety in women and autism and ADHD in men.

It was not until the fourth paragraph of the article that we learned that this study was actually carried out on hamsters. They may indeed suffer from hamster versions of post-traumatic stress disorder or ADHD, but their relevance to the human condition is debatable at best.

This kind of problem is also illustrated by the journey from journal to online resource. One journal article, catchily titled 'Esr1$^+$ Cells in the Ventromedial Hypothalamus Control of Female Aggression', was an investigation into the brain bases of aggression in female mice.[6] The findings suggested that these might be different from those in male mice (which were not tested). A journalist had contacted a top neuroscience researcher about the potential 'human' significance of the study. In response, the researcher wrote a careful and thoughtful reply, copied in to colleagues to check that her cautious view was representative of opinion in the field.[7] Two key points she made were that the study was only carried out on females (so talking about sex differences was stretching a point) and that the participants were mice, so the human significance might be limited. So far so good.

It was therefore somewhat surprising a few weeks later to see the headline 'Science explains why some people are into BDSM and some aren't', helpfully illustrated with an eye-catching image of a scantily clad (human) couple linked by what was possibly a leather belt (I lead a sheltered life), and with an accompanying introductory line: 'Do you like the rough stuff even in the bedroom? Well, recent studies claim that sex and aggression may go hand-in-hand in the human brain!'[8] Tracking back through the tortuous chain of provenance, it emerged that this referred to the very same mice and hypothalamus study that had been carefully commented on earlier. Yet another example of how science can be scrambled on its journey to the public domain.

Neurotrash and the Whac-a-Mole problem

Worse still, even resources that have clearly been identified as neurotrash are still being hijacked in the cause of the female–male brain debate. Just when you think it is safe to have well-researched and well-informed discussions about where and why sex differences in the brain might be found, and what they might mean for the brains' owners, up pops an old piece of neurotrash.

You will recall my less than charitable comments on Louann Brizendine's book *The Female Brain*, generally identified as a rich source of inaccurate and/or untraceable assertions about sex differences, and now a film too (currently with a thirty-one per cent rating on Rotten Tomatoes).

Some colleagues and I were contacted by a *Newsweek* journalist for comment in relation to the film, with a list of some of the neuroscience claims in the film for confirmation. A few of these 'facts' proved particularly puzzling. For example, 'Gossip is critical for building social bonds, so women's brains have a "hardwired" dopamine-reward system for gossiping' (this appears to link back to Paleolithic versions of *Hello!* magazine, helpfully supported, one assumes, by findings from Paleolithic endocrinologists). And I'm afraid I gave up the fight with this claim: 'So, I know I said women seek consensus, but, if the amygdala is activated, her adrenaline can give her enough confidence to override the instinct to be cooperative.' I googled this to try and work out what on earth it was referring to, and was taken to one site on the 'psychopharmacology of pictorial pornography' and another one on horse behaviour. Which I felt said it all (both sounding more intriguing than the film!).[9]

This film is not going to single-handedly undermine all serious neuroscience attempts to get at the truth, but it is yet another little echo to add to those uncritically circulating in the ether. It seems we have not yet managed to take out the neurotrash once and for all.

Neurosexism lives?

You will recall that Cordelia Fine coined the term 'neurosexism' to draw attention to problematic practices in neuroscience itself which might be contributing to the sustaining of stereotypes and belief in hard-wiring.[10] How are we doing on that front?

Some early brain imaging studies focussed on sex differences in the size of particular structures, such as the corpus callosum or the hippocampus, as the potential source of given differences in behaviour and ability (echoing, in fact, the early 'missing five ounces' approach way back in the nineteenth century). But more sophisticated approaches to calculating size-related aspects of the brain, such as its volume as a function of the head size of its owner, revealed that, simply speaking, it was the size of the brain, not its sex, which determined the size of different structures within it.[11] More recently, this has been shown to be true of the pathways between different structures. Again, simply put, bigger brains have longer (and possibly stronger) pathways to cope with additional distances. If you compare big brains (from males or females) with little brains (ditto) you'll find it is size not sex that is most important.[12] So neuroimagers who are interested in comparing males and females need to factor in additional calculations into their analyses and, importantly, to demonstrate that they have done so.

I've said before that we are still not clear what the relationship is between structure and function in the brain. Does having a bigger amygdala make you more aggressive? Does having a higher ratio of grey to white matter make you more intelligent? If we don't know the answers to these questions, is it worth continuing to use our ever more sophisticated brain imaging techniques for looking at the size of different bits of the brain, in a kind of neo-craniology mission?

On the one hand, old size-related claims are disappearing, particularly in the face of detailed brain or head size corrections (although, as we've seen, there are still ongoing arguments about which brain size correction to use). Two recent meta-analyses showed that previously reliable female–male differences in both the amygdala and the hippocampus, two key structures in the brain, were eliminated once such corrections had been made.[13] On the other hand, new versions of such claims appear to be emerging, partly stimulated by access to the large brain imaging data sets that are now available.

A recent paper by a team headed by Stuart Ritchie, a psychologist from the University of Edinburgh, reports on sex differences in a cohort comprising 2,750 females and 2,466 males.[14] One of the things that is interesting to note about this study is how the findings are reported, both within the paper and in subsequent public comment. The paper's abstract makes reference to males having higher raw (that is, uncorrected) volumes, raw surface areas, and white matter connectivity, whereas females have higher raw cortical thickness and more complex white matter tracts. These differences are illustrated in the text with distinctive pink and blue bell curves, annotated with fairly large effect size data. The total brain volume, grey matter volume and white matter volume differences are particularly eye-catching. However, once the authors had corrected these measures for brain size, many of these differences disappeared, and those that were left were significantly reduced. This was fully acknowledged in the text, but the initial impression for a potentially unwary journalist could well be that these differences were highly significant (in both the popular and the statistical sense of the word).

Indeed, the paper was commented on in an article excitedly entitled 'Why can't a woman be more like a man?', which described it as 'something of a reality check' for those who believe that there are no differences between men's and women's brains.[15] The author also threw into the mix a 'well-established connection between brain volume and IQ', rather oddly quoting a paper in support whose authors actually claim as one of its highlight findings that 'brain size is not a necessary cause for human IQ differences'.[16] So, what one might call a lack of caution in the original paper resulted in a hot-topic headline for an ill-informed journalist, hailing it as a neural reality check.

This kind of rapid pouncing on confirmatory evidence of sex differences in the brain was also evident in a rather more worrying recent event. A paper from the University of Wisconsin–Madison reported the results of a study of brain structures in 143 (73 female, 70 male) one-month-old infants.[17] This was a large-scale study on healthy full-term infants using a high-resolution scanner, so an important data set for this area of research. We know that in the past it has been claimed that female–male differences are evident at birth and that this is powerful support for a biological determinist viewpoint, but that, crucially, there was a lack of large-scale studies on typically developing infants to support this. Perhaps this study would be the decider?

The authors of this paper reported marked sex differences in total brain volume, grey matter and white matter. Again, this was quickly disseminated into the public realm, this time by an online research summary source that pitched this report as an important breakthrough in the search for explanations of female–male differences in behaviour. The source concluded that 'pretending these early sex differences in the brain don't exist will not help us make society fairer'.[18] The trouble was that the reported findings were actually wrong. Although the researchers claimed to have corrected for brain size, an eagle-eyed neuroscientist noted that the data in the paper weren't consistent with this claim. The authors were contacted, rapid checking and reanalysis followed and all the claimed significant differences disappeared.

A correction was quickly issued, published on both the journal's and the research digest's websites.[19] But there was a two-month gap between these events, and social media had already pounced. Reference to the paper had already appeared on Facebook with one telling comment: 'I actually had an argument about this with someone who claimed to have a degree in the field very recently. Rubbing her face in this will make me so happy.' It can still be found on Pinterest too.[20]

In these days of ideological echo chambers it is the fake news, or in this case the fake neuronews, that sticks around, even if later disproved.

The iceberg problem

There is still evidence that the 'file drawer problem' or 'iceberg problem' is alive and well too. As mentioned in Chapter 3, this is the old issue of a reporting bias, where only those studies that find differences get published and those that don't get put in the file drawer.[21] One way of checking this out is to calculate the expected proportions of significant and non-significant differences in a research field, given the known effect sizes of the differences you are investigating. Then you can compare this with what you actually get.

In the battle against the iceberg problem, John P. A. Ioannidis, a professor of medicine, health research and policy and statistics at Stanford University, has been the scourge of poor statistical practice in clinical research for the last decade or so, and has shown how science needs to be more aware of the need to self-correct, to keep an ever-watchful brief on non-reproducible findings or on anomalies in published data sets. In 2018, he and his team turned their attention to neuroimaging studies of sex differences. They looked at the proportion of those that reported differences as opposed to those that reported similarities or no differences.[22] Of the 179 papers they looked at, only two highlighted in their title the fact that they had found no difference. Overall, eighty-eight per cent reported significant differences of some kind. As the authors pointed out, this 'success rate' is implausibly high. The team also looked at the relationship between sample size (number of participants in a study) and the number of brain areas in each study where sex differences had been identified. There should be a correlation between these two factors, as smaller, underpowered studies would normally be expected to find fewer areas of significant activation. However hard they tried, though, the researchers could not find this expected statistical relationship. It looked like there were many more 'positive' findings than you might have expected in the small-scale studies, which were reporting just about the same number of significant areas as the larger-scale studies. This could be for a variety of reasons, ranging from the researchers only submitting papers that had positive findings, or journals only publishing papers with significant findings, or researchers underreporting negative findings.

Given how small the differences between the sexes are, fully acknowledged even by the most fervent defenders of the biological determinist position, research seeking sex differences just should not be showing this enormously high 'success rate'. It has been shown that belief in the 'hard-wiring' approach to sex differences is strongly reinforced by 'brain difference' reports such as those examined by Ioannides' group, so it is concerning for all of us when this body of work is shown to be biased in this way.[23] Neuroscience-type evidence is a powerful outside influence in the brain-changing effects of the world, in the sustaining of stereotypes and in the cataloguing of self and other profiles by our rule-gathering social brains. So if we are getting a distorted view, perhaps the truth but not the whole truth, then we – and our brains – are being misled.

Plasticity, plasticity, plasticity – and the rigid problem of sex[24]

We've seen how early brain imaging assumed that in a healthy adult human, brain structures and functions typically became 'hard-wired' into the brain and were stable and fixed. This meant that whenever you tried out a language task or a visual paradigm or a decision-making exercise on a participant you should get similar if not identical activation patterns and images, measurable at any time point and easily replicable if necessary. So, if you were going to be comparing males and females, apart from ensuring that your participants didn't have an unusual neurological history, weren't taking any kind of brain-altering drugs and were broadly speaking within a similar age range, then all you really needed to know about your male and female participants was just that, whether they were male or female.

And you assumed that all your female participants would be representative of the group you labelled 'female' and the males of the group labelled 'male'. If, for example, you were testing language skills in females, you would assume that the 'opportunity' sample you picked one year (very often from among your undergraduate or graduate students) would be pretty much the same as a similar sample you might pick the following year, if you decided to repeat your study. And you would then explain any group differences that you found in terms of this maleness or femaleness. You had picked these two groups based on their 'natural' differences and if they performed differently or their brains looked different, then that had to be because males were different from females.

The discovery of life-long experience-dependent plasticity in the human brain means that, in studying sex/gender differences in the brain, we have to pay attention to more than just the sex and age of our participants. But it seems that the 'spotlight of plasticity', the increasingly powerful evidence of how mouldable our brains are by their life-long experiences, is rarely turned onto the sex-differences-in-the-brain debate.

We now know that acquiring different types of expertise, playing videogames, even being exposed to different kinds of expectations about what we might achieve, can change our brains. For example, if you are interested in differences in spatial cognition, you might need to know what kind of relevant experiences your participants have had. Do they play videogames a lot? Do they play sports, have hobbies that involve some kind of spatial skill? Does their job involve some kind of spatial awareness? As we saw in our look at the gendered world to which our brains are exposed, it is more than possible that this will intersect with whether we are female or male. So neuroimaging research needs to factor this in when designing studies and analysing and interpreting results; we need to acknowledge that our brains are irrevocably entangled with the worlds in which they operate, so to understand these brains we need to look at their worlds as well.

This is particularly true when researchers are interrogating the very large neuroimaging data sets that are now available. Labs across the world are collaborating to share the measures they have collected in the course of their own studies, to ensure there are large central collections of brain structure and function measurements to which all brain imagers can have access, to test out their own theories or to check out the generalisability of their own findings. Instead of participant numbers in the tens or twenties, we are now looking at hundreds, even thousands of brain scans.

One paper reported on the analysis of resting-state data (that is, data from brain activity when participants were just lying in a scanner without having to carry out any particular task) from over 1,400 participants.[25] By looking at measures of connectivity in these brains, the researchers reported that age and sex were key distinguishing factors in various ways of comparing brain connectivity, helpfully illustrating these with bell-shaped pink and blue graphs. These actually served to indicate how closely overlapping the data from females and males were – but no effect sizes were given. Although demographic data such as years in education or occupation were available from the central data set, the authors didn't take these into account in making their comparisons. So this paper looked like impressive support for the biological determinist view from an enormous data set. Yet key plasticity-related features were not considered. All the participants were aged between eighteen and sixty, so they'd all had plenty of time for gendered life experiences to affect their brains and behaviour.

If we are still asking the same questions, with the same mindset, then the answers aren't necessarily going to be any better, even if we do have better technology and better data sets. More and more comparisons of bigger and bigger data sets will not get us any nearer to understanding our brains if we only focus on binary biological characteristics and continue to ignore psychological, social and cultural factors. There may not be too many juggling taxi drivers, or even violin-playing slackliners, among the people being studied but you can bet that there would be a pretty wide variety of educational experience, of occupations, even of sporting or other hobbies among a group of 1,400 or more individuals.

And it seems that it is not just brains themselves which reflect the world in which they are working; emerging evidence shows that we also need to acknowledge how entangled the activities of hormones are with the world in which we humans function.[26] Alongside the discovery that brain development is not a unidirectional unrolling of a predetermined template, but a dynamic process of change reflecting interactions with the environment, it has become clear that fluctuations in hormone levels similarly reflect what is going on around us. Far from the 'biology in the driving seat' characterisation of hormones such as testosterone, it's clear that hormone levels can be driven by engagement in social activities.

An astonishing example of this is that testosterone levels in fathers will vary as a function of how much time they spend caring for their children. And this can reflect cultural expectations as well. In one study of two different groups in Tanzania, in the group where it was normal for fathers to care for their children, testosterone levels were lower than in the group where it was not.[27]

This 'smart' testosterone effect was neatly demonstrated by the social neuroendocrinologist Sari van Anders, using a crying baby doll and three unwary groups of men.[28] (This is one of those studies when I'd really like to have been on the other side of one of those one-way mirrors common in developmental psychology labs or in television crime drama interview rooms.) One group had to just listen to the baby cry with no possibility of intervening; one group was allowed to interact with the doll, which, however, was programmed to cry no matter what you did (I am familiar with human babies who displayed the same characteristics); and the lucky third group had a doll that was programmed to eventually respond to one of the several 'nurturant' activities on offer (feeding, nappy change, burping, etc.). Salivary testosterone levels were measured before and after the doll experience. The 'successful calming' group showed a significant decrease in testosterone, while the 'just listening' group showed a significant increase. The group who had interacted unsuccessfully with the doll showed little change in before and after levels. Van Anders suggests that as the stimuli were the same for each group, the variations in testosterone levels were reflecting the social context, the availability or not of some action that would 'solve the problem'. So, like our ever-plastic brains, our hormonal levels are not as fixed as previously thought.

Are there any other aspects of our human condition which we can no longer assume as fixed? It turns out that our personality profiles may well change over time too. Even accepting that it is often clear what a personality questionnaire is trying to measure or the 'social desirability' effect of coming up with answers to any kind of personal profile inventory that will paint you in the most positive light, it was generally assumed that individual measures of, for example, what are called the 'Big Five' personality traits (openness, conscientiousness, extraversion, agreeableness and neuroticism) were pretty stable. The nineteenth-century thinker William James, known as the 'father of American psychology', even described personality as being 'set like plaster' after the age of about thirty.[29]

This tied in nicely with the model that personality characteristics, certainly in adults, were a reflection of our (fixed) biological characteristics. But a recent study, combining data from fourteen longitudinal studies, where measures that had been taken on at least four different occasions were available for nearly 50,000 people, showed that the 'plaster-like' nature of personalities is anything but.[30] Across all studies, all traits but agreeableness showed significant reductions over time (with the latter showing increasing crankiness in some studies and increasing charm in others). Explanations included a kind of pragmatic 'best face forward' effect, where, as a young person(ality), you may sell yourself as 'optimally' conscientious and extravert, but you will calm down a bit as you age (this is the delightfully named Dolce Vita effect). There was also clear evidence that not everyone changes at the same rate or in the same direction.

Overall, then, it would seem that our personality, our outward-facing profile, is not a steady fixed point in our journey through life, but can vary quite significantly. This finding could, of course, just reflect the vicissitudes of the various ways of assessing personality, but it could equally reflect the way in which who we want people to think we are is entangled with social factors, such as the 'who is asking', 'why are you asking', or 'when are you being asked' aspects. So we have plastic, flexible personalities in the same way that we have plastic, flexible biologies.

Diminishing differences?

The assessment of personality characteristics was just one of the contributions of psychology to the sex/gender differences debate that we looked at in Chapter 3. Another core offering was detailed cataloguing of the kind of cognitive skills that it was claimed reliably distinguished females from males. Has this go-to list stood the test of time or should we be revisiting it?

The psychological study of sex differences in behaviour has attracted a fair degree of criticism, from the trenchant scorn of Helen Thompson Woolley's observations of psychology's contributions at the beginning of the twentieth century, to Cordelia Fine's forensic scrutiny of decades of misinterpreted, misunderstood or misrepresented research at the beginning of this century. Neither had fundamental objections to the research being done, rather with how it was done: both felt that the area was characterised by poor scientific practice, which had to throw doubt on many of the conclusions.

With respect to cognitive skills, as we saw in Chapter 3, Eleanor Maccoby and Carol Jacklin did a neat job of tidying up the field in the early 1970s, leaving us with verbal ability, visuospatial ability, mathematical ability and aggression as the reliable characteristics that could differentiate men from women. At this stage, little attention was paid to any contributory factors other than biological sex – it was assumed that as long as you knew whether your participants ticked the 'female' or the 'male' box then everything else (apart, perhaps, from age) was irrelevant.

But this gradually changed as it became clear that environmental variables needed to be considered alongside biological variables, not as alternatives but as part of the same process. It is even possible to track the emergence of this kind of thinking by comparing the prefaces to the four editions of Diane Halpern's excellent book *Sex Differences in Cognitive Abilities*, published between 1987 and 2012.[31] Halpern noted both the increasing input from cognitive neuroscience techniques, including evidence of the brain-changing nature of environmental events, and the growing politicisation of the research field. She also headed the group of psychologists that, following the notorious Larry Summers speech, produced an authoritative summary of the current state of research into sex differences in science and mathematics.[32] So as someone with a comprehensive overview of the landscape of this type of research, one particular feature she noted was that these differences were actually diminishing or disappearing, or were even reversed, in various cultures. Evidence like this makes it harder and harder to maintain that the differences are biologically determined, by genetics or hormones or both.

In 2005, Janet Hyde (who is in fact the Helen Thompson Woolley Professor of Psychology and Women's Studies at the University of Wisconsin–Madison) reviewed forty-six meta-analyses of such studies, together with the outcomes of many social and personality investigations into some of the 'psychological well-being' measures such as 'self-esteem' and 'life satisfaction', with a few motor behaviours such as throwing or jumping.[33] As you will know by now, each meta-analysis in itself will have reviewed dozens if not hundreds of different research papers, so it is clear that the 'psychology of sex differences' industry was hugely productive.

Hyde came up with the startling conclusion that, as opposed to the current 'differences model', stressing the nearly dimorphic distinctions between males and females, the data were showing that males and females are similar on most, but not all, psychological variables. Of the 124 effect sizes that the meta-analyses had revealed, seventy-eight per cent were small or close to zero, including some relating to old favourites such as mathematic ability (+0.16) and helping behaviour (+0.13). Very few could be considered large (more than 0.6).

Bearing in mind that much of the study of sex differences has been to justify why men are to be found in positions of power and influence (and women aren't), the characteristics that showed the greatest differences between men and women will probably not be found in too many job descriptions for future captains of industry. They included masturbation (a 'social and personality variable', apparently, with a whopping effect size of +0.96) and throwing velocity (+2.18), as well as throwing distance (+1.98).

If you thought Hyde's collection of forty-six meta-analyses was an impressive index of psychology's output on this question, only ten years later Ethan Zell and colleagues put together 106 meta-analyses to carry out a higher-level form of effect size assessment (known as a metasynthesis).[34] This was, in fact, specifically aimed to be an evaluation of Hyde's gender similarities hypothesis. As they calculated they had data from over 20,000 individual studies and over twelve million participants, it was an impressively detailed test.

What did they find? The overall effect size, across all the different characteristics they included, was +0.21, with eighty-five per cent of the male–female differences being very small or small. The largest difference they found, in masculine versus feminine traits, was an effect size of +0.73, so not extreme even given the nature of the characteristic being measured. The conclusion was that their metasynthesis offered 'compelling' support for the gender similarities hypothesis.

These two examples illustrate that, on closer inspection of better-conducted studies on larger groups of people, it looks like items from psychology's go-to list of differences in cognitive skills and personality profiles are rapidly disappearing. Those reliable differences in maths performance identified by Maccoby and Jacklin? Gone. And the reliable female superiority in verbal skills? Vanishingly small across many different measures including vocabulary, reading comprehension and essay writing, with verbal fluency the only potential candidate for a difference, but with an effect size of only −0.33 not a hugely predictive variable. In fairness, it is the case that studies of mental rotation ability did throw up an average, but only moderate, effect size of 0.57 – but, as we have seen, this can vary as a function of the type of test used to measure it and can also disappear with training.

So psychology's reliable index of how men differ from women, which has not only supported a centuries-old belief system but also, in some instances, has informed the research agenda of cutting-edge laboratories, appears to be in need of a radical update.

Could we, in fact, have been wrong about sex all along?

Chapter 14:
Mars, Venus or Earth? Have we been wrong about sex all along?

The more we learn about sex and gender, the more these attributes appear to exist on a spectrum.

Amanda Montañez[1]

As we have seen, the hunt for differences between the brains of men and women has been vigorously pursued down the ages with all the techniques that science could muster. It has been a certainty as old as life itself that men and women are different. The empathic, emotionally and verbally fluent females (brilliant at remembering birthdays) could almost belong to a different tribe from the systemising, rational, spatially skillful males (great with a map).

The claim that we have been looking at so far is that there are two distinct groups of people, who think, behave and achieve differently. Where might those differences come from? We have looked at old arguments about the 'essence' of males and females, and the biologically determined, innate, fixed, hard-wired processes that underpin their evolutionarily adaptive differences. We have looked at more recent claims that these differences are socially constructed, that men and women learn to be different, shaped from birth by the specific gendered attitudes, expectations and role-determining opportunities on offer in their environment. And we've mused on even more recent versions that acknowledge the entangled nature of the relationship between brains and the culture in which they function, an understanding that our brain characteristics can be just as much a social construction as the printout of a genetic blueprint.

But, whatever the cause, the basic premise is that there are differences that need explaining. So, whether we are filling empty skulls with bird seed or tracking the passage of radio-active isotopes through the corridors of the brain, or testing empathy or spatial cognition, we *will* find these differences. Separately and together, and through the centuries, psychologists and neuroscientists have pursued the question, what makes men and women different? Answers have been extensively researched, widely reported, enthusiastically believed or heavily criticised.

But in the twenty-first century, psychologists and neuroscientists are beginning to question the question. Just how different *are* men and women, not only at the behavioural level but even at the more fundamental brain level? Have we spent all this effort looking at two separate groups who aren't actually that different, and may not even be distinct groups?

Sex redefined

In Daphna Joel's terms, we have long assumed that sorting individuals into 'female' or 'male' is based on the 3G model, that human beings can be classified into two neat categories according to their genetic, gonadal and genital make-up.[2] An XX individual will have ovaries and a vagina; an XY individual will have testes and a penis. Exceptions to this rule, for example in individuals born with ambiguous genitalia or who later develop secondary sexual characteristics at odds with their assigned gender, were viewed as intersex anomalies or disorders of sex development (DSDs) requiring medical management, possibly including very early surgical interventions.[3]

In 2015, an article in *Nature* by Claire Ainsworth, a science journalist, drew attention to the fact that 'sex can be more complicated than it at first seems'.[4] She reported case histories showing that individuals could have mixed sets of chromosomes (some cells XY, some XX), with emerging techniques in DNA sequencing and cell biology revealing that this was by no means a rare occurrence. And evidence that expression of the gonad-determining genes could continue postnatally undermined the concept of core physical sex differences being hard-wired. Perhaps there should be a wider definition of different types of sex development, including variations in sperm production, wide variations in hormone levels, or more subtle anatomical differences in penile structure? This could reveal that the manifestations of biological sex occur on a spectrum, which would include both subtle and moderate variations, rather than as the 'binary divide' which had held sway to date. This approach would therefore include, rather than exclude, DSDs, no longer labelling them as exceptions to the rule.[5]

However, this wasn't quite the ground-breaking piece of news that it seemed, as Vanessa Heggie, a *Guardian* journalist, pointed out.[6] In a 1993 article, 'The five sexes', Anne Fausto-Sterling had already suggested (as the title of her paper indicated) that we needed at least five categories of sex to cover intersex occurrences.[7] She felt that this grouping should include males with testes and some female characteristics, and females with ovaries and some male characteristics, as well as 'true' hermaphrodites, with one testis and one ovary. Fausto-Sterling's observations had a political context and she felt that society needed to move away from 'the assumption that in a sex-divided culture people can realize their greatest potential for happiness and productivity only if they are sure they belong to one of only two acknowledged sexes'.

When she revisited these thoughts in 2000, she noted that although they had proved controversial at the time, much thinking about 'intersex' individuals had changed in the following few years to the extent that the medical profession were taking a much more cautious attitude to apparently anomalous sex development.[8]

There was even a suggestion that gender should not be determined by genitals, but certainly that the existence of more than two categories (however defined) should be acknowledged.

So, we have a challenge at the most fundamental layer of our chain of arguments. Can human beings reliably be clearly differentiated into two categories, male and female, with membership of each category determined by genes, gonads and genitals, and with the differences in these clearly defined and easily identifiable? It would appear that a genotype may be heterogeneous and variable, and that it is possible to divert the emerging phenotype from its original destination. Neurobiologist Professor Art Arnold has shown that you can separate out the influence of chromosomes from gonads and that these can vary independently, with quite different effects on physical characteristics and on behaviour.[9] Hormone levels can fluctuate widely within as well as between groups, and as a function of different contexts and different lifestyles. Genitals, even where clearly identifiable as labia or penises, can present in a startling variety of forms. There is a wonderfully illustrated *Scientific American* article on the extraordinary complexity of sex determination that makes you wonder that we ever arrived at an end product that looks even remotely classifiable into just two categories.[10]

What about the brain?

The next line of argument has been that, just as men and women could be anatomically segregated, so too could their brains. Be it size, structure or function, it must be possible to find those characteristics that would distinguish the brain of a man from the brain of a woman. As we have seen, the search for such differences has been a centuries-long crusade, from the reading of skull bumps to the measurement of brain blood flow, and it's certainly not been a story of linear progression. Back in 1966 the only brain region identified as relevant for understanding sex differences was the hypothalamus.[11] Things have certainly changed since then; there have been nearly 300 studies on imaging of sex or gender differences in the human brain in the last ten years, with hundreds of reports of sex differences in dozens of different brain characteristics.

As we saw in Chapter 4, although the techniques involved with examining the brain are clearly more sophisticated than bumps and buckshot, many of the arguments have remained the same. Establishing differences in the brain first involves an agreement on how different structures should be measured, and that has still not been fully reached today. For example, there is a consensus that it is necessary to undertake some kind of size correction in order to compare brains from men with brains from women; but as we have seen there are still arguments about what the corrections should be based on. Should it be total brain volume or intracranial brain volume, height or weight or head size, all of the above or just some of them? We know there has been a long-standing list of those brain areas where 'reliable' sex differences have been found. This includes the suggestion that two key structures in the brain, the amygdala and the hippocampus, are larger in males. This was apparently confirmed in a meta-analysis in 2014, which looked at over 150 studies.[12]

But, in the brain imaging world, updates come thick and fast, and more and more subtle techniques are allowing us to revisit past certainties about brain differences. Yesterday's candidate region *du jour* for understanding the sex difference can be today's hippocampal rethink. Even in the four years since the above review, new research has challenged some of those conclusions. Lise Eliot's team from the Rosalind Franklin University in Chicago have carried out meta-analyses of structural data on both the hippocampus and the amygdala, in 2016 and 2017 respectively.[13] In both cases, they demonstrated that the initial claims that these structures were bigger in males than in females were unfounded, and that the differences diminished markedly or disappeared when the measures were corrected for intracranial brain volume.

It is becoming increasingly clear that structures within the brain are carefully scaled as a function of brain size, quite possibly to optimise metabolic needs or inter-cellular communication. This means that *any* reports of sex differences in brain structures that have not made some form of volume adjustment are really not giving an accurate picture.

Rather than focussing on the size of different structures, there is now a greater interest in patterns of connections in the understanding of brain–behaviour links. In Chapter 4, we looked at a study from Ruben and Raquel Gur's lab, which was one of the first to apply structural connectivity measures to the study of sex differences, and which claimed to find greater within-hemisphere connectivity for males and between-hemisphere connectivity for females.[14] But, as we discussed in our journey through neurononsense and neurosexism, there were difficulties with this study, particularly with respect to the authors' (or the authors of their press releases') overemphasis on the significance of their findings, which did not accurately reflect the extent of these differences. What has also become clear is that the 'size matters' issue applies here too: findings from Lutz Jäncke's lab in Zurich have shown that the larger the brain, the stronger the connections *within* its hemispheres, and importantly that this is independent of the sex of the brain's owner (although, of course, most of the bigger brains do belong to men).[15] Again, we could look on this as a scaling issue. As brains grow in size, the distances between key processing hubs will increase so there needs to be a mechanism to ensure that processing speed isn't compromised. Bigger countries need better roads.

The other key matter to take into account is that of brain plasticity. As we have seen, life's experiences and attitudes can shape and reshape brains, so attempting to measure structures in brains as though they were fixed endpoints, without taking into account the kind of brain-changing experiences they might have had, is likely to be of limited value at best. Researchers who found size/sex differences in the amygdala and the hippocampus, as mentioned above, do acknowledge this, and point out that both these structures are renowned for the extent to which they can be influenced by experience and lifestyle. We need to know what kind of lives these brains have lived – their owners will probably have had varying amounts of education, different occupations or kinds of life experiences arising from their socio-economic status.

You might wonder if, after all this time and effort, it might be time to call a halt to this decades-long attempt to generate a catalogue of differences in brain structures and pathways. Daphna Joel and Margaret McCarthy have proposed an improved framework for interpreting sex differences which may take the field forward.[16] They suggest four additional questions that we might need to ask of any differences that are found. Firstly, are they persistent or transient across the lifespan – are we talking about ever-present differences or differences which, say, come and go with different hormone levels, or appear in childhood but disappear after adolescence? Secondly, are they context-dependent or independent: will they only be found in certain circumstances or certain cultures, or are they universal? Thirdly, are they clearly dimorphic (non-overlapping), is there lots of overlap, or are they better characterised on a continuum? And finally, can they be *directly* attributed to biological sex (via chromosomes or hormones), or do they arise because of indirect effects, such as (in the case of humans) societal expectations and cultural norms which vary as a function of whether you are male or female? So we need to ask not just whether there is a difference, but what kind of difference it might be.

This framework would certainly offer a more nuanced answer to the perennial question 'So what *are* the sex differences in the brain?', and maybe even induce less of the customary eye-roll when you begin your answer with 'Well, it depends what you mean by different ...'

Alternatively, perhaps we should just stop looking for differences altogether?

The mosaic brain

In searching for sex differences in the brain, the underlying assumption is, of course, that brains from women will be different from brains from men. Perhaps, in the way television-series detectives seem to be able to identify discovered body parts, there is some kind of reliable set of clues that *this* is a female brain and *that* is a male brain?

In 2015, a team led by Daphna Joel from Tel Aviv University reported on the results of a long and very detailed investigation of over 1,400 brain scans from four different labs.[17] They examined the grey matter volumes in 116 regions in each brain. From a subset of their scans they identified ten features out of these 116 which showed the largest differences between the brains from females and the brains from males; the data was then, thoughtfully, colour-coded pink and blue respectively. Those features which were most consistently larger in males were called 'male-end' and those which were most consistently larger in females were called 'female-end'. When they mapped these colour-coded features onto another subset of their original data, comprising 169 brains from females and 112 brains from males, it instantly became clear that each brain exhibited a veritable mosaic of both male-end and female-end features, as well as a number of intermediate ones. Less than six per cent of the sample were consistently 'male' or 'female', that is, where the majority of the 116 features were from the male end or the female end respectively. The rest showed a wide range of variability between each brain, with a general 'pick and mix' collection of maleness and femaleness evident in the different brains. This kind of distribution was also found in other data sets, and a similar pattern of results was shown with structural pathways. The conclusion from this investigation was that we should 'shift from thinking of brains as falling into two classes, one typical of males and the other typical of females, to appreciating the variability of the human brain mosaic'.

This paper had a major impact on the sex differences research community. It provided a compelling image of the variability within brain data from males and females, and asserted that there was so little internal consistency between groups divided according to their sex that the notion of a male or a female brain should be abandoned. Although there was general consensus as to the variability in all such data (and no brain imagers in this field could really challenge that), there was some disquiet that the technique used was 'stacking the odds' against finding neat, clear-cut categories. For example, one paper applied a similar technique to the very dissimilar faces of different types of monkey and said that it couldn't distinguish these.[18] So the reason that Joel couldn't group her data into two neat piles was not a function of the data, but of how she was trying to sort it. Others applied automatic pattern recognition techniques and reported they could correctly identify a brain's 'sex category' between sixty-five and ninety per cent of the time. Joel defended her approach by emphasising that the key message was that the range of the variability within the brain data sets (which wasn't quite as evident in the monkey faces) was so extensive that, at the individual level, it would be impossible to reliably predict the 'brain profile' of someone just on the basis of their sex. So no, you can't pick up a brain and fill in a set of 116 tick boxes and come up with the answer 'female' or 'male'.

Joel also points out that this biological mosaicism is also entangled with the plasticity issue. For example, it has been shown that supposedly typical female or male characteristics of certain nerve cells can change as a function of external stress to being more male-like or more female-like respectively,[19] so the different patterns of a brain's mosaic could well reflect the different life experiences to which it has been exposed.

At the most fundamental end of the sex story, then, it looks as if it is increasingly difficult to square the accumulating evidence with the notion of a neat binary divide. With respect to the brain, there are four emerging issues that suggest that it might just be time to move on from the simplistic 'male brain'/'female brain' divide. Decades of findings using increasingly sophisticated imaging techniques have not yet yielded anything like a consensus as to what might differentiate a 'male brain' from a 'female brain'. There are difficulties of knowing what structural differences in the brain might mean anyway in terms of the behaviour you might be interested in understanding. The plasticity issue means that a wide range of psycho-cultural factors need to be taken into account when looking at any measures of brain structure or function, and it also means that any pattern of brain activity we look at can only, at best, be considered a 'snapshot' of that brain, only reflecting its current profile. And Joel's recent findings have drawn attention to the enormous amount of variability in individual brains at the level of quite fundamental structures, to the extent that one interpretation is that the biggest myth of all is that our brains are 'male' or 'female'.[20]

Those raging hormones?

What about hormones, those chemical messengers that control most of our bodily functions, that have long been attributed a very special role in determining differences between females and males? Indeed, as we saw in Chapter 3, two of the hormone groups, the androgens and the oestrogens, are called the 'sex' hormones. Testosterone, the best-known androgen, is known as a male hormone, and oestradiol as a female hormone, despite both occurring in both females and males. A recent review has noted that average levels of oestradiol and progesterone, so-called 'female' hormones, do not differ between women and men.[21] So, just as with the brain, what looks like a neat binary measure of differences between females and males has not withstood closer scrutiny.

The issue of plasticity is relevant here, too. The long-standing impression of hormones as the drivers of behaviour (or 'stirrers into action', which, as we saw in Chapter 2, is what their name means) implies that they are the cause of all sorts of behaviours. But twenty-first-century research looking at the effects of social context on hormone levels means that we have to rethink this causal role, and acknowledge that human hormones are just as much entangled with and responsive to what is going on in their world as are human brains.[22] In the last chapter, we saw the effect on men's testosterone levels of dealing more or less successfully with a crying doll, with significant reductions in testosterone for the successful calmers. And this has been mirrored in real-world situations, with testosterone levels varying as a function of the 'hands-on-ness' of fathers.

Just as we are noting the power of society and its expectations as brain-changing variables, it is clear that the same effect is evident with respect to hormones. Social endocrinology has shown that 'androgens and estrogens are not two distinct sets of sex hormones – one set for women and one set for men – but rather hormones that are found in all humans ... Moreover, levels of these hormones are not fixed, but are dynamic and can be influenced by gendered social experiences.'[23]

But we need to recall where this all started, the *status quo ante* that males and females have such widely differing skills, temperaments, personalities, aptitudes and interests that they could be mistaken for members of different species or even denizens of different planets. And the question 'What makes men and women different?' is what this long-running scientific crusade has been about. We have been scratching our heads over the confusing answers we have been getting. But perhaps, as with brains and hormones, we should actually be looking at the question itself?

Black and white or shades of grey: diminishing differences and disappearing dichotomies

Psychology's go-to list of sex/gender differences, with cognitive factors such as verbal or spatial ability alongside empathising or systemising, had long been accepted as well-established gender differentiators. But as we saw in Chapter 13, these confident assertions were beginning to be challenged, and meta-analyses by Janet Hyde in 2005 and Ethan Zell and colleagues in 2015[24] suggested that the overwhelming message from decades of research on millions of participants was that, actually, women and men were more similar than they were different, with differences disappearing over time.

And now, let us take yet another step away from past certainties about the different planetary origins of males and females. Even the gender similarities hypothesis is based on an argument as to how different or similar the two categories of male and female actually are. But supposing we are making a fundamental mistake in thinking that there are two categories in the first place? That, with respect to all the cognitive skills or personality attributes or social behaviours that we have earnestly been meta-analysing, men and women do not fall into the two groups that their different (though more complex than initially thought) anatomies led us all to believe must exist?

Two papers with wonderful titles sparked off this line of thinking: 'Men and Women are from Earth: Examining the Latent Structure of Gender' in 2013[25] and 'Black and White or Shades of Grey: Are Gender Differences Categorical or Dimensional?' in 2014,[26] by Harry Reis from the University of Rochester in New York state and Bobbi Carothers from Washington University in St Louis. Both papers went right back to the bedrock of the 'difference' question. They pointed out that comparing two groups assumed that there were two groups in the first place. If you're going to put people (or anything) into separate groups, you'll need to know the ground rules for the 'grouping variable', the basis on which you can make your decision as to who or what belongs to which group (known as establishing a 'taxon'). For a category or taxon to be meaningful, members of it should possess a collection of recognisable characteristics that generally go together (internal consistency). Overall, these characteristics should add up to a category that is recognisably distinct from others. This then means that just knowing the label that is fixed to each category (let's call it a box for the time being) should give you a hefty clue as to what is in that box. This is, of course, exactly what stereotyping does – provides a label from which, it is assumed, all else will follow.

Carothers and Reis analysed data on 122 measures that supposedly differentiated men from women. This included many different measures of masculinity or femininity, measures of empathy, of fear of success, of interest in science, and of the Big Five personality traits such as neuroticism. They ran these data through three different types of analysis specifically designed to show whether or not the outcome measures were taxonic (belonging to discrete groups) or dimensional (belonging on a single scale). Almost all of the comparisons they looked at showed that the data fitted best along a single dimension. Clearly James Damore, the Google memo author, hadn't come across these papers!

To make sure that this wasn't just a problem with their methods of analysis and that the data would be separated into distinct groups if appropriate, they did a separate report using solely data that did divide neatly into groups, such as physical measurements and athletic achievements. The measures were also effective at reliably sorting sex-stereotypical activities such as enjoying boxing, watching porn, taking a bath and talking on the phone into the appropriate groups (I'll leave you to take a wild guess as to which activities were stereotypical of which sex). So the very origin of our hunt for sex differences in the brain, that there are clear differences in female and male behaviours, aptitudes, temperaments, likes and dislikes, would appear to be in need of a radical update.

To add to this, Daphna Joel and her team also used the 'mosaic' approach they had applied to their brain measures to look at psychological variables.[27] They took data from two large open data sets and, in addition, the data set from the Carothers and Reis study which specifically showed the biggest sex differences. They took the variables from each set that showed the largest sex differences, for example traits like 'worrying about weight', 'gambling', 'doing housework', 'having construction hobbies', 'good at communicating with their mothers' or 'keen on watching porn'. They then looked at the distribution of each of these apparently highly discriminating variables in each of the contributing participants. How many had a 'matching set' of either strongly male or strongly female characteristics? If you worried about your weight, were you also more likely to watch talk shows and do housework? If you were into gambling, did you also like boxing or have construction-type hobbies? The answer was 'no'. Of the 3,160 females and 2,533 males tested in these studies, only just over one per cent were consistently boxing-loving porn watchers, or bath-taking weight worriers, with the other ninety-nine per cent just as likely to be, say, inveterate gamblers who were good at communicating with their mothers.

So, just as with brains, there is no such thing as a typically female behavioural profile or a typically male behavioural profile – each of us is a mosaic of different skills, aptitudes and abilities, and attempting to pigeonhole us into two archaically labelled boxes will fail to capture the true essence of human variability.

The more we look at all sorts of different measures from males and females, from fundamental biology through brain characteristics to behaviour and personality profiles, the less and less likely it looks that these measures are coming from two reliably distinguishable groups of people. This obviously has implications for all of our well-established stereotypes and for all sorts of discriminatory practices based, consciously or unconsciously, on these stereotypes.

Beyond the binary – implications for gender identity

An obvious adjunct to this potential move away from simple male / female dichotomies is how this might relate to the whole concept of gender identity. As we have seen, the origins of the sex differences debate assumed that there was an unbreakable, unidirectional, causal link between your biological sex (the genes, genitals and gonads you were born with) and your social gender (how you identified yourself, what roles 'people like you' played in society). This was supposedly proved by the unassailable evidence that there were sex differences in the brain that caused sex differences in aptitudes, skills and personality and identity, which in their turn then explained sex differences in achievement, status and positions of power.

But once that chain of evidence starts to unravel then past certainties can be challenged, including the idea that the sex you are assigned at birth is in some way related to your self-identity. So the rethinking of sex in the twenty-first century has implications for more than just our understanding of brain and behaviour. Do we feel that we are male or female because we have a male or a female brain? If there is no such thing as a male or a female brain, then where will our gender identity come from?

As with so much else in this field, we need to be clear about definitions. Gender identity refers to our sense of ourselves as male or female, to whether we *identify* as a male or a female. If someone stops you in the street and asks you to complete a survey, how would you describe yourself? It is not the same as gender preference or sexual orientation, which generally refers to who you might choose as a sexual partner. The two can go together but they have been shown to vary independently.

We know that children are junior gender detectives and are on the case from early on, and by about three years old, they have sorted out what gender they feel they are, and what that means for how they should behave, how they should dress and what toys they might play with. And this gender is almost invariably linked to their perception of anatomical differences – boys have a penis and girls don't. Based on this, where they 'have the evidence' or on clues such as hair length, or name, they attribute different genders to others in their social circle, and firmly state the ground rules associated with each gender.[28] Woe betide any transgressors – children themselves are the most intransigent of gender police!

As this starts early in life, there is one school of thought that says that gender identity is due to the expression of innate biological factors. In just the same way that our genetic and hormonal sex has been identified as the cause of sex differences in brain and behaviour, biological processes are thought to underpin the emergence of gender identity.[29] You 'feel' male or female because you have a male or a female brain, organised and canalised by the action of genetic and hormonal factors. The reports that girls with congenital adrenal hyperplasia show low levels of satisfaction with their assigned female sex have been cited as evidence of the biological determination of gender identity.[30] Similarly, the case of David Reimer, the boy who was raised as a girl who we met in Chapter 2, shows how extraordinarily determined socialisation efforts were apparently insufficient to establish a gender identity at odds with the biological one.[31]

But supposing you feel a disconnect between what your biology is telling you and the gender you identify with, despite all the necessary evidence and despite the very powerful messages that your culture is giving you that the two go together? To the extent that you find yourself so unhappy that you are prepared to undergo drastic medical procedures, including surgery, to alter your biological sex to match the gender with which you identify? A perhaps understandable choice if you are immersed in a culture where biology is insistently being identified as the prime 'cause' of gender.

Gender identity is currently a hot topic. A survey of 10,000 people carried out by the Equality and Human Rights Commission in 2012 indicated that approximately one per cent of the population reports this kind of disconnect.[32] Although this does not always result in medical intervention, there is quite a dramatic increase in people taking this route. In 2017 the American Society of Plastic Surgeons reported a nineteen per cent increase from the previous year in gender reassignment surgery (3,250 operations in total).[33] In the UK, the full statistics are difficult to access as, although the NHS does carry out some surgery, much is privately provided; but a report from the House of Commons Women and Equalities Committee in 2015 reported that referrals to gender identity clinics are growing by about twenty-five to thirty per cent a year.[34]

An additional factor that has attracted comment is a dramatic increase in the number of children who declare themselves to be gender-invariant and a decrease in the age at which this is happening. A report in the *Telegraph* in 2017 said that the number of children under ten years old visiting the NHS's sole facility for transgender children had quadrupled in four years, from 36 in 2012/13 to 165 in 2016/17.[35] It was also noted that eighty-four children aged between three and seven were referred in 2016/17, compared to twenty in 2012/13. A controversial aspect of this is that one form of treatment involves the use of puberty-blocking hormones, sometimes followed by cross-gender hormones to enable the development of the secondary sexual characteristics of the gender with which the child/adolescent wishes to identify.

The revelation in 2015 that Bruce Jenner, an Olympic decathlete, was transitioning to become Caitlyn Jenner certainly put the issue in the public spotlight.[36] And her assertion that 'my brain is much more female than it is male' encapsulated a frequent claim from transgender individuals that they feel as if they have a male brain in a female body (or vice versa) or, more colloquially, that they have been born in the 'wrong box'. They feel that something has gone wrong with their biology–gender link, so they wish to realign it by changing their biology to be consistent with the gender they identify with. But perhaps we should be challenging the link? Perhaps we should be challenging the concept of any kind of prelabelled box into which human beings are being slotted?

We have seen the difficulties for women associated with an unshakable conviction that their biology determines their interests, aptitudes, personality, occupation and so forth. Perhaps these also extend to those who are questioning their gender identity. The ubiquitous and insistent nudges from gendered marketing, the unrelenting gendered bombardment from social media and entertainment outlets, the constant availability of gendered displays can add up to a much more rigid and prescriptive stereotype of what it means to be male or female than we have encountered before. So if 'none of the above' appears to be your answer to the kind of characteristics that are to be expected of you as a boy or as a girl, it may just be that there is a problem with the question of what makes a boy or a girl and not with your answers. Debunking the myth of the male brain or the female brain should have implications for the transgender community which will hopefully be seen as positive.

Sex still matters – don't shoot the messenger!

One consequence of sticking your head above the Mars/Venus parapet and pointing out that 'well-established' differences in the brains and behaviours of males and females are actually not that well established, and may in fact not even be differences, is that you can attract what might politely be called 'adverse comment'.

I treasure a cutting from Cristina Odone, writing in the *Telegraph*: 'Pity the scientist. Locked up in labs, handling vials full of toxic liquids, surrounded by white mice and white coats – no wonder she sometimes loses her common sense. This seems to be the case with Gina Rippon.'[37] I'm further taken to task for espousing a theory which 'smacks of feminism with an equality fetish'. I'll gloss over another description from a *Daily Mail* comments thread that I am 'full of carp' (I am assuming that this is a spelling mistake and not a criticism of my fish-eating habits). And add 'grumpy old harridan' and 'post-menopausal affirmative action loser' into the mix and you'll begin to get the picture.

Hopefully in our discussions so far in the area of sex differences research and its findings, two things have been clear. The first is that a *full* understanding of any sex differences that there are and, even more importantly, where they come from, is crucial, particularly with respect to anything pertaining to the brain. This is because incomplete brain-based explanations often mistakenly contribute to a belief in the fixed inevitability of a status quo, be it in who succeeds at science or who can or can't read maps. And this can lead to unhelpful stereotypes, ill-informed conscious or unconscious bias, and potentially a significant waste of human capital.

The second is that these critiques are *not* a denial of the existence of *any* sex differences. Given the necessity of getting this kind of research right, it is important that any research into sex differences is well designed, with careful choice of the dependent and independent variables that are to be examined, appropriately selected groups of participants, and thoughtful analysis and interpretation of the data. Once this is reliably in place, then we will begin to accumulate a more genuinely useful portfolio of findings. Those who refer to people like me and my colleagues as 'anti-sex difference' or 'sex difference deniers' seem to have missed the point. Similarly, the accusation that we are putting women's lives at risk by *preventing* research into sex differences is puzzling (as well as inaccurate).[38] Sex difference studies are alive and well, as we've seen throughout this book, so it strikes me that this 'feminist neuroscience' (or 'feminazi') guerrilla movement to which I apparently belong isn't being hugely successful!

There are clear indications that there are marked sex differences in the incidence of both physical and mental health problems. It is obviously essential to identify how much the sex *or gender* of an individual has contributed to this. The key issue is to look beyond the simple category of sex, not to stop there when it is clear that it can be an influential factor, but to see what other factors might be entangled with it as well. Do more women than men suffer from depression because of sex-linked genetic or hormonal factors, or because of a 'self-esteem deficit' associated with a highly gendered lifestyle? Or both? If we can get a handle on these kinds of issues we will see significant progress in the field of sex/gender differences.

The findings that we have considered in this chapter suggest that looking at biological sex in terms of the categorical differences it causes will distort the picture. Far better to think of it as a continuous dimensional variable that may exert influences – profound, moderate or maybe only trivial – on the process we are trying to understand or the problem we are trying to solve. From what we have looked at so far, the term 'influence' would seem to be a much more accurate reflection of the role that biological sex can play in our brains' journeys through life.

Sex definitely matters; this is not an 'inconvenient truth' but it is a truth that needs to be carefully revealed. We need to go 'beyond the binary', to stop thinking of 'the male brain' or 'the female brain' and see our brains as a mosaic of past events and future possibilities.

Conclusion: Raising Dauntless Daughters* (and Sympathetic Sons)

We have seen how our fantastic plastic brains are plunged into a gendered world, a world that has for hundreds of years treated the sexes differently. We may have moved on from the days of the two-headed gorilla, but even in the twenty-first century we can still find evidence of a world constructed to offer different opportunities to males and to females, based on long-standing stereotypes of differences in abilities, temperaments and preferences. From the moment of birth (and even before) our rule-hungry brains are confronted with the different expectations of families, teachers, employers, the media and, eventually, ourselves. Even with the emergence of amazing brain imaging technology and evidence of diminishing and disappearing differences, men and women are still being sold the concept of *the* male brain and *the* female brain, which will determine what they can and can't do, what they will and won't achieve. All washed down with a hefty dose of neurononsense.

But brain scientists can move and are moving the debate forward. The question of sex differences in the brain is being

---

* See www.dauntlessdaughters.co.uk.

challenged. If there are sex differences, where do they come from? And what do they mean for the brain's owner? We have seen that our brains are rule-seeking systems, generating predictions based on the world in which they are functioning in order to guide us through that world. So, in order to understand how different brains arrive at different destinations (because they do), we have realised that we need to be much more aware of exactly what social rules (right or wrong) are out there to be absorbed. An emerging understanding of the life-long plasticity of the brain means we also have to factor in what kind of brain-changing experiences our brains will encounter en route.

Brain plasticity is not just about taxi driving and juggling (intriguing as the insights these provide really are) but about the impact that attitudes and beliefs can have on our flexible brains. Understanding our brains as deep-learning systems means we can see how, just like Microsoft's unfortunate chatbot, Tay, a biased world will produce a biased brain. We need to register the gendered bombardment that is coming from social and cultural media, as well as from family, friends, employers, teachers (and ourselves), and understand the very real impact it is having on our brains.

Developmental cognitive neuroscience is showing us just how sophisticated babies and their brains are. We used to think of them rather sneerily as merely 'reactive' and 'subcortical' beings. But from day one (and possibly even before) these tiny social sponges, fully equipped with 'cortical start-up kits', are getting themselves embedded into their social networks, and searching out the rules of social engagement in their world. So we need to be on the alert as to just what they might be encountering out there.

We are just starting to realise that we have a second 'window of opportunity' to watch the construction and deconstruction of brain networks and the (eventual) emergence of an adult social being. Adolescence marks a period of dynamic reorganisation in brain networks, a 'system-level rewiring' which sees a shift from local, within-system connections to more widespread global connections between different parts of the brain.[1] These changes are almost as dramatic as those we have been tracking in baby brains. As adolescents are (generally) a more accessible and compliant cohort than newborn babies, there is a possibility that neuroscientists will be able to track these changes in brain organisation alongside accompanying changes in behaviour.

You don't have to be a neuroscientist to know that adolescents have difficulty with emotion regulation and impulse inhibition, as well as appearing to be inordinately susceptible to peer pressure and social rejection.[2] All of these processes, as we know, are core features of the activities of the social brain, and can be modelled for investigation in the scanner. As an understanding of the activities of the social brain appears to be at the core of understanding how a brain interacts with its world, of how an emerging sense of self can be reflected in both brain and behaviour, then a focus on these processes in adolescence could bring insights into how, for example, social rules and influential others can determine cortical processes.[3]

Social cognitive neuroscience is putting the self centre stage, making us realise that the construction of ourselves as social beings is perhaps the most powerful triumph of the brain's evolution. It is clear that understanding the social brain could offer us a hugely effective lens to investigate how a gendered world can produce a gendered brain, how gender stereotypes are a very real brain-based threat that can divert brains from the endpoint they deserve. Understanding the importance of self-esteem and processes such as self-silencing will give us a much better handle on the brain bases of gender gaps and underachievement.[4] If we know how an 'inner limiter' is constructed we have a better chance of recalibrating it to be a useful component of our social brain.

Now that we know that explanations for all kinds of gender gaps are a tangle of brain-based and world-based processes, we must realise that solving the problem will involve untangling each of the threads to see if we can come up with a better version.

Brain matters – don't blame the water?

There is a saying that if you have a leaky pipeline then don't blame the water. Replacing the whole pipeline can be a long-term solution, but sometimes finding ways of stopping leaks can help.

We looked at the continuing underrepresentation of women in science as a case study of exactly the sort of tangled web of different factors that can underpin this kind of issue. This can include a worldview of science as a masculine institution, with scientists almost invariably being men, offering a chilly climate to female would-be scientists, or a worldview of women (commonly shared by the women themselves) as lacking the necessary aptitude and temperament, disempowered, disconnected, disenchanted. Cue the notorious Google memo incident of summer 2017.[5]

Confronted with this conjunction of exclusionary characteristics, women can succumb to the self-fulfilling prophecy of stereotype threat. And so the cycle continues. Do the 'paradoxical' findings that gender gaps in science are greatest in countries with the greatest gender equality really support the argument that where women have more freedom of choice they will naturally gravitate to non-scientific careers?[6] Or could it be that where they have more freedom of choice they naturally gravitate away from unwelcoming workplaces, particularly where their behavioural and cortical training has instilled in them a belief that such workplaces are not for them?

There are clearly steps that need to be taken within the culture of science in order to improve its attractiveness to those who don't currently engage or who over time disappear off the books.[7] Big strides are being taken with, for example, more family-friendly policies, but the ongoing gender gaps in the higher echelons indicate there is still some way to go.[8]

An additional approach could be to find ways of empowering those whose inner limiter may be set too low (or may be reflecting a lifetime of low expectations). The issue of self-silencing and disengagement needs to be tackled.[9] As we have seen, the problem of maths anxiety offers useful insights into the many entangled causes of underperformance, showing how emotion regulation processes can interfere with ongoing processing.[10] But we can also illustrate exactly what is going wrong – how potent and attention-grabbing negative feedback is for the anxiety sufferer, revealed by clearly marked activation in her brain's 'error-evaluate system', how her attention is diverted away from possible sources of support and drives her to throw in the towel all too soon.[11]

There are ways to make our brains more resilient and to deactivate the negative inhibitory processes that can lead to defeat and disengagement. Psychologists Katie van Loo and Robert Rydell from Indiana University have demonstrated that a very simple 'empowerment' process inoculated girls against the effect of stereotype threat.[12] They showed that allowing girls to imagine themselves in positions of power, for example as a CEO ranking her employees in order of usefulness, could moderate the effect of stereotype threat on a subsequent maths test. This effect has been shown in the brain as well, where 'high-power priming' can reduce the levels of activation in those parts of the brain associated with cognitive interference.[13]

Sometimes 'empowerment' can be manipulated in very easy ways – social psychologists have shown that something as simple as having pictures of powerful women like Hillary Clinton or Angela Merkel in the background can aid nervous female speakers.[14] The importance of role models in establishing self-identity or overcoming challenges to it has been well established in both social psychology and social cognitive neuroscience.[15] Role models can also play a vital part in bolstering negative self-images and flagging self-esteem at all ages, and in many situations.

Similarly, evidence of the strong drive to belong to an in-group could be harnessed in the development of initiatives to encourage girls (or indeed any individuals who are under-represented). A great example is the initiative run by the campaigning group Women in Science and Engineering (WISE). 'People Like Me' taps into the 'fitting-in' agenda in career choice, and shows how any type of personality can find a match in different types of scientific careers.[16] This recruitment campaign targets not only girls, to show how they can match their personal characteristics to different types of scientists, but also parents, teachers and employers, to ensure they are aware of such individual differences and tailor their science encouragement accordingly.

Culture-based problems clearly need to be solved by fixing the culture, but empowering those who engage with the culture can buttress the solutions offered and speed the process of change. And understanding what drives engagement or disengagement could provide answers to paradoxical gender gaps, where evidently capable individuals appear to turn away from the opportunities on offer.

It's not all about sex

A key message in this book is that a persistent focus of the brain research agenda, effectively since the early nineteenth century, has been driven by the perceived need to explain differences between two groups divided by their biological sex. As we saw in the last chapter, we are only just starting to wake up to the idea that there is currently no good evidence of there actually being any relevant differences either in the brains of these two groups or in the behaviours these brains support. For sure, there are average differences between the groups, but the effect sizes are characteristically very small, and averaging will wash out any interesting individual differences. Even the notion of sex differences at the fundamental biological level is being challenged.[17]

Perhaps it is time for a rethink of how brain science might come to understand the strengths and weaknesses, the abilities and aptitudes, the 'brain histories' of individuals by doing just that – looking at individuals. Both the techniques and the aims of early brain imaging concerned generic descriptions of *the* language areas, *the* semantic memory stores, *the* pattern recognition areas, in all of us. Individual differences were treated as noise, data averaged across participants to remove such variability. From personal experience, intriguing findings of significant differences in single participants would disappear once a group average was generated. Early forms of data and data analysis were too crude to offer any insights into individual differences, but we have moved on from then. It is now possible to generate functional connectivity profiles, patterns of task- or rest-related synchronised activity in the brain which, it is claimed, are like a fingerprint, unique to each individual, sufficiently distinct that they could be linked to their owners with up to ninety-nine per cent accuracy.[18] So we *can* look at brains at the individual level. And the evidence about what can affect brains, and when, indicates that we *should*. We need to really understand the external factors that shape such individual differences, with social variables such as level of engagement in social networks and self-esteem, and opportunity variables such as sport, hobbies or videogame experience alongside more standard measures such as education and occupation. Each of these can alter the brain – sometimes independently of sex and sometimes very much entangled with it, but they will contribute to the almost unique mosaic that we now know characterises each and every brain.[19]

Cognitive neuroscientists Lucy Foulkes and Sarah-Jayne Blakemore, writing about the adolescent brain, also urged that we should look at individual differences here as well.[20] They noted that social factors such as socio-economic status, culture and 'peer environment' factors including size of social networks or experience of bullying, where there are marked individual variations, have been shown to have significant impact on brain activity profiles.

With the availability of sufficiently large data sets from much more powerful brain imaging systems and the powerful analytical protocols now being developed, as well as machine-based pattern recognition programs, the possibility of looking at the influences of multiple variables, of which biological sex could indeed be one, is well within our reach.[21]

This proposal is in no way a denial that sex differences might matter. We know that there are gender imbalances in mental health conditions such as depression and autism, as well as physical health conditions such as Alzheimer's disease and immune disorders.[22] But to understand these, we need to acknowledge that in order to unpack the reasons for any gender imbalance we should not assume that just focussing on biological sex will provide the answers.

A recent mandate by the National Institutes of Health in the United States has insisted that all fundamental and preclinical research should include sex as a biological variable in trials, specifically to shed light on gender imbalances in pathological conditions.[23] This will provide valuable additional data on the influence of biological sex on such conditions. But the data on mosaic brains and dimensionality in behavioural measures reveals that using sex as a catch-all category may miss key contributory influences and paint a misleading picture.

Perils and pitfalls – neurosexism and neurotrash

Cognitive neuroscience is rightly seen as a key player in building up a picture of human behaviour at all levels of scrutiny, from genetic to cultural. Its outputs can be more accessible than some of the more complex epigenetic or neurobiochemical research and can often relate more closely to the public's everyday experiences. But with this accessibility should come responsibility. As Donna Maney, a professor of psychology at Emory University in Atlanta, has indicated, there are perils and pitfalls associated with the reporting of sex differences; overemphasising the essentialist aspects of findings can reinforce unfounded biological determinist views.[24] Using terms like 'profound' or 'fundamental' when the effects sizes are tiny is irresponsible; ignoring the contributions of variables other than sex is misleading. A 'belief in biology' brings with it a particular mindset regarding the fixed and unchangeable nature of human activity, and overlooks the possibilities offered by our emerging understanding of the extent to which our flexible brain and its adjustable world are inextricably entangled (and could lead misinformed individuals such as James Damore to write misguided memos to their employers).

It's also worth keeping an eye on the neural Looney Tunes that characterise some of the worst of popular science writing, particularly in the self-help genre. However flattering it is to be thought of as a mind reader who can solve all relationship problems, neuroscientists must be wary of the headlines that can seep into the public consciousness and mislead and misinform. Neurotrash can discredit the genuine and important work that is emerging from neuroscience labs. And when there are already so many sources of brain-changing bias in our world, cutting one out can only help (or only allowing it into the public domain with 'brain health' warnings – reading this rubbish can damage your brain).

Baby matters – look after the little humans

Neuroscientists have long identified the early years as the most plastic of all stages of brain development. The new revelations from developmental cognitive neuroscientists as to just how early the world can impact on these tiny brains should give us pause for thought about that world. So grassroots campaigns such as Let Toys Be Toys[25] really do matter – and we know that our little gender detective will root out any 'hidden truths' so we may need to stand firmer than we like on 'princessification'. If we want to raise dauntless daughters and sympathetic sons, perhaps she can have a pink fairy castle – but she has to build it herself. And hold fire on the 'man up'-type dialogues. These things do matter.

A 2017 BBC programme called *No More Boys and Girls* investigated the extent of sex/gender-stereotyped beliefs in seven-year-old girls and boys, and, following them up over a six-week period, saw what happened when they tried to remove as many stereotypical influences as possible from their classroom.[26] Did it change their own self-belief or their behaviour? The opening credits were sobering, with little girls emphasising the importance of being pretty, and boys reckoning they could 'get to be president'. There were plenty of other thought-provoking moments; the girls' level of self-esteem (at *seven years old*) was much lower than the boys'; the (male) teacher was unaware that he gender-labelled his pupils ('mate' for boys, 'sweet pea' for girls) but gamely joined in on a self-improvement regime;

girls hugely underestimated their skill at a game of strength (and cried if they got a top score) whereas boys overestimated theirs (and had a full-blown tantrum if they came last). We met a mother who had allowed her daughter to accumulate a wardrobe full of princess dressing-up clothes; another mother agreeing that she probably wouldn't let her daughter wear a pink T-shirt proclaiming 'Born to Be Underpaid' (although there were several examples of the 'Born to Be a Footballer's Wife' genre). There were some changes, even over only six weeks: the girls grew in confidence and the idea of mixed-sex football proved to be an eventual hit.

But it was perhaps the initial status quo that was revealed at the outset which was most concerning. As we now know, by seven years old, our junior gender detectives (both they and their brains) will already have been out gender-questing for more than half their lives, sifting through the gender-labelled detritus for affirmation as to their identity and what it means, not only for now but for their future. Schools can play a major part in spotting and, if necessary, trying to undo the effects of gender stereotyping, particularly with respect to the kind of low expectations they can create.

Don't Let It Go – the gendered waters in which we swim

There is a joke about Fish One and Fish Two swimming along when they meet Fish Three en route. 'How's the water?' Fish Three asks. 'Er … great,' says Fish One. A little further on, Fish Two turns to his companion and says, 'What water?' The moral of this story is that we can be blissfully unaware of the world we are moving through. In the twenty-first century, gender stereotypes are more ubiquitous than ever, with the bombardment so constant that we may well tune out, claim it is not relevant to how we live our lives, assume that it is sorted, or dismiss attempts to address the problem as mere political correctness.

We have to remember stereotypes serve a purpose – they are cognitive shortcuts which make negotiating the world that much quicker. They can be self-reinforcing, either because they prove to be useful and all the little girls do sit quietly completing the sticker books while the boys race around playing football outside, or because they contain an element of self-fulfilment: 'Women are bad at maths; here's a maths task, girls; didn't you all do badly.' And they serve the purpose of our predictive brain, providing input for the establishment of a prior, being rarely associated with a prediction error and reflecting faithfully the culture in which the brain is functioning.

Where stereotypes are linked to self-identity they become firmly embedded in the workings of the social brain, with even a suggestion that they have their own separate cortical store.[27] This is surely true of gender stereotypes. Attacks on this class of stereotypes can be equivalent to an attack on one's own self-image, and so will be fiercely defended. Even allowing for the lunacy of the Twittersphere, some of the nastiest tweets I have received followed my involvement in the No More Boys and Girls programme, with accusations of supporting the BBC's social engineering agenda and even more unpleasant references to 'interfering with children'.

We need to persistently challenge gender stereotypes. We can see how they are shaping the lives of young children, how they are serving as gatekeepers to the higher echelons of power, politics, business, science, as well as possibly contributing to mental health conditions such as depression or eating disorders.

Neuroscience can play a role here. It can help bridge the gap between the old nature versus nurture arguments and show how our world can affect our brains. Neuroscientists can lead people away from the fixed mindset that you are stuck with the biology that nature has dealt you. We can ensure that brain owners are aware of just how flexible and malleable an asset they have in their heads, but also make our society aware of the brain-changing nature of negative stereotypes (of any kind), which can lead to self-silencing, self-blame, self-criticism and plummeting self-esteem. Despite earlier waves of neurobunk and neurobaloney, neuroscience explanations are not always seductive nonsense.

Challenging stereotypical views on sex and gender may not be as straightforward as it seems. Calling attention to evidence of *racial* bias has been shown to fairly easily induce guilt and future-based determination to reduce bias; evidence of *gender* bias can have quite a different response. The 'accused' may deny bias ('I think women are wonderful'), justify the bias ('women don't belong in science anyway') or criticise the complainants as over-sensitive or trying to ignore 'inconvenient truths'.

How important are such challenges? Aren't we just talking about a bit of marketing froth? Twitter-based echo chambers which we can loftily ignore? But there are still problems to solve. Gender gaps still abound; attempts to address the lack of women in science and technology have had limited success, resulting in a waste of badly needed human capital; the greater incidence of depression and social anxiety and eating disorders in women can be a waste of human life.

Another strand of concern is the possibility, probability even, that stereotypes may serve as some kind of biosocial strait-jackets, a form of 'brainbinding'. Advances in evolutionary theory could have much significance on our thinking about the limiting characteristics of stereotypes.[28] An oft-repeated claim is that gender gaps reflect firmly rooted, genetically determined differences which hold fast in the face of well-meaning but ultimately fruitless attempts to level the playing field. But maybe social and cultural factors have a much greater role to play in what look like biologically fixed differences. Maybe these differences appear fixed because they reflect the determinedly strati-fied requirements of that environment. Perhaps the source of stability (or lack of change depending on your viewpoint) comes from a 'fixed environment'. As we have seen in this book, the long, intensive socialisation that human infants experience is full of emphasis on differences between the sexes, via stereo-typical toys, clothing, names, expectations and role models. And as we know, our brains will reflect this input. Stereotypes could be straitjacketing our flexible, plastic brains. So, yes, challenging them does matter.

*

Bias in, bias out. Let's finish by recalling Tay, the deep-learning system chatbot which was enthusiastically launched into the Twittersphere to see if she could learn some 'casual and playful conversation' by interacting with Twitter users. Tay went from tweeting about how 'humans are supercool' to becoming a 'sexist, racist asshole'[29] within sixteen hours. The world of strait-jacketing stereotypes that our brains are exposed to can have the same effect.

'I'm better than this,' tweeted poor Tay before they closed her down.[30]

And the same is true of our brains.

Acknowledgements

If you set out to write a book about sex differences in the brain, you cannot but fail to be aware that, Whac-a-Mole-like, many have fought this battle before you. Of these, several in particular stand out for their contributions to this argument and for the insights they offered me into what was often unfamiliar territory. Principal among these are Cordelia Fine's *Delusions of Gender* (Icon Books, 2010) and *Testosterone Rex* (Icon Books, 2017) as prime examples of how to be rigorous and accessible (and even funny) at the same time, and of how to track the backstory behind neuroscience research findings. I also benefited from Rebecca Jordan-Young's *Brain Storm* (HUP, 2011), which is full of detailed insights into unexplored stories behind research into hormones and the brain. Of course, Anne Fausto-Sterling's various texts have been an inspiration for many years, but *Myths of Gender: Biological Theories about Women and Men* (Basic Books, 1992), *Sexing the Body: Gender Politics and the Construction of Sexuality* (Basic Books, 2000) and *Sex/Gender: Biology in a Social World* (Routledge, 2012) have all provided sounding boards for my thoughts, particularly about rethinking the nature/nurture divide. Londa Schiebinger's *The Mind has No Sex?* (HUP, 1989) opened my eyes to the hidden and forgotten heritage of women in early science, providing a framework for the parallels that, sadly, still exist today. Lise Eliot's *Pink Brain, Blue Brain* (Oneworld, 2010) is a model of how we need to pay attention to the human brain's early years, and Eliot's trenchant critiques of single-sex schooling are a forceful exemplar of the need to keep an eye on how the research we neuroscientists are doing may be translated into educational practice. As a neuroscientist who has spent much of my time at the more technical end of brain imaging data, Matthew Lieberman's *Social* (OUP, 2013) was a real eye-opener as an introduction to the social implications of cognitive neuroscience research – it provided many of the missing links in the story I was struggling to put together.

Individual researchers have also swiftly and thoughtfully answered my cold-calling emails about aspects of their work and even carried out further analyses to provide more detailed input. Simon Baron-Cohen, Sarah-Jayne Blakemore, Paul Bloom, Karen Wynn and Tim Dalgleish were particularly helpful in this respect. Chris and Uta Frith also tirelessly answered repeated queries and even found time to give feedback on draft chapters – their work on the social brain and its implications for typical and atypical behaviour has been a continued inspiration to me. Uta's work as Chair of the Royal Society Diversity committee has also shown how research in this area can (and should) be put into practice.

I have also benefited hugely from the input and support of members of the NeuroGenderings Network, whose thought-provoking writings and discussions on issues in sex/gender research certainly widened my restricted horizons. Their selfless sharing of key materials and helpful comments on early drafts have been invaluable. Input into the cover design was also much appreciated. Particular thanks go to Cordelia Fine, Rebecca Jordan-Young, Daphna Joel, Anelis Kaiser and Giordana Grossi for sharing their critical thinking and writing skills with me, and, en route, becoming friends as well as colleagues. Needless to say, any errors or omissions in *The Gendered Brain* are entirely my responsibility.

Although this book has a single author, there are many, many others who have contributed in their various ways. Kate Barker, my indefatigable agent, 'found' me and stuck with me and has been more supportive (and tirelessly optimistic) than I probably deserve – I realise how lucky I am to be published by The Bodley

Head and know that Kate was instrumental in making this happen. Anna-Sophia Watts took on the daunting task of editing a first-time popular science author and the end product is a reflection of her diligent input. Jonathan Wadman's copy-editing was hugely insightful and, in parts, very funny, so made a daunting task ultimately achievable. I'm also grateful to Alison Davies, Sophie Painter and the rest of the excellent team at The Bodley Head, and to Maria Goldverg at Pantheon.

My own particular foray into the sex/gender research and commentary sphere began with the support and encouragement of Professor Dame Julia King, then Aston's vice-chancellor, and now Baroness Brown of Cambridge. Her tireless support for diversity agendas provided the space and opportunities for developing this aspect of my academic life. Her encouragement still continues, despite an awesomely busy schedule, via chirpy postcards and messages. The support of the British Science Association has been instrumental in paving my way to producing this book and I'd particularly like to thank Kath Mathieson, Ivvet Modinou, Amy MacLaren and Louise Ogden, though I'm aware there are many more dedicated behind-the-scenes helpers in this amazing organisation. Thanks also to Jon Wood and Anna Zecharia, via their involvement in ScienceGrrl, and Martin Davies, from the Royal Institution, who are actually responsible for starting this whole journey.

The support that Aston University provided for the development of the Aston Brain Centre has resulted not only in a facility with state-of-the-art brain imaging equipment but also in a community of colleagues and friends who have been core to my academic career. They are an amazing example of how multidisciplinary teams can work together and I am endlessly grateful for the patience and support shown not only to me but also to my research students, associates and visitors. To name but a few, Gareth Barnes, Adrian Burgess, Paul Furlong, Arjan Hillebrand, Ian Holliday, Klaus Kessler, Brian Roberts, Stefano Seri, Krish Singh, Joel Talcott and Caroline Witton have all played various roles in my development as a brain imager and critical neuroscientist. I would also like to single out Andrea

Scott, who, in the later stages of my time at Aston, always cheerfully supported the various forays of documentary-makers and film crews into my lab, way beyond the call of duty and, all too often, her official working hours.

My thanks are also due to family and friends outside my academic circle whose cautious enquiries as to progress – and patient listening to the often lengthy and gloomy answers – helped me out more than they may have realised. My daughters, Anna and Eleanor, put up with a mother who was often mentally if not physically absent and certainly perpetually distracted. Their input into this book, both through how they have lived their lives so far, as well as via real-time feedback into its content, has been immeasurably helpful. Grandson Luke suffered from my unavailability as goalie, baseball pitcher and lemon drizzle baker without (much) complaint, but he was always right when he said it was time for a break. My horse-riding friends very helpfully distracted me and made sure that thinking about brains didn't quite take over my life. Non-human companions also helped, and thanks are due to Just Joseph, Nick, and other members of our horsey herd, as well as Bob and his various predecessors who ensured that, come rain, shine, hail or snow, I always got a hefty dose of fresh air.

Almost invariably, living with Someone Who Is Writing A Book takes its toll on those who have to deal with that Someone. Now I have emerged blinking into the daylight, the most important acknowledgement of all is due to Dennis, who has taken over just about everything that keeps our show on the road. Not only has he continued in the role of Head Gardener and Chief Cook (I still do the Bottle Washing), but his patience, advice and support has appeared limitless, and his timing with the Best Gin and Tonic in the World has invariably been impeccable. Without him, there would be no book, so all the thanks in the world are due to him.

Notes

CHAPTER 1: INSIDE HER PRETTY LITTLE HEAD –
THE HUNT BEGINS

1. F. Poullain de la Barre, *De l'égalité des deux sexes, discours physique et moral où l'on voit l'importance de se défaire des préjugés* (Paris, Jean Dupuis, 1673), translated by D. M. Clarke as *The Equality of the Sexes* (Manchester, Manchester University Press, 1990). The importance of Poullain de la Barre has been detailed in Londa Schiebinger's wonderfully comprehensive history of women and science, *The Mind Has No Sex? Women in the Origins of Modern Science* (Cambridge, MA, Harvard University Press, 1991). • **2.** F. Poullain de la Barre, *De l'éducation des dames pour la conduite de l'esprit, dans les sciences et dans les moeurs: entretiens* (Paris, Jean Dupuis, 1674). • **3.** 'Si l'on y fait attention, l'on trouvera que chaque science de raisonnement demande moins d'esprit de temps qu'il n'en faut pour bien apprendre le point ou la tapisserie.' ('If one paid attention, one would see that every rational science requires less intelligence and less time than is necessary for learning embroidery or needlework well.' (Poullain de la Barre, *The Equality of the Sexes*, p. 86.) • **4.** 'L'anatomie la plus exacte ne nous fait remarquer aucune différence dans cette partie entre les hommes et les femmes; le cerveau de celles-si est entièrement semblable au notre.' (Poullain de la Barre, *The Equality of the Sexes*, p. 88.) • **5.** 'Il est aisé de remarquer que les différences des sexe ne regardent que le corps … l'esprit … n'a point de sexe.' ('It is easy to see that sexual differences apply only to the body … the mind … has no sex.' (Poullain de la Barre, *The Equality of the Sexes*, p. 87.) • **6.** L. K. Kerber, 'Separate Spheres, Female Worlds, Woman's Place: The Rhetoric of Women's History', *Journal of American History* 75:1 (1988), pp. 9–39. • **7.** E. M. Aveling, 'The Woman Question', *Westminster Review* 125:249 (1886), pp. 207–22. • **8.** C. Darwin, *The Descent of Man and Selection in Relation to Sex*, 2nd edn (London, John Murray, 1888), vol. 1. • **9.** G. Le Bon (1879) cited in S. J. Gould, *The Panda's Thumb: More Reflections in Natural History* (New York, W. W. Norton, 1980). • **10.** G. Le Bon (1879) cited in Gould, *The Panda's Thumb.* • **11.** S. J. Morton, *Crania Americana; or, a comparative view of the skulls of various aboriginal nations of North and South America: to which is prefixed an essay on the varieties of the human species* (Philadelphia, J. Dobson, 1839). • **12.** G. J. Romanes, 'Mental Differences of Men and Women', *Popular Science Monthly* 31 (1887), pp. 383–401; J. S. Mill, *The Subjection of Women* (London, Transaction, [1869] 2001). • **13.** T. Deacon, *The Symbolic Species: The Co-evolution of Language and the Human Brain* (Allen Lane, London, 1997). • **14.** E. Fee, 'Nineteenth-Century Craniology: The Study of the Female Skull', *Bulletin of the History of Medicine* 53:3 (1979), pp. 415–33. • **15.** A. Ecker, 'On a Characteristic Peculiarity in the Form of the Female Skull, and Its Significance for Comparative Anthropology', *Anthropological Review* 6:23 (1868), pp. 350–56. • **16.** J. Cleland, 'VIII. An Inquiry into the Variations of the Human Skull, Particularly the Antero-posterior Direction', *Philosophical Transactions of the Royal Society*, 160 (1870), pp. 117–74. • **17.** J. Barzan, *Race: A Study in Superstition* (New York, Harper & Row, 1965). • **18.** A. Lee, 'V. Data for the Problem of Evolution in Man – VI. A First Study of the Correlation of the Human Skull', *Philosophical Transactions of the Royal Society A* 196:274–86 (1901), pp. 225–64. • **19.** K. Pearson, 'On the Relationship of Intelligence to Size and Shape of Head, and to Other Physical and Mental Characters', *Biometrika* 5:1–2 (1906), pp. 105–46. • **20.** F. J. Gall, *On the Functions of the Brain and of Each of Its Parts: with observations on the possibility of determining the instincts, propensities, and talents, or the moral and intellectual dispositions of men and animals, by the configuration of the brain and head* (Boston, Marsh, Capen & Lyon, 1835), vol. 1. • **21.** J. G. Spurzheim, *The Physiognomical System of Drs Gall and Spurzheim: founded on an anatomical and physiological examination of the nervous system in general, and of the brain in particular; and indicating the dispositions and manifestations of the mind* (London, Baldwin, Cradock & Joy, 1815). • **22.** C. Bittel, 'Woman, Know Thyself: Producing and Using Phrenological Knowledge in 19th-Century America', *Centaurus* 55:2 (2013), pp. 104–30. • **23.** P. Flourens, *Phrenology Examined* (Philadelphia, Hogan & Thompson, 1846). • **24.** P. Broca, 'Sur le siège de la faculté du langage articulé (15 juin)', *Bulletins de la Société d'Anthropologie de Paris* 6 (1865), pp. 377–93; E. A. Berker, A. H. Berker and A. Smith, 'Translation of Broca's 1865 Report: Localization of Speech in the Third Left Frontal Convolution', *Archives of Neurology* 43:10 (1986), pp. 1065–72. • **25.** J. M. Harlow, 'Passage of an Iron Rod through the Head', *Boston Medical and Surgical Journal* 39:20 (1848), pp. 389–93; J. M. Harlow, 'Recovery from the Passage of an Iron Bar through the Head', *History of Psychiatry* 4:14 (1993), pp. 274–81. • **26.** H. Ellis, *Man and Woman: A Study of Secondary and Tertiary Sexual Characteristics*, 8th edn

(London, Heinemann, 1934), cited in S. Shields, 'Functionalism, Darwinism, and the Psychology of Women', *American Psychologist* 30:7 (1975), p. 739. • **27.** G. T. W. Patrick, 'The Psychology of Women', *Popular Science Monthly*, June 1895, pp. 209–25, cited in S. Shields, 'Functionalism, Darwinism, and the Psychology of Women', *American Psychologist* 30:7, (1975), p. 739. • **28.** Schiebinger, *The Mind Has No Sex?*, p. 217. • **29.** J-J. Rousseau, *Émile, ou de l'éducation* (Paris, Firmin Didot, [1762] 1844). • **30.** J. McGrigor Allan, 'On the Real Differences in the Minds of Men and Women', *Journal of the Anthropological Society of London*, 7 (1869), pp. cxcv–ccxix, at p. cxcvii. • **31.** McGrigor Allan, 'On the Real Differences in the Minds of Men and Women', p. cxcviii. • **32.** W. Moore, 'President's Address, Delivered at the Fifty-Fourth Annual Meeting of the British Medical Association, Held in Brighton, August 10th, 11th, 12th, and 13th, 1886', *British Medical Journal*, 2:295 (1886), pp. 295–9. • **33.** R. Malane, *Sex in Mind: The Gendered Brain in Nineteenth-Century Literature and Mental Sciences* (New York, Peter Lang, 2005). • **34.** H. Berger, 'Über das Elektrenkephalogramm des Menschen', *Archiv für Psychiatrie und Nervenkrankheiten*, 87 (1929), pp. 527–70; D. Millett, 'Hans Berger: From Psychic Energy to the EEG', *Perspectives in Biology and Medicine*, 44:4 (2001), pp. 522–42. • **35.** D. Millet, 'The Origins of EEG', 7th Annual Meeting of the International Society for the History of the Neurosciences, Los Angeles, 2 June 2002. • **36.** R. S. J. Frackowiak, K. J. Friston, C. D. Frith, R. J. Dolan, C. J. Price, S. Zeki, J. T. Ashburner and W. D. Penny (eds), *Human Brain Function*, 2nd edn (San Diego, Academic Press, 2004). • **37.** Friston et al., *Human Brain Function*. • **38.** A. Fausto-Sterling, *Sexing the Body: Gender Politics and the Construction of Sexuality* (New York, Basic, 2000). • **39.** R. L. Holloway, 'In the Trenches with the Corpus Callosum: Some Redux of Redux', *Journal of Neuroscience Research* 95:1–2 (2017), pp. 21–3. • **40.** E. Zaidel and M. Iacoboni, *The Parallel Brain: The Cognitive Neuroscience of the Corpus Callosum* (Cambridge, MA, MIT Press, 2003). • **41.** C. DeLacoste-Utamsing and R. L. Holloway, 'Sexual Dimorphism in the Human Corpus Callosum', *Science*, 216:4553 (1982), pp. 1431–2. • **42.** N. R. Driesen and N. Raz, 'The Influence of Sex, Age, and Handedness on Corpus Callosum Morphology: A Meta-analysis', *Psychobiology* 23:3 (1995), pp. 240–47. • **43.** Cleland, 'VIII. An Inquiry into the Variations of the Human Skull'. • **44.** W. Men, D. Falk, T. Sun, W. Chen, J. Li, D. Yin, L. Zang and M. Fan, 'The Corpus Callosum of Albert Einstein's Brain: Another Clue to His High Intelligence?', *Brain* 137:4 (2014), p. e268. • **45.** R. J. Smith, 'Relative Size versus Controlling for Size: Interpretation of Ratios in Research on Sexual Dimorphism in the Human Corpus Callosum', *Current Anthropology* 46:2 (2005), pp. 249–73. • **46.** Ibid, p. 264. • **47.** S. P. Springer and G. Deutsch, *Left Brain, Right Brain: Perspectives from Cognitive Neuroscience*, 5th edn (New York, W. H. Freeman, 1998).

• **48.** G. D. Schott, 'Penfield's Homunculus: A Note on Cerebral Cartography', *Journal of Neurology, Neurosurgery and Psychiatry* 56:4 (1993), p. 329. • **49.** K. Woollett, H. J. Spiers and E. A. Maguire, 'Talent in the Taxi: a Model System for Exploring Expertise', *Philosophical Transactions of the Royal Society B: Biological Sciences* 364:1522 (2009), pp. 1407–16. • **50.** H. Vollmann, P. Ragert, V. Conde, A. Villringer, J. Classen, O. W. Witte and C. J. Steele, 'Instrument Specific Use-Dependent Plasticity Shapes the Anatomical Properties of the Corpus Callosum: A Comparison between Musicians and Non-musicians', *Frontiers in Behavioral Neuroscience* 8 (2014), p. 245. • **51.** L. Eliot, 'Single-Sex Education and the Brain', *Sex Roles* 69:7–8 (2013), pp. 363–81. • **52.** R. C. Gur, B. I. Turetsky, M. Matsui, M. Yan, W. Bilker, P. Hughett and R. E. Gur, 'Sex Differences in Brain Gray and White Matter in Healthy Young Adults: Correlations with Cognitive Performance', *Journal of Neuroscience* 19:10 (1999), pp. 4065–72. • **53.** J. S. Allen, H. Damasio, T. J. Grabowski, J. Bruss and W. Zhang, 'Sexual Dimorphism and Asymmetries in the Gray–White Composition of the Human Cerebrum, *NeuroImage* 18:4 (2003), pp. 880–94; M. D. De Bellis, M. S. Keshavan, S. R. Beers, J. Hall, K. Frustaci, A. Masalehdan, J. Noll and A. M. Boring, 'Sex Differences in Brain Maturation during Childhood and Adolescence', *Cerebral Cortex* 11:6 (2001), pp. 552–7; J. M. Goldstein, L. J. Seidman, N. J. Horton, N. Makris, D. N. Kennedy, V. S. Caviness Jr, S. V. Faraone and M. T. Tsuang, 'Normal Sexual Dimorphism of the Adult Human Brain Assessed by In Vivo Magnetic Resonance Imaging', *Cerebral Cortex* 11:6 (2001), pp. 490–97; C. D. Good, I. S. Johnsrude, J. Ashburner, R. N. A. Henson, K. J. Friston and R. S. Frackowiak, 'A Voxel-Based Morphometric Study of Ageing in 465 Normal Adult Human Brains', *NeuroImage* 14:1 (2001), pp. 21–36. • **54.** A. N. Ruigrok, G. Salimi-Khorshidi, M. C. Lai, S. Baron-Cohen, M. V. Lombardo, R. J. Tait and J. Suckling, 'A Meta-analysis of Sex Differences in Human Brain Structure', *Neuroscience & Biobehavioral Reviews* 39 (2014), pp. 34–50. • **55.** R. J. Haier, R. E. Jung, R. A. Yeo, K. Head and M. T. Alkire, 'The Neuroanatomy of General Intelligence: Sex Matters', *NeuroImage* 25:1 (2005), pp. 320–27.

CHAPTER 2: HER RAGING HORMONES

1. C. Fine, *Testosterone Rex: Unmaking the Myths of our Gendered Minds* (London, Icon, 2017); G. Breuer, *Sociobiology and the Human Dimension* (Cambridge, Cambridge University Press, 1983). • **2.** C. H. Phoenix, R. W. Goy, A. A. Gerall and W. C. Young, 'Organizing Action of Prenatally Administered Testosterone Propionate on the Tissues Mediating Mating Behavior in the Female Guinea Pig', *Endocrinology* 65:3 (1959), pp. 369–82; K. Wallen, 'The Organizational Hypothesis: Reflections on the 50th

Anniversary of the Publication of Phoenix, Goy, Gerall, and Young (1959)', *Hormones and Behavior* 55:5 (2009), pp. 561–5; M. Hines, *Brain Gender* (Oxford, Oxford University Press, 2005); R. M. Jordan-Young, *Brain Storm: The Flaws in the Science of Sex Differences* (Cambridge, MA, Harvard University Press, 2011). • **3.** J. D. Wilson, 'Charles-Edouard Brown-Sequard and the Centennial of Endocrinology', *Journal of Clinical Endocrinology and Metabolism* 71:6 (1990), pp. 1403–9. • **4.** J. Henderson, 'Ernest Starling and "Hormones": An Historical Commentary', *Journal of Endocrinology* 184:1 (2005), pp. 5–10. • **5.** B. P. Setchell, 'The Testis and Tissue Transplantation: Historical Aspects', *Journal of Reproductive Immunology* 18:1 (1990), pp. 1–8. • **6.** M. L. Stefanick, 'Estrogens and Progestins: Background and History, Trends in Use, and Guidelines and Regimens Approved by the US Food and Drug Administration', *American Journal of Medicine* 118:12 (2005), pp. 64–73. • **7.** 'Origins of Testosterone Replacement', Urological Sciences Research Foundation website, https://www.usrf.org/news/000908-origins.html (accessed 4 November 2018). • **8.** J. Schwarcz, 'Getting "Steinached" was all the rage in roaring '20s', 20 March 2017, McGill Office for Science and Security website, https://www.mcgill.ca/oss/article/health-history-science-science-everywhere/getting-steinached-was-all-rage-roaring-20s (accessed 4 November 2018). • **9.** A. Carrel and C. C. Guthrie, 'Technique de la transplantation homoplastique de l'ovaire', *Comptes rendus des séances de la Société de biologie* 6 (1906), pp. 466–8, cited in E. Torrents, I. Boiso, P. N. Barri and A. Veiga, 'Applications of Ovarian Tissue Transplantation in Experimental Biology and Medicine', *Human Reproduction Update* 9:5 (2003), pp. 471–81; J. Woods, 'The history of estrogen', menoPAUSE blog, February 2016, https://www.urmc.rochester.edu/ob-gyn/gynecology/menopause-blog/february-2016/the-history-of-estrogen.aspx (accessed 4 November 2018). • **10.** J. M. Davidson and P. A. Allinson, 'Effects of Estrogen on the Sexual Behavior of Male Rats', *Endocrinology* 84:6 (1969), pp. 1365–72. • **11.** R. H. Epstein, *Aroused: The History of Hormones and How They Control Just About Everything* (New York, W. W. Norton, 2018). • **12.** R. T. Frank, 'The Hormonal Causes of Premenstrual Tension', *Archives of Neurology and Psychiatry* 26:5 (1931), pp. 1053–7. • **13.** R. Greene and K. Dalton, 'The Premenstrual Syndrome', *British Medical Journal* 1:4818 (1953), p. 1007. • **14.** C. A. Boyle, G. S. Berkowitz and J. L. Kelsey, 'Epidemiology of Premenstrual Symptoms', *American Journal of Public Health* 77:3 (1987), pp. 349–50. • **15.** J. C. Chrisler and P. Caplan, 'The Strange Case of Dr Jekyll and Ms Hyde: How PMS Became a Cultural Phenomenon and a Psychiatric Disorder', *Annual Review of Sex Research* 13:1 (2002), pp. 274–306. • **16.** J. T. E. Richardson, 'The Premenstrual Syndrome: A Brief History', *Social Science and Medicine* 41:6 (1995), pp. 761–7. • **17.** 'Raging hormones', *New York Times*, 11 January 1982, http://www.nytimes.com/1982/01/11/

opinion/raging-hormones.html (accessed 4 November 2018). • **18.** K. L. Ryan, J. A. Loeppky and D. E. Kilgore Jr, 'A Forgotten Moment in Physiology: The Lovelace Woman in Space Program (1960–1962)', *Advances in Physiology Education* 33:3 (2009), pp. 157–64. • **19.** R. K. Koeske and G. F. Koeske, 'An Attributional Approach to Moods and the Menstrual Cycle', *Journal of Personality and Social Psychology* 31:3 (1975), p. 473. • **20.** D. N. Ruble, 'Premenstrual Symptoms: A Reinterpretation', *Science* 197:4300 (1977), pp. 291–2. • **21.** Chrisler and Caplan, 'The Strange Case of Dr Jekyll and Ms Hyde'. • **22.** R. H. Moos, 'The Development of a Menstrual Distress Questionnaire', *Psychosomatic Medicine* 30:6 (1968), pp. 853–67. • **23.** J. Brooks-Gunn and D. N. Ruble, 'The Development of Menstrual-Related Beliefs and Behaviors during Early Adolescence', *Child Development* 53:6 (1982), pp. 1567–77. • **24.** S. Toffoletto, R. Lanzenberger, M. Gingnell, I. Sundström-Poromaa and E. Comasco, 'Emotional and Cognitive Functional Imaging of Estrogen and Progesterone Effects in the Female Human Brain: A Systematic Review', *Psychoneuroendocrinology* 50 (2014), pp. 28–52. • **25.** D. B. Kelley and D. W Pfaff, 'Generalizations from Comparative Studies on Neuroanatomical and Endocrine Mechanisms of Sexual Behaviour', in J. B. Hutchison (ed.), *Biological Determinants of Sexual Behaviour* (Chichester, John Wiley, 1978), pp. 225–54. • **26.** M. Hines, 'Gender Development and the Human Brain', *Annual Review of Neuroscience* 34 (2011), pp. 69–88. • **27.** Phoenix et al., 'Organizing Action of Prenatally Administered Testosterone Propionate'. • **28.** M. Hines and F. R. Kaufman, 'Androgen and the Development of Human Sex-Typical Behavior: Rough-and-Tumble Play and Sex of Preferred Playmates in Children with Congenital Adrenal Hyperplasia (CAH)', *Child Development* 65:4 (1994), pp. 1042–53; C. van de Beek, S. H. van Goozen, J. K. Buitelaar and P. T. Cohen-Kettenis, 'Prenatal Sex Hormones (Maternal and Amniotic Fluid) and Gender-Related Play Behavior in 13-Month-Old Infants', *Archives of Sexual Behavior* 38:1 (2009), pp. 6–15. • **29.** J. B. Watson, 'Psychology as the Behaviorist Views It', *Psychological Review* 20:2 (1913), pp. 158–77. • **30.** G. Kaplan and L. J. Rogers, 'Parental Care in Marmosets (*Callithrix jacchus jacchus*): Development and Effect of Anogenital Licking on Exploration', *Journal of Comparative Psychology* 113:3 (1999), p. 269. • **31.** S. W. Bottjer, S. L. Glaessner and A. P. Arnold, 'Ontogeny of Brain Nuclei Controlling Song Learning and Behavior in Zebra Finches', *Journal of Neuroscience* 5:6 (1985), pp. 1556–62. • **32.** D. W. Bayless and N. M. Shah, 'Genetic Dissection of Neural Circuits Underlying Sexually Dimorphic Social Behaviours', *Philosophical Transactions of the Royal Society B: Biological Sciences*, 371:1688 (2016), 20150109. • **33.** R. M. Young and E. Balaban, 'Psychoneuroindoctrinology', *Nature* 443:7112 (2006), p. 634. • **34.** A. Fausto-Sterling, *Sexing the Body*. • **35.** D. P. Merke and S. R. Bornstein, 'Congenital Adrenal Hyperplasia', *Lancet* 365:9477 (2005), pp. 2125–36. • **36.** Jordan-Young, *Brain Storm*. • **37.** Hines and Kaufman, 'Androgen and the Development of Human Sex-Typical Behavior'. • **38.** P. Plumb and G. Cowan, 'A Developmental Study of Destereotyping and Androgynous Activity Preferences of Tomboys, Nontomboys, and Males', *Sex Roles* 10:9–10 (1984), pp. 703–12. • **39.** J. Money and A. A. Ehrhardt, *Man and Woman, Boy and Girl: The Differentiation and Dimorphism of Gender Identity from Conception to Maturity* (Baltimore, Johns Hopkins University Press, 1972). • **40.** M. Hines, *Brain Gender*. • **41.** D. A. Puts, M. A. McDaniel, C. L. Jordan and S. M. Breedlove, 'Spatial Ability and Prenatal Androgens: Meta-analyses of Congenital Adrenal Hyperplasia and Digit Ratio (2D:4D) Studies', *Archives of Sexual Behavior* 37:1 (2008), p. 100. • **42.** Jordan-Young, *Brain Storm*. • **43.** Ibid., p. 289. • **44.** J. Colapinto, *As Nature Made Him: The Boy Who Was Raised as a Girl* (New York, HarperCollins, 2001). • **45.** J. Colapinto, 'The True Story of John/Joan', *Rolling Stone*, 11 December 1997, pp. 54–97. • **46.** M. V. Lombardo, E. Ashwin, B. Auyeung, B. Chakrabarti, K. Taylor, G. Hackett, E. T. Bullmore and S. Baron-Cohen, 'Fetal Testosterone Influences Sexually Dimorphic Gray Matter in the Human Brain', *Journal of Neuroscience* 32:2 (2012), pp. 674–80. • **47.** S. Baron-Cohen, S. Lutchmaya and R. Knickmeyer, *Prenatal Testosterone in Mind: Amniotic Fluid Studies* (Cambridge, MA, MIT Press, 2004). • **48.** R. Knickmeyer, S. Baron-Cohen, P. Raggatt and K. Taylor, 'Foetal Testosterone, Social Relationships, and Restricted Interests in Children', *Journal of Child Psychology and Psychiatry* 46:2 (2005), pp. 198–210; E. Chapman, S. Baron-Cohen, B. Auyeung, R. Knickmeyer, K. Taylor and G. Hackett, 'Fetal Testosterone and Empathy: Evidence from the Empathy Quotient (EQ) and the "Reading the Mind in the Eyes" Test', *Social Neuroscience* 1:2 (2006), pp. 135–48. • **49.** S. Lutchmaya, S. Baron-Cohen, P. Raggatt, R. Knickmeyer and J. T. Manning, '2nd to 4th Digit Ratios, Fetal Testosterone and Estradiol', *Early Human Development* 77:1–2 (2004), pp. 23–8. • **50.** J. Hönekopp and C. Thierfelder, 'Relationships between Digit Ratio (2D:4D) and Sex-Typed Play Behavior in Pre-school Children', *Personality and Individual Differences* 47:7 (2009), pp. 706–10; D. A. Putz, S. J. Gaulin, R. J. Sporter and D. H. McBurney, 'Sex Hormones and Finger Length: What Does 2D:4D Indicate?', *Evolution and Human Behavior* 25:3 (2004), pp. 182–99. • **51.** J. M. Valla and S. J. Ceci, 'Can Sex Differences in Science Be Tied to the Long Reach of Prenatal Hormones? Brain Organization Theory, Digit Ratio (2D/4D), and Sex Differences in Preferences and Cognition', *Perspectives on Psychological Science*, 6:2 (2011), pp. 134–46. • **52.** S. M. Van Anders, K. L. Goldey and P. X. Kuo, 'The Steroid/Peptide Theory of Social Bonds: Integrating Testosterone and Peptide Responses for Classifying Social Behavioral Contexts', *Psychoneuroendocrinology* 36:9 (2011), pp. 1265–75.

CHAPTER 3: THE RISE OF PSYCHOBABBLE

1. H. T. Woolley, 'A Review of Recent Literature on the Psychology of Sex', *Psychological Bulletin* 7:10 (1910), pp. 335–42. • **2.** C. Fine, *Delusions of Gender: How Our Minds, Society, and Neurosexism Create Difference* (New York, W. W. Norton, 2010), p. xxvii. • **3.** C. Darwin, *On the Origin of Species by Means of Natural Selection* (London, John Murray, 1859); C. Darwin, *The Descent of Man and Selection in Relation to Sex* (London, John Murray, 1871). • **4.** S. A. Shields, *Speaking from the Heart: Gender and the Social Meaning of Emotion* (Cambridge, Cambridge University Press, 2002), p. 77. • **5.** Darwin, *The Descent of Man*, p. 361. • **6.** S. A. Shields, 'Passionate Men, Emotional Women: Psychology Constructs Gender Difference in the Late 19th Century', *History of Psychology* 10:2 (2007), pp. 92–110, at p. 93. • **7.** Shields, 'Passionate Men, Emotional Women', p. 97. • **8.** Ibid., p. 94. • **9.** L. Cosmides and J. Tooby, 'Cognitive Adaptations for Social Exchange', in J. H. Barkow, L. Cosmides and J. Tooby (eds), *The Adapted Mind: Evolutionary Psychology and the Generation of Culture* (New York, Oxford University Press, 1992). • **10.** L. Cosmides and J. Tooby, 'Beyond Intuition and Instinct Blindness: Toward an Evolutionarily Rigorous Cognitive Science', *Cognition* 50:1–3 (1994), pp. 41–77. • **11.** A. C. Hurlbert and Y. Ling, 'Biological Components of Sex Differences in Color Preference', *Current Biology* 17:16 (2007), pp. R623–5. • **12.** S. Baron-Cohen, *The Essential Difference* (London, Penguin, 2004). • **13.** Ibid., p. 26. • **14.** Ibid., p. 63. • **15.** Ibid., p. 127. • **16.** Ibid., p. 185. • **17.** Ibid., p. 123. • **18.** Ibid, p. 185. • **19.** S. Baron-Cohen, J. Richler, D. Bisarya, N. Gurunathan and S. Wheelwright, 'The Systemizing Quotient: An Investigation of Adults with Asperger Syndrome or High-Functioning Autism, and Normal Sex Differences', *Philosophical Transactions of the Royal Society B: Biological Sciences* 358:1430 (2003), pp. 361–74; S. Baron-Cohen and S. Wheelwright, 'The Empathy Quotient: An Investigation of Adults with Asperger Syndrome or High Functioning Autism, and Normal Sex Differences', *Journal of Autism and Developmental Disorders* 34:2 (2004), pp. 163–75; A. Wakabayashi, S. Baron-Cohen, S. Wheelwright, N. Goldenfeld, J. Delaney, D. Fine, R. Smith and L. Weil, 'Development of Short Forms of the Empathy Quotient (EQ-Short) and the Systemizing Quotient (SQ-Short)', *Personality and Individual Differences* 41:5 (2006), pp. 929–40. • **20.** B. Auyeung, S. Baron-Cohen, F. Chapman, R. Knickmeyer, K. Taylor and G. Hackett, 'Foetal Testosterone and the Child Systemizing Quotient', *European Journal of Endocrinology* 155:Supplement 1 (2006), pp. S123–30; E. Chapman, S. Baron-Cohen, B. Auyeung, R. Knickmeyer, K. Taylor and G. Hackett, 'Fetal Testosterone and Empathy: Evidence from the Empathy Quotient (EQ) and the "Reading the Mind in the Eyes" Test', *Social Neuroscience* 1:2 (2006), pp. 135–48. • **21.** S. Baron-Cohen, S. Wheelwright, J. Hill, Y. Raste and I. Plumb, 'The "Reading the Mind in the Eyes" Test

Revised Version: A Study with Normal Adults, and Adults with Asperger Syndrome or High-Functioning Autism', *Journal of Child Psychology and Psychiatry* 42:2 (2001), pp. 241–51. • **22.** J. Billington, S. Baron-Cohen and S. Wheelwright, 'Cognitive Style Predicts Entry into Physical Sciences and Humanities: Questionnaire and Performance Tests of Empathy and Systemizing', *Learning and Individual Differences* 17:3 (2007), pp. 260–68. • **23.** Baron-Cohen, *The Essential Difference*, pp. 185, 1. • **24.** Ibid., pp. 185, 8. • **25.** E. B. Titchener, 'Wilhelm Wundt', *American Journal of Psychology* 32:2 (1921), pp. 161–78; W. C. Wong, 'Retracing the Footsteps of Wilhelm Wundt: Explorations in the Disciplinary Frontiers of Psychology and in *Völkerpsychologie*', *History of Psychology* 12:4, (2009), p. 229. • **26.** R. W. Kamphaus, M. D. Petoskey and A. W. Morgan, 'A History of Intelligence Test Interpretation', in D. P. Flanagan, J. L. Genshaft and P. L Harrison (eds), *Contemporary Intellectual Assessment: Theories, Tests, and Issues* (New York, Guilford, 1997), pp. 3–16. • **27.** R. E. Gibby and M. J. Zickar, 'A History of the Early Days of Personality Testing in American Industry: An Obsession with Adjustment', *History of Psychology* 11:3 (2008), p. 164. • **28.** Woodworth Psychoneurotic Inventory, https://openpsychometrics.org/tests/WPI.php (accessed 4 November 2018). • **29.** J. Jastrow, 'A Study of Mental Statistics', *New Review* 5 (1891), pp. 559–68. • **30.** Woolley, 'A Review of the Recent Literature on the Psychology of Sex', p. 335. • **31.** N. Weisstein, 'Psychology Constructs the Female; or the Fantasy Life of the Male Psychologist (with Some Attention to the Fantasies of his Friends, the Male Biologist and the Male Anthropologist)', *Feminism and Psychology* 3:2 (1993), pp. 194–210. • **32.** S. Schachter and J. Singer, 'Cognitive, Social, and Physiological Determinants of Emotional State', *Psychological Review* 69:5 (1962), p. 379. • **33.** E. E. Maccoby and C. N. Jacklin, *The Psychology of Sex Differences, Vol. 1: Text* (Stanford, CA, Stanford University Press, 1974). • **34.** J. Cohen, *Statistical Power Analysis for the Behavioral Sciences*, 2nd edn (Hillsdale, NJ, Laurence Erlbaum Associates, 1988); K. Magnusson, 'Interpreting Cohen's *d* effect size: an interactive visualisation', R Psychologist blog, 13 January 2014, http://rpsychologist.com/d3/cohend (accessed 4 November 2018); SexDifference website, https://sexdifference.org (accessed 4 November 2018). • **35.** K. Magnusson, 'Interpreting Cohen's *d* effect size'; SexDifference website. • **36.** SexDifference website. • **37.** T. D. Satterthwaite, D. H. Wolf, D. R. Roalf, K. Ruparel, G. Erus, S. Vandekar, E. D. Gennatas, M. A. Elliott, A. Smith, H. Hakonarson and R. Verma, 'Linked Sex Differences in Cognition and Functional Connectivity in Youth', *Cerebral Cortex* 25:9 (2014), pp. 2383–94, at p. 2383. • **38.** A. Kaiser, S. Haller, S. Schmitz and C. Nitsch, 'On Sex/Gender Related Similarities and Differences in fMRI Language Research', *Brain Research Reviews* 61:2 (2009), pp. 49–59. • **39.** R. Rosenthal, 'The File Drawer Problem and Tolerance for Null Results', *Psychological Bulletin* 86:3

(1979), p. 638. • **40.** D. J. Prediger, 'Dimensions Underlying Holland's Hexagon: Missing Link between Interests and Occupations?', *Journal of Vocational Behavior* 21:3 (1982), pp. 259–87. • **41.** Ibid, p. 261. • **42.** United States Bureau of the Census, *1980 Census of the Population: Detailed Population Characteristics* (US Department of Commerce, Bureau of the Census, 1984). • **43.** B. R. Little, 'Psychospecialization: Functions of Differential Orientation towards Persons and Things', *Bulletin of the British Psychological Society* 21 (1968), p. 113. • **44.** P. I. Armstrong, W. Allison and J. Rounds, 'Development and Initial Validation of Brief Public Domain RIASEC Marker Scales', *Journal of Vocational Behavior* 73:2 (2008), pp. 287–99. • **45.** V. Valian, 'Interests, Gender, and Science', *Perspectives on Psychological Science* 9:2 (2014), pp. 225–30. • **46.** R. Su, J. Rounds and P. I. Armstrong, 'Men and Things, Women and People: A Meta-analysis of Sex Differences in Interests', *Psychological Bulletin* 135:6 (2009), p. 859. • **47.** M. T. Orne, 'Demand Characteristics and the Concept of Quasi-controls', in R. Rosenthal and R. L. Rosnow, *Artifacts in Behavioral Research* (Oxford, Oxford University Press, 2009), pp. 110–37. • **48.** J. C. Chrisler, I. K. Johnston, N. M Champagne and K. E. Preston, 'Menstrual Joy: The Construct and Its Consequences', *Psychology of Women Quarterly* 18:3 (1994), pp. 375–87. • **49.** J. L. Hilton and W. Von Hippel, 'Stereotypes', *Annual Review of Psychology* 47:1 (1996), pp. 237–71. • **50.** N. Eisenberg and R. Lennon, 'Sex Differences in Empathy and Related Capacities', *Psychological Bulletin* 94:1 (1983), p. 100. • **51.** C. M. Steele and J. Aronson, 'Stereotype Threat and the Intellectual Test Performance of African Americans', *Journal of Personality and Social Psychology* 69:5 (1995), p. 797; S. J. Spencer, C. Logel and P. G. Davies, 'Stereotype Threat', *Annual Review of Psychology* 67 (2016), pp. 415–37. • **52.** S. J. Spencer, C. M. Steele and D. M. Quinn, 'Stereotype Threat and Women's Math Performance', *Journal of Experimental Social Psychology* 35:1 (1999), pp. 4–28. • **53.** M. A. Pavlova, S. Weber, E. Simoes and A. N. Sokolov, 'Gender Stereotype Susceptibility', *PLoS One* 9:12 (2014), e114802. • **54.** Fine, *Delusions of Gender*. • **55.** D. Carnegie, *How to Win Friends and Influence People* (New York, Simon & Schuster, 1936).

CHAPTER 4: BRAIN MYTHS, NEUROTRASH
AND NEUROSEXISM

1. N. K. Logothetis, 'What We Can Do and What We Cannot Do with fMRI', *Nature* 453:7197 (2008), p. 869. • **2.** R. S. J. Frackowiak, K. J. Friston, C. D. Frith, R. J. Dolan, C. J. Price, S. Zeki, J. T. Ashburner and W. D. Penny (eds), *Human Brain Function*, 2nd edn (San Diego and London, Academic Press, 2004). • **3.** A. L. Roskies, 'Are Neuroimages like Photographs of the Brain?', *Philosophy of Science* 74:5 (2007), pp. 860–72. • **4.** R. A. Poldrack,

'Can Cognitive Processes Be Inferred from Neuroimaging Data?', *Trends in Cognitive Sciences* 10:2 (2006), pp. 59–63. • **5.** J. B. Meixner and J. P. Rosenfeld, 'A Mock Terrorism Application of the P300-Based Concealed Information Test', *Psychophysiology* 48:2 (2011), pp. 149–54. • **6.** A. Linden and J. Fenn, 'Understanding Gartner's Hype Cycles', Strategic Analysis Report R-20–1971 (Stamford, CT, Gartner, 2003). • **7.** J. Devlin and G. de Ternay, 'Can neuromarketing really offer you useful customer insights?', *Medium*, 8 October 2016, https://medium.com/@GuerricdeTernay/can-neuromarketing-really-offer-you-useful-customer-insights-e4d0f515f1ec (accessed 13 November 2018). • **8.** A. Orlowski, 'The Great Brain Scan Scandal: It isn't just boffins who should be ashamed', *Register*, 7 July 2016, https://www.theregister.co.uk/2016/07/07/the_great_brain_scan_scandal_it_isnt_just_boffins_who_should_be_ashamed (accessed 13 November 2018). • **9.** S. Ogawa, D. W. Tank, R. Menon, J. M. Ellermann, S. G. Kim, H. Merkle and K. Ugurbil, 'Intrinsic Signal Changes Accompanying Sensory Stimulation: Functional Brain Mapping with Magnetic Resonance Imaging', *Proceedings of the National Academy of Sciences* 89:13 (1992), pp. 5951–5. • **10.** K. K. Kwong, J. W. Belliveau, D. A. Chesler, I. E. Goldberg, R. M. Weisskoff, B. P. Poncelet, D. N. Kennedy, B. E. Hoppel, M. S. Cohen and R. Turner, 'Dynamic Magnetic Resonance Imaging of Human Brain Activity during Primary Sensory Stimulation', *Proceedings of the National Academy of Sciences* 89:12 (1992), pp. 5675–9. • **11.** K. Smith, 'fMRI 2.0', *Nature* 484:7392 (2012), p. 24. • **12.** Presidential Proclamation 6158, 17 July 1990, Project on the Decade of the Brain, https://www.loc.gov/loc/brain/proclaim.html (accessed 4 November 2018); E. G. Jones and L. M. Mendell, 'Assessing the Decade of the Brain', *Science*, 30 April 1999, p. 739. • **13.** 'Neurosociety Conference: What Is It with the Brain These Days?', Oxford Martin School website, https://www.oxfordmartin.ox.ac.uk/event/895 (accessed 4 November 2018). • **14.** B. Carey, 'A neuroscientific look at speaking in tongues', *New York Times*, 7 November 2006, https://www.nytimes.com/2006/11/07/health/07brain.html (accessed 4 November 2018); M. Shermer, 'The political brain', *Scientific American*, 1 July 2006, https://www.scientificamerican.com/article/the-political-brain (accessed 4 November 2018); E. Callaway, 'Brain quirk could help explain financial crisis', *New Scientist*, 24 March 2009, https://www.newscientist.com/article/dn16826-brain-quirk-could-help-explain-financial-crisis (accessed 4 November 2018). • **15.** '"Beliebers" suffer a real fever: How fans of the pop sensation have brains hard wired to be obsessed with him', *Mail Online*, 1 July 2012, https://www.dailymail.co.uk/sciencetech/article-2167108/Beliebers-suffer-real-fever-How-fans-Justin-Bieber-brains-hard-wired-obsessed-him.html (accessed 4 November 2018). • **16.** J. Lehrer, 'The neuroscience of Bob Dylan's genius', *Guardian*, 6 April 2012, https://www.theguardian.com/music/2012/apr/06/neuroscience-bob-dylan-genius-creativity (accessed 4 November 2018). • **17.** 'The neuroscience of kitchen cabinetry', The Neurocritic blog, 5 December 2010, https://neurocritic. blogspot.com/2010/12/neuroscience-of-kitchen-cabinetry.html (accessed 4 November 2018). • **18.** 'Spanner or sex object?', Neurocritic blog, 20 February 2009, https://neurocritic.blogspot.com/2009/02/spanner-or-sex-object.html (accessed 4 November 2018). • **19.** I. Sample, 'Sex objects: pictures shift men's view of women', *Guardian*, 16 February 2009, https://www.theguardian.com/science/2009/feb/16/sex-object-photograph (accessed 4 November 2018). • **20.** E. Rossini, 'Princeton study: "Men view half-naked women as objects"', *Illusionists* website, 18 February 2009, https://theillusionists.org/2009/02/princeton-objectification (accessed 4 November 2018). • **21.** C. dell'Amore, 'Bikinis make men see women as objects, scans confirm', *National Geographic*, 16 February 2009, https://www.national-geographic.com/science/2009/02/bikinis-women-men-objects-science (accessed 4 November 2018). • **22.** E. Landau, 'Men see bikini-clad women as objects, psychologists say', CNN website, 2 April 2009, http://edition.cnn.com/2009/HEALTH/02/19/women.bikinis.objects (accessed 4 November 2018). • **23.** C. O'Connor, G. Rees and H. Joffe, 'Neuroscience in the Public Sphere', *Neuron* 74:2 (2012), pp. 220–26. • **24.** J. Dumit, *Picturing Personhood: Brain Scans and Biomedical Identity* (Princeton, NJ, Princeton University Press, 2004). • **25.** http://www.sandsresearch.com/coke-heist.html (accessed 4 November 2018). • **26.** D. P. McCabe and A. D. Castel, 'Seeing Is Believing: The Effect of Brain Images on Judgments of Scientific Reasoning', *Cognition* 107:1 (2008), pp. 343–52; D. S. Weisberg, J. C. V. Taylor and E. J. Hopkins, 'Deconstructing the Seductive Allure of Neuroscience Explanations', *Judgment and Decision Making*, 10:5 (2015), p. 429. • **27.** K. A. Joyce, 'From Numbers to Pictures: The Development of Magnetic Resonance Imaging and the Visual Turn in Medicine', *Science as Culture*, 15:01 (2006), pp. 1–22. • **28.** M. J. Farah and C. J. Hook, 'The Seductive Allure of "Seductive Allure"', *Perspectives on Psychological Science* 8:1 (2013), pp. 88–90. • **29.** R. B. Michael, E. J. Newman, M. Vuorre, G. Cumming and M. Garry, 'On the (Non) Persuasive Power of a Brain Image', *Psychonomic Bulletin and Review* 20:4 (2013), pp. 720–25. • **30.** D. Blum, 'Winter of Discontent: Is the Hot Affair between Neuroscience and Science Journalism Cooling Down?', *Undark*, 3 December 2012, https://undark.org/2012/12/03/winter-discontent-hot-affair-between-neu (accessed 4 November 2018). • **31.** A. Quart, 'Neuroscience: under attack', *New York Times*, 23 November 2012, https://www.nytimes. com/2012/11/25/opinion/sunday/neuroscience-under-attack.html (accessed 4 November 2018). • **32.** S. Poole, 'Your brain on pseudoscience: the rise of popular neurobollocks', *New Statesman*, 6 September 2012, https://www. newstatesman.com/culture/books/2012/09/your-brain-pseudoscience-rise-popular-neurobollocks (accessed 4 November 2018). • **33.** E. Racine,

O. Bar-Ilan and J. Illes, 'fMRI in the Public Eye', *Nature Reviews Neuroscience* 6:2 (2005), p. 159. • **34.** 'Welcome to the Neuro-Journalism Mill', James S. McDonnell Foundation website, https://www.jsmf.org/neuromill/about. htm (accessed 4 November 2018). • **35.** E. Vul, C. Harris, P. Winkielman and H. Pashler, 'Puzzlingly High Correlations in fMRI Studies of Emotion, Personality, and Social Cognition', *Perspectives on Psychological Science* 4:3 (2009), pp. 274–90. • **36.** C. M. Bennett, M. B. Miller and G. L. Wolford, 'Neural Correlates of Interspecies Perspective Taking in the Post-Mortem Atlantic Salmon: An Argument for Multiple Comparisons Correction', *NeuroImage* 47:Supplement 1 (2009), p. S125. • **37.** A. Madrigal, 'Scanning dead salmon in fMRI machine highlights risk of red herrings', *Wired*, 18 September 2009, https://www.wired.com/2009/09/fmrisalmon (accessed 4 November 2018); Neuroskeptic, 'fMRI gets slap in the face with a dead fish', *Discover*, 16 September 2009, http://blogs.discovermagazine.com/ neuroskeptic/2009/09/16/fmri-gets-slap-in-the-face-with-a-dead-fish (accessed 4 November 2018). • **38.** Scicurious, 'IgNobel Prize in Neuroscience: the dead salmon study', *Scientific American*, 25 September 2012, https://blogs. scientificamerican.com/scicurious-brain/ignobel-prize-in-neuroscience-the- dead-salmon-study (accessed 4 November 2018). • **39.** S. Dekker, N. C. Lee, P. Howard-Jones and J. Jolles, 'Neuromyths in Education: Prevalence and Predictors of Misconceptions among Teachers', *Frontiers in Psychology* 3 (2012), p. 429. • **40.** Human Brain Project website, https://www.human- brainproject.eu/en/; H. Markram, 'The human brain project', *Scientific American*, June 2012, pp. 50–55. • **41.** UK Biobank website, https://www. ukbiobank.ac.uk (accessed 4 November 2018); C. Sudlow, J. Gallacher, N. Allen, V. Beral, P. Burton, J. Danesh, P. Downey, P. Elliott, J. Green, M. Landray and B. Liu, 'UK Biobank: An Open Access Resource for Identifying the Causes of a Wide Range of Complex Diseases of Middle and Old Age', *PLoS Medicine* 12:3 (2015), e1001779. • **42.** BRAIN Initiative website, https://www.braininitiative.nih.gov (accessed 4 November 2018); T. R. Insel, S. C. Landis and F. S. Collins, 'The NIH Brain Initiative', *Science* 340:6133 (2013), pp. 687–8. • **43.** Human Connectome Project website, http://www.humanconnectomeproject.org (accessed 4 November 2018); D. C. Van Essen, S. M. Smith, D. M. Barch, T. E. Behrens, E. Yacoub, K. Ugurbil and WU-Minn HCP Consortium, 'The WU-Minn Human Connectome Project: An Overview', *NeuroImage* 80 (2013), pp. 62–79. • **44.** R. A. Poldrack and K. J. Gorgolewski, 'Making Big Data Open: Data Sharing in Neuroimaging', *Nature Neuroscience* 17:11 (2014), p. 1510. • **45.** J. Gray, *Men Are from Mars, Women Are from Venus* (New York, HarperCollins, 1992). • **46.** L. Brizendine, *The Female Brain* (New York: Morgan Road, 2006). • **47.** Young and Balaban, 'Psychoneuroindoctrinology', p. 634. • **48.** 'Sex-linked lexical budgets', Language Log, 6 August 2006, http://

itre.cis.upenn.edu/~myl/languagelog/archives/003420.html (accessed 4 November 2018). • 49. 'Neuroscience in the service of sexual stereotypes', Language Log, 6 August 2006, http://itre.cis.upenn.edu/~myl/languagelog/archives/003419.html (accessed 4 November 2018). • 50. Fine, *Delusions of Gender*, p. 161. • 51. M. Liberman, 'The Female Brain movie', Language Log, 21 August 2016, http://languagelog.ldc.upenn.edu/nll/?p=27641 (accessed 4 November 2018). • 52. V. Brescoll and M. LaFrance, 'The Correlates and Consequences of Newspaper Reports of Research on Sex Differences', *Psychological Science* 15:8 (2004), pp. 515–20. • 53. Fine, *Delusions of Gender*, pp. 154–75; C. Fine, 'Is There Neurosexism in Functional Neuroimaging Investigations of Sex Differences?', *Neuroethics* 6:2 (2013), pp. 369–409. • 54. R. Bluhm, 'New Research, Old Problems: Methodological and Ethical Issues in fMRI Research Examining Sex/Gender Differences in Emotion Processing', *Neuroethics*, 6:2 (2013), pp. 319–30. • 55. K. McRae, K. N. Ochsner, I. B. Mauss, J. J Gabrieli and J. J. Gross, 'Gender Differences in Emotion Regulation: An fMRI Study of Cognitive Reappraisal', *Group Processes and Intergroup Relations*, 11:2 (2008), pp. 143–62; R. Bluhm, 'Self-Fulfilling Prophecies: The Influence of Gender Stereotypes on Functional Neuroimaging Research on Emotion', *Hypatia* 28:4 (2013), pp. 870–86. • 56. B. A. Shaywitz, S. E. Shaywitz, K. R. Pugh, R. T. Constable, P. Skudlarski, R. K. Fulbright, R. A. Bronen, J. M. Fletcher, D. P. Shankweiler, L. Katz and J. C. Gore, 'Sex Differences in the Functional Organization of the Brain for Language', *Nature* 373:6515 (1995), p. 607. • 57. G. Kolata, 'Men and women use brain differently, study discovers', *New York Times*, 16 February 1995, https://www.nytimes.com/1995/02/16/us/men-and-women-use-brain-differently-study-discovers.html (accessed 4 November 2018). • 58. Fine, 'Is There Neurosexism'. • 59. Ibid., p. 379. • 60. I. E. C. Sommer, A. Aleman, A. Bouma and R. S. Kahn, 'Do Women Really Have More Bilateral Language Representation than Men? A Meta-analysis of Functional Imaging Studies', *Brain* 127:8 (2004), pp. 1845–52. • 61. M. Wallentin, 'Putative Sex Differences in Verbal Abilities and Language Cortex: A Critical Review', *Brain and Language* 108:3 (2009), pp. 175–83. • 62. M. Ingalhalikar, A. Smith, D. Parker, T. D. Satterthwaite, M. A. Elliott, K. Ruparel, H. Hakonarson, R. E. Gur, R. C. Gur and R. Verma, 'Sex Differences in the Structural Connectome of the Human Brain', *Proceedings of the National Academy of Sciences* 111:2 (2014), pp. 823–8. • 63. Ibid, p. 823, abstract. • 64. 'Brain connectivity study reveals striking differences between men and women', Penn Medicine press release, 2 December 2013, https://www.pennmedicine.org/news/news-releases/2013/december/brain-connectivity-study-revea (accessed 4 November 2018). • 65. D. Joel and R. Tarrasch, 'On the Mis-presentation and Misinterpretation of Gender-Related Data: The Case of Ingalhalikar's

Human Connectome Study', *Proceedings of the National Academy of Sciences* 111:6 (2014), p. E637; M. Ingalhalikar, A. Smith, D. Parker, T. D. Satterthwaite, M. A. Elliott, K. Ruparel, H. Hakonarson, R. E. Gur, R. C. Gur and R. Verma, 'Reply to Joel and Tarrasch: On Misreading and Shooting the Messenger', *Proceedings of the National Academy of Sciences* 111:6 (2014), 201323601; 'Expert reaction to study on gender differences in brains', Science Media Centre, 3 December 2013, http://www.sciencemediacentre.org/expert-reaction-to-study-on-gender-differences-in-brains (accessed 4 November 2018); Neuroskeptic, 'Men, women and big PNAS papers', *Discover*, 3 December 2013, http://blogs.discovermagazine.com/neuroskeptic/2013/12/03/men-women-big-pnas-papers/#.W69vxltyKpo (accessed 4 November 2018); 'Men are map readers and women are intuitive, but bloggers are fast', The Neurocritic blog, 5 December 2013, https://neurocritic.blogspot.com/2013/12/men-are-map-readers-and-women-are.html (accessed 4 November 2018); https://blogs.biomedcentral.com/on-biology/2013/12/12/lets-talk-about-sex/ • **66.** G. Ridgway, 'Illustrative effect sizes for sex differences', Figshare, 3 December 2013, https://figshare.com/articles/Illustrative_effect_sizes_for_sex_differences/866802 (accessed 4 November 2018). • **67.** S. Connor, 'The hardwired difference between male and female brains could explain why men are "better at map reading"', *Independent*, 3 December 2013, https://www.independent.co.uk/life-style/the-hardwired-difference-between-male-and-female-brains-could-explain-why-men-are-better-at-map-8978248.html (accessed 4 November 2018); J. Naish, 'Men's and women's brains: the truth!', *Mail Online*, 5 December 2013, https://www.dailymail.co.uk/femail/article-2518327/Mens-womens-brains-truth-As-research-proves-sexes-brains-ARE-wired-differently-womens-cleverer-ounce-ounce--men-read-female-feelings.html (accessed 4 November 2018). • **68.** C. O'Connor and H. Joffe, 'Gender on the Brain: A Case Study of Science Communication in the New Media Environment', *PLoS One* 9:10 (2014), e110830.

CHAPTER 5: THE TWENTY-FIRST-CENTURY BRAIN

1. K. J. Friston, 'The Fantastic Organ', *Brain* 136:4 (2013), pp. 1328–32. • **2.** N. K. Logothetis, 'The Ins and Outs of fMRI Signals', *Nature Neuroscience* 10:10 (2007), p. 1230. • **3.** K. J. Friston, 'Functional and Effective Connectivity: A Review', *Brain Connectivity* 1:1 (2011), pp. 13–36. • **4.** Y. Assaf and O. Pasternak, 'Diffusion Tensor Imaging (DTI)-Based White Matter Mapping in Brain Research: A Review', *Journal of Molecular Neuroscience* 34:1 (2008), pp. 51–61. • **5.** A. Holtmaat and K. Svoboda, 'Experience-Dependent Structural Synaptic Plasticity in the Mammalian Brain', *Nature Reviews Neuroscience* 10:9 (2009), p. 647. • **6.** A. Razi and K. J. Friston, 'The Connected

Brain: Causality, Models, and Intrinsic Dynamics', *IEEE Signal Processing Magazine* 33:3 (2016), pp. 14–35. • **7.** A. von Stein and J. Sarnthein, 'Different Frequencies for Different Scales of Cortical Integration: From Local Gamma to Long Range Alpha/Theta Synchronization', *International Journal of Psychophysiology* 38:3 (2000), pp. 301–13. • **8.** S. Baillet, 'Magnetoencephalography for Brain Electrophysiology and Imaging', *Nature Neuroscience* 20:3 (2017), p. 327. • **9.** W. D. Penny, S. J. Kiebel, J. M. Kilner and M. D. Rugg, 'Event-Related Brain Dynamics', *Trends in Neurosciences* 25:8 (2002), pp. 387–9. • **10.** K. Kessler, R. A. Seymour and G. Rippon, 'Brain Oscillations and Connectivity in Autism Spectrum Disorders (ASD): New Approaches to Methodology, Measurement and Modelling', *Neuroscience and Biobehavioral Reviews* 71 (2016), pp. 601–20. • **11.** S. E. Fisher, 'Translating the Genome in Human Neuroscience', in G. Marcus and J. Freeman (eds), *The Future of the Brain: Essays by the World's Leading Neuroscientists* (Princeton, NJ, Princeton University Press, 2015), pp. 149–58. • **12.** S. R. Chamberlain, U. Müller, A. D. Blackwell, L. Clark, T. W. Robbins and B. J. Sahakian, 'Neurochemical Modulation of Response Inhibition and Probabilistic Learning in Humans', *Science* 311:5762 (2006), pp. 861–3. • **13.** C. Eliasmith, 'Building a Behaving Brain', in Marcus and Freeman (eds), *The Future of the Brain*, pp. 125–36. • **14.** A. Zador, 'The Connectome as a DNA Sequencing Problem', in Marcus and Freeman (eds), *The Future of the Brain*, 2015), pp. 40–49, at p. 46. • **15.** J. W. Lichtman, J. Livet and J. R. Sanes, 'A Technicolour Approach to the Connectome', *Nature Reviews Neuroscience* 9:6 (2008), p. 417. • **16.** G. Bush, P. Luu and M. I. Posner, 'Cognitive and Emotional Influences in Anterior Cingulate Cortex', *Trends in Cognitive Sciences* 4:6 (2000), pp. 215–22. • **17.** M. Alper, 'The "God" Part of the Brain: A Scientific Interpretation of Human Spirituality and God' (Naperville, IL, Sourcebooks, 2008). • **18.** J. H. Barkow, L. Cosmides and J. Tooby (eds), *The Adapted Mind: Evolutionary Psychology and the Generation of Culture* (New York, Oxford University Press, 1992). • **19.** Penny et al., 'Event-Related Brain Dynamics'. • **20.** G. Shen, T. Horikawa, K. Majima and Y. Kamitani, 'Deep Image Reconstruction from Human Brain Activity', *bioRxiv* (2017), 240317. • **21.** R. A. Thompson and C. A. Nelson, 'Developmental Science and the Media: Early Brain Development', *American Psychologist* 56:1 (2001), pp. 5–15. • **22.** Thompson and Nelson, 'Developmental Science and the Media', p. 5. • **23.** A. May, 'Experience-Dependent Structural Plasticity in the Adult Human Brain', *Trends in Cognitive Sciences* 15:10 (2011), pp. 475–82. • **24.** Y. Chang, 'Reorganization and Plastic Changes of the Human Brain Associated with Skill Learning and Expertise', *Frontiers in Human Neuroscience* 8 (2014), art. 35. • **25.** B. Draganski and A. May, 'Training-Induced Structural Changes in the Adult Human Brain', *Behavioural Brain Research* 192:1 (2008), pp. 137–42. • **26.**

E. A. Maguire, D. G. Gadian, I. S. Johnsrude, C. D. Good, J. Ashburner, R. S. Frackowiak and C. D. Frith, 'Navigation-Related Structural Change in the Hippocampi of Taxi Drivers', *Proceedings of the National Academy of Sciences* 97:8 (2000), pp. 4398–403; K. Woollett, H. J. Spiers and E. A. Maguire, 'Talent in the Taxi: A Model System for Exploring Expertise', *Philosophical Transactions of the Royal Society B: Biological Sciences* 364:1522 (2009), pp. 1407–16. • **27.** M. S. Terlecki and N. S. Newcombe, 'How Important Is the Digital Divide? The Relation of Computer and Videogame Usage to Gender Differences in Mental Rotation Ability', *Sex Roles* 53:5–6 (2005), pp. 433–41. • **28.** R. J. Haier, S. Karama, L. Leyba and R. E. Jung, 'MRI Assessment of Cortical Thickness and Functional Activity Changes in Adolescent Girls Following Three Months of Practice on a Visual-Spatial Task', *BMC Research Notes* 2:1 (2009), p. 174. • **29.** S. Kühn, T. Gleich, R. C. Lorenz, U. Lindenberger and J. Gallinat, 'Playing Super Mario Induces Structural Brain Plasticity: Gray Matter Changes Resulting from Training with a Commercial Video Game', *Molecular Psychiatry* 19:2 (2014), p. 265. • **30.** N. Jaušovec and K. Jaušovec, 'Sex Differences in Mental Rotation and Cortical Activation Patterns: Can Training Change Them?', *Intelligence* 40:2 (2012), pp. 151–62. • **31.** A. Clark, 'Whatever Next? Predictive Brains, Situated Agents, and the Future of Cognitive Science', *Behavioral and Brain Sciences* 36:3 (2013), pp. 181–204; E. Pellicano and D. Burr, 'When the World Becomes "Too Real": A Bayesian Explanation of Autistic Perception', *Trends in Cognitive Sciences* 16:10 (2012), pp. 504–10. • **32.** D. I. Tamir and M. A. Thornton, 'Modeling the Predictive Social Mind', *Trends in Cognitive Sciences* 22:3 (2018), pp. 201–12. • **33.** A. Clark, *Surfing Uncertainty: Prediction, Action, and the Embodied Mind* (New York, Oxford University Press, 2015); Clark, 'Whatever Next?'; D. D. Hutto, 'Getting into Predictive Processing's Great Guessing Game: Bootstrap Heaven or Hell?', *Synthese* 195:6 (2018), pp. 2445–8. • **34.** The Invisible Gorilla, http://www.theinvisiblegorilla.com/videos.html (accessed 4 November 2018). • **35.** L. F. Barrett and J. Wormwood, 'When a gun is not a gun', *New York Times*, 17 April 2015, https://www.nytimes.com/2015/04/19/opinion/sunday/when-a-gun-is-not-a-gun.html (accessed 4 November 2018). • **36.** Kessler et al., 'Brain Oscillations and Connectivity in Autism Spectrum Disorders (ASD)'. • **37.** E. Hunt, 'Tay, Microsoft's AI chatbot, gets a crash course in racism from Twitter', *Guardian*, 24 March 2016, https://www.theguardian.com/technology/2016/mar/24/tay-microsofts-ai-chatbot-gets-a-crash-course-in-racism-from-twitter (accessed 4 November 2018); I. Johnston, 'AI robots learning racism, sexism and other prejudices from humans, study finds', *Independent*, 13 April 2017, https://www.independent.co.uk/life-style/gadgets-and-tech/news/ai-robots-artificial-intelligence-racism-sexism-prejudice-bias-language-learn-from-humans-a7683161.html (accessed 4 November 2018). • **38.** Y. LeCun, Y. Bengio and G. Hinton,

'Deep Learning', *Nature* 521:7553 (2015), p. 436; R. D. Hof, 'Deep learning', *MIT Technology Review*, https://www.technologyreview.com/s/513696/deep-learning (accessed 4 November 2018). • **39.** T. Simonite, 'Machines taught by photos learn a sexist view of women', *Wired*, 21 August 2017, https://www.wired.com/story/machines-taught-by-photos-learn-a-sexist-view-of-women (accessed 4 November 2018). • **40.** J. Zhao, T. Wang, M. Yatskar, V. Ordonez and K. W. Chang, 'Men Also Like Shopping: Reducing Gender Bias Amplification Using Corpus-Level Constraints', *arXiv*:1707.09457, 29 July 2017. • **41.** R. I. Dunbar, 'The Social Brain Hypothesis', *Evolutionary Anthropology: Issues, News, and Reviews* 6:5 (1998), pp. 178–90. • **42.** U. Frith and C. Frith, 'The Social Brain: Allowing Humans to Boldly Go Where No Other Species Has Been', *Philosophical Transactions of the Royal Society B: Biological Sciences* 365:1537 (2010), pp. 165–76.

CHAPTER 6: YOUR SOCIAL BRAIN

1. M. D. Lieberman, *Social: Why Our Brains Are Wired to Connect* (Oxford, Oxford University Press, 2013). • **2.** R. Adolphs, 'Investigating the Cognitive Neuroscience of Human Social Behavior', *Neuropsychologia* 41:2 (2003), pp. 119–26; D. M. Amodio, E. Harman-Jones, P. G. Devine, J. J. Curtin, S. L. Hartley and A. E. Covert, 'Neural Signals for the Detection of Unintentional Race Bias', *Psychological Science* 15:2 (2004), pp. 88–93. • **3.** D. I. Tamir and M. A. Thornton, 'Modeling the Predictive Social Mind', *Trends in Cognitive Sciences* 22:3 (2018), pp. 201–12; P. Hinton, 'Implicit Stereotypes and the Predictive Brain: Cognition and Culture in "Biased" Person Perception', *Palgrave Communications* 3 (2017), 17086. • **4.** Frith and Frith, 'The Social Brain: Allowing Humans to Boldly Go Where No Other Species Has Been', pp. 165–76. • **5.** P. Adjamian, A. Hadjipapas, G. R. Barnes, A. Hillebrand and I. E. Holliday, 'Induced Gamma Activity in Primary Visual Cortex Is Related to Luminance and Not Color Contrast: An MEG Study', *Journal of Vision* 8:7 (2008), art. 4. • **6.** M. V. Lombardo, J. L. Barnes, S. J. Wheelwright and S. Baron-Cohen, 'Self-Referential Cognition and Empathy in Autism', *PLoS One* 2:9 (2007), e883. • **7.** T. Singer, 'The Neuronal Basis and Ontogeny of Empathy and Mind Reading: Review of Literature and Implications for Future Research', *Neuroscience and Biobehavioral Reviews* 30:6 (2006), pp. 855–63. • **8.** C. D. Frith, 'The Social Brain?', *Philosophical Transactions of the Royal Society B: Biological Sciences*, 362:1480 (2007), pp. 671–8. • **9.** R. Adolphs, D. Tranel and A. R. Damasio, 'The Human Amygdala in Social Judgment', *Nature* 393:6684 (1998), p. 470. • **10.** A. J. Hart, P. J. Whalen, L. M. Shin, S. C. McInerney, H. Fischer and S. L. Rauch, 'Differential Response in the Human Amygdala to Racial Outgroup vs Ingroup Face Stimuli', *Neuroreport* 11:11 (2000), pp. 2351–4. • **11.**

D. M. Amodio and C. D. Frith, 'Meeting of Minds: The Medial Frontal Cortex and Social Cognition', *Nature Reviews Neuroscience* 7:4 (2006), p. 268. • **12.** Ibid. • **13.** Ibid. • **14.** S. J. Gillihan and M. J. Farah, 'Is Self Special? A Critical Review of Evidence from Experimental Psychology and Cognitive Neuroscience', *Psychological Bulletin* 131:1 (2005), p. 76. • **15.** D. A. Gusnard, E. Akbudak, G. L. Shulman and M. E. Raichle, 'Medial Prefrontal Cortex and Self-Referential Mental Activity: Relation to a Default Mode of Brain Function', *Proceedings of the National Academy of Sciences* 98:7 (2001), pp. 4259–64; R. B. Mars, F. X. Neubert, M. P. Noonan, J. Sallet, I. Toni and M. F. Rushworth, 'On the Relationship between the "Default Mode Network" and the "Social Brain"', *Frontiers in Human Neuroscience* 6 (2012), p. 189. • **16.** N. I. Eisenberger, M. D. Lieberman and K. D. Williams, 'Does Rejection Hurt? An fMRI Study of Social Exclusion', *Science* 302:5643 (2003), pp. 290–92. • **17.** N. I. Eisenberger, T. K. Inagaki, K. A. Muscatell, K. E. Byrne Haltom and M. R. Leary, 'The Neural Sociometer: Brain Mechanisms Underlying State Self-Esteem', *Journal of Cognitive Neuroscience* 23:11 (2011), pp. 3448–55. • **18.** L. H. Somerville, T. F. Heatherton and W. M. Kelley, 'Anterior Cingulate Cortex Responds Differentially to Expectancy Violation and Social Rejection', *Nature Neuroscience* 9:8 (2006), p. 1007. • **19.** T. Dalgleish, N. D. Walsh, D. Mobbs, S. Schweizer, A-L. van Harmelen, B. Dunn, V. Dunn, I. Goodyer and J. Stretton, 'Social Pain and Social Gain in the Adolescent Brain: A Common Neural Circuitry Underlying Both Positive and Negative Social Evaluation', *Scientific Reports* 7 (2017), 42010. • **20.** N. I. Eisenberger and M. D. Lieberman, 'Why Rejection Hurts: A Common Neural Alarm System for Physical and Social Pain', *Trends in Cognitive Sciences* 8:7 (2004), pp. 294–300. • **21.** M. R. Leary, E. S. Tambor, S. K. Terdal and D. L. Downs, 'Self-Esteem as an Interpersonal Monitor: The Sociometer Hypothesis', *Journal of Personality and Social Psychology* 68:3 (1995), p. 518. • **22.** M. M. Botvinick, J. D. Cohen and C. S. Carter, 'Conflict Monitoring and Anterior Cingulate Cortex: An Update', *Trends in Cognitive Sciences* 8:12 (2004), pp. 539–46. • **23.** Botvinick et al., 'Conflict Monitoring and Anterior Cingulate Cortex'. • **24.** A. D. Craig, 'How Do You Feel – Now? The Anterior Insula and Human Awareness', *Nature Reviews Neuroscience*, 10:1 (2009), pp. 59–70. • **25.** Ibid. • **26.** Eisenberger et al., 'The Neural Sociometer'. • **27.** K. Onoda, Y. Okamoto, K. I. Nakashima, H. Nittono, S. Yoshimura, S. Yamawaki, S. Yamaguchi and M. Ura, 'Does Low Self-Esteem Enhance Social Pain? The Relationship between Trait Self-Esteem and Anterior Cingulate Cortex Activation Induced by Ostracism', *Social Cognitive and Affective Neuroscience* 5:4 (2010), pp. 385–91. • **28.** J. P. Bhanji and M. R. Delgado, 'The Social Brain and Reward: Social Information Processing in the Human Striatum', *Wiley Interdisciplinary Reviews: Cognitive Science* 5:1 (2014), pp. 61–73. • **29.** S. Bray and J. O'Doherty, 'Neural Coding of Reward-Prediction Error Signals during Classical Conditioning with Attractive Faces', *Journal of Neurophysiology* 97:4 (2007), pp. 3036–45. • **30.** D. A. Hackman and M. J. Farah, 'Socioeconomic Status and the Developing Brain', *Trends in Cognitive Sciences* 13:2 (2009), pp. 65–73. • **31.** P. J. Gianaros, J. A. Horenstein, S. Cohen, K. A. Matthews, S. M. Brown, J. D. Flory, H. D. Critchley, S. B. Manuck and A. R. Hariri, 'Perigenual Anterior Cingulate Morphology Covaries with Perceived Social Standing', *Social Cognitive and Affective Neuroscience* 2:3 (2007), pp. 161–73. • **32.** O. Longe, F. A. Maratos, P. Gilbert, G. Evans, F. Volker, H. Rockliff and G. Rippon, 'Having a Word with Yourself: Neural Correlates of Self-Criticism and Self-Reassurance', *NeuroImage* 49:2 (2010), pp. 1849–56. • **33.** B. T. Denny, H. Kober, T. D. Wager and K. N. Ochsner, 'A Meta-analysis of Functional Neuroimaging Studies of Self- and Other Judgments Reveals a Spatial Gradient for Mentalizing in Medial Prefrontal Cortex', *Journal of Cognitive Neuroscience* 24:8 (2012), pp. 1742–52. • **34.** H. Tajfel, 'Social Psychology of Intergroup Relations', *Annual Review of Psychology* 33 (1982), pp. 1–39. • **35.** P. Molenberghs, 'The Neuroscience of In-Group Bias', *Neuroscience and Biobehavioral Reviews* 37:8 (2013), pp. 1530–36. • **36.** J. K. Rilling, J. E. Dagenais, D. R. Goldsmith, A. L. Glenn and G. Pagnoni, 'Social Cognitive Neural Networks during In-Group and Out-Group Interactions', *NeuroImage* 41:4 (2008), pp. 1447–61. • **37.** C. Frith and U. Frith, 'Theory of Mind', *Current Biology* 15:17 (2005), pp. R644–5; D. Premack and G. Woodruff, 'Does the Chimpanzee Have a Theory of Mind?', *Behavioral and Brain Sciences* 1:4 (1978), pp. 515–26. • **38.** Amodio and Frith, 'Meeting of Minds'. • **39.** V. Gallese and A. Goldman, 'Mirror Neurons and the Simulation Theory of Mind-Reading', *Trends in Cognitive Sciences* 2:12 (1998), pp. 493–501. • **40.** M. Schulte-Rüther, H. J. Markowitsch, G. R. Fink and M. Piefke, 'Mirror Neuron and Theory of Mind Mechanisms Involved in Face-to-Face Interactions: A Functional Magnetic Resonance Imaging Approach to Empathy', *Journal of Cognitive Neuroscience* 19:8 (2007), pp. 1354–72. • **41.** S. G. Shamay-Tsoory, J. Aharon-Peretz and D. Perry, 'Two Systems for Empathy: A Double Dissociation between Emotional and Cognitive Empathy in Inferior Frontal Gyrus versus Ventromedial Prefrontal Lesions', *Brain* 132:3 (2009), pp. 617–27. • **42.** M. Iacoboni and J. C. Mazziotta, 'Mirror Neuron System: Basic Findings and Clinical Applications', *Annals of Neurology* 62:3 (2007), pp. 213–18; M. Iacoboni, 'Imitation, Empathy, and Mirror Neurons', *Annual Review of Psychology* 60 (2009), pp. 653–70. • **43.** J. M. Contreras, M. R. Banaji and J. P. Mitchell, 'Dissociable Neural Correlates of Stereotypes and Other Forms of Semantic Knowledge', *Social Cognitive and Affective Neuroscience* 7:7 (2011), pp. 764–70. • **44.** S. J. Spencer, C. M. Steele and D. M. Quinn, 'Stereotype Threat and Women's Math Performance', *Journal of Experimental Social Psychology* 35:1 (1999), pp. 4–28;

T. Schmader, 'Gender Identification Moderates Stereotype Threat Effects on Women's Math Performance', *Journal of Experimental Social Psychology* 38:2 (2002), pp. 194–201. • **45.** T. Schmader, M. Johns and C. Forbes, 'An Integrated Process Model of Stereotype Threat Effects on Performance', *Psychological Review* 115:2 (2008), p. 336. • **46.** M. Wraga, M. Helt, E. Jacobs and K. Sullivan, 'Neural Basis of Stereotype-Induced Shifts in Women's Mental Rotation Performance', *Social Cognitive and Affective Neuroscience* 2:1 (2007), pp. 12–19. • **47.** M. Wraga, L. Duncan, E. C. Jacobs, M. Helt and J. Church, 'Stereotype Susceptibility Narrows the Gender Gap in Imagined Self-Rotation Performance', *Psychonomic Bulletin and Review* 13:5 (2006), pp. 813–19. • **48.** Wraga et al., 'Neural Basis of Stereotype-Induced Shifts'. • **49.** H. J. Spiers, B. C. Love, M. E. Le Pelley, C. E. Gibb and R. A. Murphy, 'Anterior Temporal Lobe Tracks the Formation of Prejudice', *Journal of Cognitive Neuroscience* 29:3 (2017), pp. 530–44; R. I. Dunbar, 'The Social Brain Hypothesis', *Evolutionary Anthropology: Issues, News, and Reviews* 6:5 (1998), pp. 178–90. • **50.** Dunbar, 'The Social Brain Hypothesis'. • **51.** J. Stiles, 'Neural Plasticity and Cognitive Development', *Developmental Neuropsychology* 18:2 (2000), pp. 237–72.

CHAPTER 7: BABY MATTERS – TO BEGIN AT THE BEGINNING (OR EVEN A BIT BEFORE)

1. J. Connellan, S. Baron-Cohen, S. Wheelwright, A. Batki and J. Ahluwalia, 'Sex Differences in Human Neonatal Social Perception', *Infant Behavior and Development* 23:1 (2000), pp. 113–18. • **2.** Y. Minagawa-Kawai, K. Mori, J. C. Hebden and E. Dupoux, 'Optical Imaging of Infants' Neurocognitive Development: Recent Advances and Perspectives', *Developmental Neurobiology* 68:6 (2008), pp. 712–28. • **3.** C. Clouchoux, N. Guizard, A. C. Evans, A. J. du Plessis and C. Limperopoulos, 'Normative Fetal Brain Growth by Quantitative In Vivo Magnetic Resonance Imaging', *American Journal of Obstetrics and Gynecology* 206:2 (2012), pp. 173.e1–8. • **4.** J. Dubois, G. Dehaene-Lambertz, S. Kulikova, C. Poupon, P. S. Hüppi and L. Hertz-Pannier, 'The Early Development of Brain White Matter: A Review of Imaging Studies in Fetuses, Newborns and Infants', *Neuroscience* 276 (2014), pp. 48–71. • **5.** M. I. van den Heuvel and M. E. Thomason, 'Functional Connectivity of the Human Brain In Utero', *Trends in Cognitive Sciences* 20:12 (2016), pp. 931–9. • **6.** J. Dubois, M. Benders, C. Borradori-Tolsa, A. Cachia, F. Lazeyras, R. Ha-Vinh Leuchter, S. V. Sizonenko, S. K. Warfield, J. F. Mangin and P. S. Hüppi, 'Primary Cortical Folding in the Human Newborn: An Early Marker of Later Functional Development', *Brain* 131:8 (2008), pp. 2028–41. • **7.** D. Holland, L. Chang, T. M. Ernst, M. Curran, S. D. Buchthal, D. Alicata, J. Skranes, H. Johansen, A. Hernandez,

R. Yamakawa and J. M. Kuperman, 'Structural Growth Trajectories and Rates of Change in the First 3 Months of Infant Brain Development', *JAMA Neurology* 71:10 (2014), pp. 1266–74. • **8.** G. M. Innocenti and D. J. Price, 'Exuberance in the Development of Cortical Networks', *Nature Reviews Neuroscience* 6:12 (2005), p. 955. • **9.** Holland et al., 'Structural Growth Trajectories'. • **10.** J. Stiles and T. L. Jernigan, 'The Basics of Brain Development', *Neuropsychology Review* 20:4 (2010), pp. 327–48. • **11.** S. Jessberger and F. H. Gage, 'Adult Neurogenesis: Bridging the Gap between Mice and Humans', *Trends in Cell Biology* 24:10 (2014), pp. 558–63. • **12.** W. Gao, S. Alcauter, J. K. Smith, J. H. Gilmore and W. Lin, 'Development of Human Brain Cortical Network Architecture during Infancy', *Brain Structure and Function* 220:2 (2015), pp. 1173–86. • **13.** Dubois et al., 'The Early Development of Brain White Matter'. • **14.** B. J. Casey, N. Tottenham, C. Liston and S. Durston, 'Imaging the Developing Brain: What Have We Learned about Cognitive Development?', *Trends in Cognitive Sciences* 9:3 (2005), pp. 104–10. • **15.** Holland et al., 'Structural Growth Trajectories'. • **16.** J. H. Gilmore, W. Lin, M. W. Prastawa, C. B. Looney, Y. S. K. Vetsa, R. C. Knickmeyer, D. D. Evans, J. K. Smith, R. M. Hamer, J. A. Lieberman and G. Gerig, 'Regional Gray Matter Growth, Sexual Dimorphism, and Cerebral Asymmetry in the Neonatal Brain', *Journal of Neuroscience* 27:6 (2007), pp. 1255–60. • **17.** R. C. Knickmeyer, J. Wang, H. Zhu, X. Geng, S. Woolson, R. M. Hamer, T. Konneker, M. Styner and J. H. Gilmore, 'Impact of Sex and Gonadal Steroids on Neonatal Brain Structure', *Cerebral Cortex* 24:10 (2013), pp. 2721–31. • **18.** R. K. Lenroot and J. N. Giedd, 'Brain Development in Children and Adolescents: Insights from Anatomical Magnetic Resonance Imaging', *Neuroscience and Biobehavioral Reviews* 30:6 (2006), pp. 718–29. • **19.** D. F. Halpern, L. Eliot, R. S. Bigler, R. A. Fabes, L. D. Hanish, J. Hyde, L. S. Liben and C. L. Martin, 'The Pseudoscience of Single-Sex Schooling', *Science* 333:6050 (2011), pp. 1706–7. • **20.** G. Dehaene-Lambertz and E. S. Spelke, 'The Infancy of the Human Brain', *Neuron* 88:1 (2015), pp. 93–109. • **21.** Gilmore et al., 'Regional Gray Matter Growth'. • **22.** G. Li, J. Nie, L. Wang, F. Shi, A. E. Lyall, W. Lin, J. H. Gilmore and D. Shen, 'Mapping Longitudinal Hemispheric Structural Asymmetries of the Human Cerebral Cortex from Birth to 2 Years of Age', *Cerebral Cortex* 24:5 (2013), pp. 1289–300. • **23.** Ibid., p. 1298. • **24.** N. Geschwind and A. M. Galaburda, 'Cerebral Lateralization: Biological Mechanisms, Associations, and Pathology – I. A Hypothesis and a Program for Research', *Archives of Neurology* 42:5 (1985), pp. 428–59. • **25.** Knickmeyer et al., 'Impact of Sex and Gonadal Steroids', p. 2721. • **26.** Van den Heuvel and Thomason, 'Functional Connectivity of the Human Brain In Utero'. • **27.** Gao et al., 'Development of Human Brain Cortical Network Architecture'. • **28.** H. T. Chugani,

M. E. Behen, O. Muzik, C. Juhász, F. Nagy and D. C. Chugani, 'Local Brain Functional Activity Following Early Deprivation: A Study of Postinstitutionalized Romanian Orphans', *NeuroImage* 14:6 (2001), pp. 1290–301. • **29.** C. H. Zeanah, C. A. Nelson, N. A. Fox, A. T. Smyke, P. Marshall, S. W. Parker and S. Koga, 'Designing Research to Study the Effects of Institutionalization on Brain and Behavioral Development: The Bucharest Early Intervention Project', *Development and Psychopathology* 15:4 (2003), pp. 885–907. • **30.** K. Chisholm, M. C. Carter, E. W. Ames and S. J. Morison, 'Attachment Security and Indiscriminately Friendly Behavior in Children Adopted from Romanian Orphanages', *Development and Psychopathology* 7:2 (1995), pp. 283–94. • **31.** Chugani et al., 'Local Brain Functional Activity'; T. J. Eluvathingal, H. T Chugani, M. E. Behen, C. Juhász, O. Muzik, M. Maqbool, D. C. Chugani and M. Makki, 'Abnormal Brain Connectivity in Children after Early Severe Socioemotional Deprivation: A Diffusion Tensor Imaging Study', *Pediatrics* 117:6 (2006), pp. 2093–100. • **32.** M. A. Sheridan, N. A. Fox, C. H. Zeanah, K. A. McLaughlin and C. A. Nelson, 'Variation in Neural Development as a Result of Exposure to Institutionalization Early in Childhood', *Proceedings of the National Academy of Sciences* 109:32 (2012), pp. 12927–32. • **33.** N. Tottenham, T. A. Hare, B. T. Quinn, T. W. McCarry, M. Nurse, T. Gilhooly, A. Millner, A. Galvan, M. C. Davidson, I. M. Eigsti, K. M. Thomas, P. J. Freed, E. S. Booma, M. R. Gunnar, M. Altemus, J. Aronson and B. J. Casey, 'Prolonged Institutional Rearing Is Associated with Atypically Large Amygdala Volume and Difficulties in Emotion Regulation', *Developmental Science* 13:1 (2010), pp. 46–61. • **34.** N. D. Walsh, T. Dalgleish, M. V. Lombardo, V. J. Dunn, A. L. Van Harmelen, M. Ban and I. M. Goodyer, 'General and Specific Effects of Early-Life Psychosocial Adversities on Adolescent Grey Matter Volume', *NeuroImage: Clinical* 4 (2014), pp. 308–18; P. Tomalski and M. H. Johnson, 'The Effects of Early Adversity on the Adult and Developing Brain', *Current Opinion in Psychiatry* 23:3 (2010), pp. 233–8. • **35.** M. H. Johnson and M. de Haan, *Developmental Cognitive Neuroscience: An Introduction*, 4th edn (Chichester, Wiley-Blackwell, 2015). • **36.** G. A. Ferrari, Y. Nicolini, E. Demuru, C. Tosato, M. Hussain, E. Scesa, L. Romei, M. Boerci, E. Iappini, G. Dalla Rosa Prati and E. Palagi, 'Ultrasonographic Investigation of Human Fetus Responses to Maternal Communicative and Non-communicative Stimuli', *Frontiers in Psychology* 7 (2016), p. 354. • **37.** M. Huotilainen, A. Kujala, M. Hotakainen, A. Shestakova, E. Kushnerenko, L. Parkkonen, V. Fellman and R. Näätänen, 'Auditory Magnetic Responses of Healthy Newborns', *Neuroreport* 14:14 (2003), pp. 1871–5. • **38.** A. R. Webb, H. T. Heller, C. B. Benson and A. Lahav, 'Mother's Voice and Heartbeat Sounds Elicit Auditory Plasticity in the Human Brain before Full Gestation', *Proceedings of the National Academy of Sciences* 112:10 (2015), 201414924. • **39.** A. J. DeCasper and W. P. Fifer, 'Of Human

Bonding: Newborns Prefer Their Mothers' Voices', *Science* 208:4448 (1980), pp. 1174–6. • **40.** M. Mahmoudzadeh, F. Wallois, G. Kongolo, S. Goudjil and G. Dehaene-Lambertz, 'Functional Maps at the Onset of Auditory Inputs in Very Early Preterm Human Neonates', *Cerebral Cortex* 27:4 (2017), pp. 2500–12. • **41.** P. Vannasing, O. Florea, B. González-Frankenberger, J. Tremblay, N. Paquette, D. Safi, F. Wallois, F. Lepore, R. Béland, M. Lassonde and A. Gallagher, 'Distinct Hemispheric Specializations for Native and Non-native Languages in One-Day-Old Newborns Identified by fNIRS', *Neuropsychologia* 84 (2016), pp. 63–9. • **42.** Y. Cheng, S. Y. Lee, H. Y. Chen, P. Y. Wang and J. Decety, 'Voice and Emotion Processing in the Human Neonatal Brain', *Journal of Cognitive Neuroscience* 24:6 (2012), pp. 1411–19. • **43.** A. Schirmer and S. A. Kotz, 'Beyond the Right Hemisphere: Brain Mechanisms Mediating Vocal Emotional Processing', *Trends in Cognitive Sciences* 10:1 (2006), pp. 24–30. • **44.** E. V. Kushnerenko, B. R. Van den Bergh and I. Winkler, 'Separating Acoustic Deviance from Novelty during the First Year of Life: A Review of Event-Related Potential Evidence', *Frontiers in Psychology* 4 (2013), p. 595. • **45.** M. Rivera-Gaxiola, G. Csibra, M. H. Johnson and A. Karmiloff-Smith, 'Electrophysiological Correlates of Cross-linguistic Speech Perception in Native English Speakers', *Behavioural Brain Research* 111:1–2 (2000), pp. 13–23. • **46.** M. Rivera-Gaxiola, J. Silva-Pereyra and P. K. Kuhl, 'Brain Potentials to Native and Non-native Speech Contrasts in 7- and 11-Month-Old American Infants', *Developmental Science* 8:2 (2005), pp. 162–72. • **47.** K. R. Dobkins, R. G. Bosworth and J. P. McCleery, 'Effects of Gestational Length, Gender, Postnatal Age, and Birth Order on Visual Contrast Sensitivity in Infants', *Journal of Vision* 9:10 (2009), art. 19. • **48.** F. Thorn, J. Gwiazda, A. A. Cruz, J. A. Bauer and R. Held, 'The Development of Eye Alignment, Convergence, and Sensory Binocularity in Young Infants', *Investigative Ophthalmology and Visual Science* 35:2 (1994), pp. 544–53. • **49.** Dobkins et al., 'Effects of Gestational Length'. • **50.** T. Farroni, E. Valenza, F. Simion and C. Umiltà, 'Configural Processing at Birth: Evidence for Perceptual Organisation', *Perception* 29:3 (2000), pp. 355–72; Thorn et al., 'The Development of Eye Alignment'. • **51.** Thorn et al., 'The Development of Eye Alignment'. • **52.** R. Held, F. Thorn, J. Gwiazda and J. Bauer, 'Development of Binocularity and Its Sexual Differentiation', in F. Vital-Durand, J. Atkinson and O. J. Braddick (eds), *Infant Vision* (Oxford, Oxford University Press, 1996), pp. 265–74. • **53.** M. C. Morrone, C. D. Burr and A. Fiorentini, 'Development of Contrast Sensitivity and Acuity of the Infant Colour System', *Proceedings of the Royal Society B: Biological Sciences* 242:1304 (1990), pp. 134–9. • **54.** T. Farroni, G. Csibra, F. Simion and M. H. Johnson, 'Eye Contact Detection in Humans from Birth', *Proceedings of the National Academy of Sciences* 99:14 (2002), pp. 9602–5. • **55.** A. Frischen, A. P. Bayliss and S. P. Tipper, 'Gaze Cueing of

Attention: Visual Attention, Social Cognition, and Individual Differences', *Psychological Bulletin* 133:4 (2007), p. 694. • **56.** S. Hoehl and T. Striano, 'Neural Processing of Eye Gaze and Threat-Related Emotional Facial Expressions in Infancy', *Child Development* 79:6 (2008), pp. 1752–60. • **57.** T. Grossmann and M. H. Johnson, 'Selective Prefrontal Cortex Responses to Joint Attention in Early Infancy', *Biology Letters* 6:4 (2010), pp. 540–43. • **58.** T. Grossmann, 'The Role of Medial Prefrontal Cortex in Early Social Cognition', *Frontiers in Human Neuroscience* 7 (2013), p. 340. • **59.** E. Nagy, 'The Newborn Infant: A Missing Stage in Developmental Psychology', *Infant and Child Development*, 20:1 (2011) pp. 3–19. • **60.** J. N. Constantino, S. Kennon-McGill, C. Weichselbaum, N. Marrus, A. Haider, A. L. Glowinski, S. Gillespie, C. Klaiman, A. Klin and W. Jones, 'Infant Viewing of Social Scenes Is under Genetic Control and Is Atypical in Autism', *Nature* 547:7663 (2017), p. 340. • **61.** J. H. Hittelman and R. Dickes, 'Sex Differences in Neonatal Eye Contact Time', *Merrill-Palmer Quarterly of Behavior and Development* 25:3 (1979), pp. 171–84. • **62.** R. T. Leeb and F. G. Rejskind, 'Here's Looking at You, Kid! A Longitudinal Study of Perceived Gender Differences in Mutual Gaze Behavior in Young Infants', *Sex Roles* 50:1–2 (2004), pp. 1–14. • **63.** S. Lutchmaya, S. Baron-Cohen and P. Raggatt, 'Foetal Testosterone and Eye Contact in 12-Month-Old Human Infants', *Infant Behavior and Development* 25:3 (2002), pp. 327–35. • **64.** A. Fausto-Sterling, D. Crews, J. Sung, C. García-Coll and R. Seifer, 'Multimodal Sex-Related Differences in Infant and in Infant-Directed Maternal Behaviors during Months Three through Twelve of Development', *Developmental Psychology* 51:10 (2015), p. 1351.

CHAPTER 8: LET'S HEAR IT FOR THE BABIES

1. D. Joel, 'Genetic-Gonadal-Genitals Sex (3G-Sex) and the Misconception of Brain and Gender, or, Why 3G-Males and 3G-Females Have Intersex Brain and Intersex Gender', *Biology of Sex Differences* 3:1 (2012), p. 27. • **2.** C. Cummings and K. Trang, 'Sex/Gender, Part I: Why Now?', *Somatosphere*, 10 March 2016, http://somatosphere.net/2016/03/sexgender-part-1-why-now.html (accessed 7 November 2018). • **3.** A. Fausto-Sterling, C. G. Coll and M. Lamarre, 'Sexing the Baby, Part 2: Applying Dynamic Systems Theory to the Emergences of Sex-Related Differences in Infants and Toddlers', *Social Science and Medicine* 74:11 (2012), pp. 1693–702. • **4.** C. Smith and B. Lloyd, 'Maternal Behavior and Perceived Sex of Infant: Revisited', *Child Development* 49:4 (1978), pp. 1263–5; E. R. Mondschein, K. E. Adolph and C. S. Tamis-LeMonda, 'Gender Bias in Mothers' Expectations about Infant Crawling', *Journal of Experimental Child Psychology* 77:4 (2000), pp. 304–16. • **5.** Holland et al., 'Structural Growth Trajectories'. • **6.** M. Pena, A. Maki, D. Kovačić,

G. Dehaene-Lambertz, H. Koizumi, F. Bouquet and J. Mehler, 'Sounds and Silence: An Optical Topography Study of Language Recognition at Birth', *Proceedings of the National Academy of Sciences* 100:20 (2003), pp. 11702–5. • **7.** P. Vannasing, O. Florea, B. González-Frankenberger, J. Tremblay, N. Paquette, D. Safi, F. Wallois, F. Lepore, R. Béland, M. Lassonde and A. Gallagher, 'Distinct Hemispheric Specializations for Native and Non-native Languages in One-Day-Old Newborns Identified by fNIRS', *Neuropsychologia* 84 (2016), pp. 63–9. • **8.** T. Nazzi, J. Bertoncini and J. Mehler, 'Language Discrimination by Newborns: Toward an Understanding of the Role of Rhythm', *Journal of Experimental Psychology: Human Perception and Performance* 24:3 (1998), p. 756. • **9.** M. H. Bornstein, C-S. Hahn and O. M. Haynes, 'Specific and General Language Performance across Early Childhood: Stability and Gender Considerations', *First Language* 24:3 (2004), pp. 267–304. • **10.** K. Johnson, M. Caskey, K. Rand, R. Tucker and B. Vohr, 'Gender Differences in Adult–Infant Communication in the First Months of Life', *Pediatrics* 134:6 (2014), pp. e1603–10. • **11.** A. D. Friederici, M. Friedrich and A. Christophe, 'Brain Responses in 4-Month-Old Infants Are Already Language Specific', *Current Biology* 17:14 (2007), pp. 1208–11. • **12.** Fausto-Sterling et al., 'Sexing the Baby, Part 2'. • **13.** V. Izard, C. Sann, E. S. Spelke and A. Streri, 'Newborn Infants Perceive Abstract Numbers', *Proceedings of the National Academy of Sciences* 106:25 (2009), pp. 10382–5. • **14.** R. Baillargeon, 'Infants' Reasoning about Hidden Objects: Evidence for Event-General and Event-Specific Expectations', *Developmental Science* 7:4 (2004), pp. 391–414. • **15.** S. J. Hespos and K. vanMarle, 'Physics for Infants: Characterizing the Origins of Knowledge about Objects, Substances, and Number', *Wiley Interdisciplinary Reviews: Cognitive Science* 3:1 (2012), pp. 19–27. • **16.** J. Connellan, S. Baron-Cohen, S. Wheelwright, A. Batki and J. Ahluwalia, 'Sex Differences in Human Neonatal Social Perception', *Infant Behavior and Development* 23:1 (2000), pp. 113–18. • **17.** A. Nash and G. Grossi, 'Picking Barbie™'s Brain: Inherent Sex Differences in Scientific Ability?', *Journal of Interdisciplinary Feminist Thought* 2:1 (2007), p. 5. • **18.** P. Escudero, R. A. Robbins and S. P. Johnson, 'Sex-Related Preferences for Real and Doll Faces versus Real and Toy Objects in Young Infants and Adults', *Journal of Experimental Child Psychology* 116:2 (2013), pp. 367–79. • **19.** D. H. Uttal, D. I. Miller and N. S. Newcombe, 'Exploring and Enhancing Spatial Thinking: Links to Achievement in Science, Technology, Engineering, and Mathematics?', *Current Directions in Psychological Science* 22:5 (2013), pp. 367–73. • **20.** D. Voyer, S. Voyer and M. P. Bryden, 'Magnitude of Sex Differences in Spatial Abilities: A Meta-analysis and Consideration of Critical Variables', *Psychological Bulletin* 117:2 (1995), p. 250. • **21.** P. C. Quinn and L. S. Liben, 'A Sex Difference in Mental Rotation in Young Infants', *Psychological Science* 19:11 (2008), pp. 1067–70. • **22.** E. S. Spelke, 'Sex Differences in Intrinsic Aptitude for Mathematics and Science? A Critical

Review', *American Psychologist* 60:9 (2005), p. 950. • **23.** I. Gauthier and N. K. Logothetis, 'Is Face Recognition Not So Unique After All?', *Cognitive Neuropsychology* 17:1–3 (2000), pp. 125–42. • **24.** M. H. Johnson, 'Subcortical Face Processing', *Nature Reviews Neuroscience* 6:10 (2005), pp. 766–74. • **25.** M. H. Johnson, A. Senju and P. Tomalski, 'The Two-Process Theory of Face Processing: Modifications Based on Two Decades of Data from Infants and Adults', *Neuroscience and Biobehavioral Reviews* 50 (2015), pp. 169–79. • **26.** F. Simion and E. Di Giorgio, 'Face Perception and Processing in Early Infancy: Inborn Predispositions and Developmental Changes', *Frontiers in Psychology* 6 (2015), p. 969. • **27.** V. M. Reid, K. Dunn, R. J. Young, J. Amu, T. Donovan and N. Reissland, 'The Human Fetus Preferentially Engages with Face-like Visual Stimuli', *Current Biology* 27:12 (2017), pp. 1825–8. • **28.** S. J. McKelvie, 'Sex Differences in Memory for Faces', *Journal of Psychology* 107:1 (1981), pp. 109–25. • **29.** C. Lewin and A. Herlitz, 'Sex Differences in Face Recognition – Women's Faces Make the Difference', *Brain and Cognition* 50:1 (2002), pp. 121–8. • **30.** A. Herlitz and J. Lovén, 'Sex Differences and the Own-Gender Bias in Face Recognition: A Meta-analytic Review', *Visual Cognition* 21:9–10 (2013), pp. 1306–36. • **31.** J. Lovén, J. Svärd, N. C. Ebner, A. Herlitz and H. Fischer, 'Face Gender Modulates Women's Brain Activity during Face Encoding', *Social Cognitive and Affective Neuroscience* 9:7 (2013), pp. 1000–1005. • **32.** Leeb and Rejskind, 'Here's Looking at You, Kid!'. • **33.** H. Hoffmann, H. Kessler, T. Eppel, S. Rukavina and H. C. Traue, 'Expression Intensity, Gender and Facial Emotion Recognition: Women Recognize Only Subtle Facial Emotions Better than Men', *Acta Psychologica* 135:3 (2010), pp. 278–83; A. E. Thompson and D. Voyer, 'Sex Differences in the Ability to Recognise Non-verbal Displays of Emotion: A Meta-analysis', *Cognition and Emotion* 28:7 (2014), pp. 1164–95. • **34.** S. Baron-Cohen, S. Wheelwright, J. Hill, Y. Raste and I. Plumb, 'The "Reading the Mind in the Eyes" Test Revised Version: A Study with Normal Adults, and Adults with Asperger Syndrome or High-Functioning Autism', *Journal of Child Psychology and Psychiatry* 42:2 (2001), pp. 241–51. • **35.** E. B. McClure, 'A Meta-analytic Review of Sex Differences in Facial Expression Processing and Their Development in Infants, Children, and Adolescents', *Psychological Bulletin* 126:3 (2000), p. 424. • **36.** Ibid. • **37.** Ibid. • **38.** Ibid. • **39.** W. D. Rosen, L. B. Adamson and R. Bakeman, 'An Experimental Investigation of Infant Social Referencing: Mothers' Messages and Gender Differences', *Developmental Psychology* 28:6 (1992), p. 1172. • **40.** A. N. Meltzoff and M. K. Moore, 'Imitation of Facial and Manual Gestures by Human Neonates', *Science* 198:4312 (1977), pp. 75–8. • **41.** A. N. Meltzoff and M. K. Moore, 'Imitation in Newborn Infants: Exploring the Range of Gestures Imitated and the Underlying Mechanisms', *Developmental Psychology* 25:6 (1989), p. 954. • **42.** P. J. Marshall and A. N. Meltzoff, 'Neural Mirroring Mechanisms and Imitation in Human

Infants', *Philosophical Transactions of the Royal Society B: Biological Sciences* 369:1644 (2014), 20130620; E. A. Simpson, L. Murray, A. Paukner and P. F. Ferrari, 'The Mirror Neuron System as Revealed through Neonatal Imitation: Presence from Birth, Predictive Power and Evidence of Plasticity', *Philosophical Transactions of the Royal Society B: Biological Sciences* 369:1644 (2014), 20130289. • **43.** E. Nagy and P. Molner, '*Homo imitans* or *Homo provocans*? Human Imprinting Model of Neonatal Imitation', *Infant Behavior and Development* 27:1 (2004), pp. 54–63. • **44.** S. S. Jones, 'Exploration or Imitation? The Effect of Music on 4-Week-Old Infants' Tongue Protrusions', *Infant Behavior and Development* 29:1 (2006), pp. 126–30. • **45.** J. Oostenbroek, T. Suddendorf, M. Nielsen, J. Redshaw, S. Kennedy-Costantini, J. Davis, S. Clark and V. Slaughter, 'Comprehensive Longitudinal Study Challenges the Existence of Neonatal Imitation in Humans', *Current Biology* 26:10 (2016), pp. 1334–8; A. N. Meltzoff, L. Murray, E. Simpson, M. Heimann, E. Nagy, J. Nadel, E. J. Pedersen, R. Brooks, D. S. Messinger, L. D. Pascalis and F. Subiaul, 'Re-examination of Oostenbroek et al. (2016): Evidence for Neonatal Imitation of Tongue Protrusion', *Developmental Science* 21:4 (2018), e12609. • **46.** Oostenbroek et al., 'Comprehensive Longitudinal Study Challenges the Existence of Neonatal Imitation in Humans'; Meltzoff et al., 'Re-examination of Oostenbroek et al. (2016)'. • **47.** Nagy and Molner, '*Homo imitans* or *Homo provocans*?'. • **48.** E. Nagy, H. Compagne, H. Orvos, A. Pal, P. Molnar, I. Janszky, K. Loveland and G. Bardos, 'Index Finger Movement Imitation by Human Neonates: Motivation, Learning, and Left-Hand Preference', *Pediatric Research* 58:4 (2005), pp. 749–53. • **49.** C. Trevarthen and K. J. Aitken, 'Infant Intersubjectivity: Research, Theory, and Clinical Applications', *Journal of Child Psychology and Psychiatry and Allied Disciplines* 42:1 (2001), pp. 3–48. • **50.** T. Farroni, G. Csibra, F. Simion and M. H. Johnson, 'Eye Contact Detection in Humans from Birth', *Proceedings of the National Academy of Sciences* 99:14 (2002), pp. 9602–5. • **51.** M. Tomasello, M. Carpenter and U. Liszkowski, 'A New Look at Infant Pointing', *Child Development* 78:3 (2007), pp. 705–22. • **52.** T. Charman, 'Why Is Joint Attention a Pivotal Skill in Autism?', *Philosophical Transactions of the Royal Society B: Biological Sciences* 358:1430 (2003), pp. 315–24. • **53.** H. L. Gallagher and C. D. Frith. 'Functional Imaging of "Theory of Mind"', *Trends in Cognitive Sciences* 7:2 (2003), pp. 77–83. • **54.** H. M. Wellman, D. Cross and J. Watson, 'Meta-analysis of Theory-of-Mind Development: The Truth about False Belief', *Child Development* 72:3 (2001), pp. 655–84. • **55.** Ibid. • **56.** 'Born good? Babies help unlock the origins of morality', CBS News/YouTube, 18 November 2012, https://youtu.be/FRvVFW85IcU (accessed 7 November 2018). • **57.** J. K. Hamlin, K. Wynn and P. Bloom, 'Social Evaluation by Preverbal Infants', *Nature* 450:7169 (2007), p. 557. • **58.** J. K. Hamlin and K. Wynn, 'Young Infants Prefer Prosocial to Antisocial Others', *Cognitive*

Development 26:1 (2011), pp. 30–39. • **59.** J. Decety and P. L. Jackson, 'The Functional Architecture of Human Empathy', *Behavioural and Cognitive Neuroscience Reviews* 3:2 (2004), pp. 71–100. • **60.** E. Geangu, O. Benga, D. Stahl and T. Striano, 'Contagious Crying beyond the First Days of Life', *Infant Behavior and Development* 33:3 (2010), pp. 279–88. • **61.** R. Roth-Hanania, M. Davidov and C. Zahn-Waxler, 'Empathy Development from 8 to 16 Months: Early Signs of Concern for Others', *Infant Behavior and Development* 34:3 (2011), pp. 447–58. • **62.** Leeb and Rejskind, 'Here's Looking at You, Kid!', p. 12. • **63.** Farroni et al., 'Eye Contact Detection in Humans from Birth'. • **64.** Ibid. • **65.** B. Auyeung, S. Wheelwright, C. Allison, M. Atkinson, N. Samarawickrema and S. Baron-Cohen, 'The Children's Empathy Quotient and Systemizing Quotient: Sex Differences in Typical Development and in Autism Spectrum Conditions', *Journal of Autism and Developmental Disorders* 39:11 (2009), p. 1509. • **66.** K. J. Michalska, K. D. Kinzler and J. Decety, 'Age-Related Sex Differences in Explicit Measures of Empathy Do Not Predict Brain Responses across Childhood and Adolescence', *Developmental Cognitive Neuroscience* 3 (2013), pp. 22–32. • **67.** Roth-Hanania et al., 'Empathy Development from 8 to 16 Months', p. 456. • **68.** Johnson, 'Subcortical Face Processing', p. 766. • **69.** D. J. Kelly, P. C. Quinn, A. M. Slater, K. Lee, L. Ge and O. Pascalis, 'The Other-Race Effect Develops during Infancy: Evidence of Perceptual Narrowing', *Psychological Science* 18:12 (2007), pp. 1084–9. • **70.** Y. Bar-Haim, T. Ziv, D. Lamy and R. M. Hodes, 'Nature and Nurture in Own-Race Face Processing', *Psychological Science* 17:2 (2006), pp. 159–63. • **71.** M. H. Johnson, 'Face Processing as a Brain Adaptation at Multiple Timescales', *Quarterly Journal of Experimental Psychology* 64:10 (2011), pp. 1873–88. • **72.** Farroni et al., 'Eye Contact Detection in Humans from Birth'; T. Farroni, M. H. Johnson and G. Csibra, 'Mechanisms of Eye Gaze Perception during Infancy', *Journal of Cognitive Neuroscience* 16:8 (2004), pp. 1320–26. • **73.** E. A. Hoffman and J. V. Haxby, 'Distinct Representations of Eye Gaze and Identity in the Distributed Human Neural System for Face Perception', *Nature Neuroscience* 3:1 (2000), p. 80. • **74.** Johnson, 'Face Processing as a Brain Adaptation'. • **75.** C. A. Nelson and M. De Haan, 'Neural Correlates of Infants' Visual Responsiveness to Facial Expressions of Emotion', *Developmental Psychobiology* 29:7 (1996), pp. 577–95; G. D. Reynolds and J. E. Richards, 'Familiarization, Attention, and Recognition Memory in Infancy: An Event-Related Potential and Cortical Source Localization Study', *Developmental Psychology* 41:4 (2005), p. 598. • **76.** T. Grossmann, T. Striano and A. D. Friederici, 'Developmental Changes in Infants' Processing of Happy and Angry Facial Expressions: A Neurobehavioral Study', *Brain and Cognition* 64:1 (2007), pp. 30–41. • **77.** T. Striano, V. M. Reid and S. Hoehl, 'Neural Mechanisms of Joint Attention in Infancy', *European Journal of Neuroscience* 23:10 (2006), pp. 2819–23. • **78.**

F. Happé and U. Frith, 'Annual Research Review: Towards a Developmental Neuroscience of Atypical Social Cognition', *Journal of Child Psychology and Psychiatry* 55:6 (2014), pp. 553–77.

CHAPTER 9: THE GENDERED WATERS IN WHICH
WE SWIM – THE PINK AND BLUE TSUNAMI

1. C. L. Martin and D. Ruble, 'Children's Search for Gender Cues: Cognitive Perspectives on Gender Development', *Current Directions in Psychological Science* 13:2 (2004), pp. 67–70. • 2. P. Rosenkrantz, S. Vogel, H. Bee, I. Broverman and D. M. Broverman, 'Sex-Role Stereotypes and Self-Concepts in College Students', *Journal of Consulting and Clinical Psychology* 32:3 (1968), p. 287. • 3. M. N. Nesbitt and N. E. Penn, 'Gender Stereotypes after Thirty Years: A Replication of Rosenkrantz, et al. (1968)', *Psychological Reports* 87:2 (2000), pp. 493–511. • 4. E. L. Haines, K. Deaux and N. Lofaro, 'The Times They Are a-Changing … Or Are They Not? A Comparison of Gender Stereotypes, 1983–2014', *Psychology of Women Quarterly* 40:3 (2016), pp. 353–63. • 5. L. A. Rudman and P. Glick, 'Prescriptive Gender Stereotypes and Backlash toward Agentic Women', *Journal of Social Issues* 57:4 (2001), pp. 743–62. • 6. C. M. Steele, *Whistling Vivaldi: And Other Clues to How Stereotypes Affect Us* (New York, W. W. Norton, 2011). • 7. C. K. Shenouda and J. H. Danovitch, 'Effects of Gender Stereotypes and Stereotype Threat on Children's Performance on a Spatial Task', *Revue internationale de psychologie sociale* 27:3 (2014), pp. 53–77. • 8. J. M. Contreras, M. R. Banaji and J. P. Mitchell, 'Dissociable Neural Correlates of Stereotypes and Other Forms of Semantic Knowledge', *Social Cognitive and Affective Neuroscience* 7:7 (2011), pp. 764–70. • 9. M. Wraga, L. Duncan, E. C. Jacobs, M. Helt and J. Church, 'Stereotype Susceptibility Narrows the Gender Gap in Imagined Self-Rotation Performance', *Psychonomic Bulletin and Review* 13:5 (2006), pp. 813–19. • 10. Shenouda and Danovitch, 'Effects of Gender Stereotypes and Stereotype Threat'. • 11. R. K. Koeske and G. F. Koeske, 'An Attributional Approach to Moods and the Menstrual Cycle', *Journal of Personality and Social Psychology* 31:3 (1975), p. 473. • 12. A. Saini, *Inferior: How Science Got Women Wrong and the New Research That's Rewriting the Story* (Boston, Beacon Press, 2017). • 13. I. K. Broverman, D. M. Broverman, F. E. Clarkson, P. S. Rosenkrantz and S. R. Vogel, 'Sex-Role Stereotypes and Clinical Judgments of Mental Health', *Journal of Consulting and Clinical Psychology* 34:1 (1970), p. 1. • 14. 'Gender stereotypes impacting behaviour of girls as young as seven', Girlguiding website, https://www.girlguiding.org.uk/what-we-do/our-stories-and-news/news/gender-stereotypes-impacting-behaviour-of-girls-as-young-as-seven (accessed 8 November 2018). • 15. S. Marsh, 'Girls as young as seven boxed in by gender stereotyping', *Guardian*, 21 September

2017, https://www.theguardian.com/world/2017/sep/21/girls-seven-uk-boxed-in-by-gender-stereotyping-equality (accessed 8 November 2018). • **16.** S. Dredge, 'Apps for children in 2014: looking for the mobile generation', *Guardian*, 10 March 2014, https://www.theguardian.com/technology/2014/mar/10/apps-children-2014-mobile-generation (accessed 8 November 2018). • **17.** 'The Common Sense Census: Media Use by Kids Age Zero to Eight 2017', Common Sense Media, https://www.commonsensemedia.org/research/the-common-sense-census-media-use-by-kids-age-zero-to-eight-2017 (accessed 8 November 2018). • **18.** Martin and Ruble, 'Children's Search for Gender Cues'. • **19.** D. Poulin-Dubois, L. A. Serbin, B. Kenyon and A. Derbyshire, 'Infants' Intermodal Knowledge about Gender', *Developmental Psychology* 30 (1994), pp. 436–42. • **20.** K. M. Zosuls, D. N. Ruble, C. S. Tamis-LeMonda, P. E. Shrout, M. H. Bornstein and F. K. Greulich, 'The Acquisition of Gender Labels in Infancy: Implications for Gender-Typed Play', *Developmental Psychology* 45:3 (2009), p. 688. • **21.** M. L. Halim, D. N. Ruble, C. S. Tamis-LeMonda, K. M. Zosuls, L. E. Lurye and F. K. Greulich, 'Pink Frilly Dresses and the Avoidance of All Things "Girly": Children's Appearance Rigidity and Cognitive Theories of Gender Development', *Developmental Psychology* 50:4 (2014), p. 1091. • **22.** L. A. Serbin, D. Poulin-Dubois and J. A. Eichstedt, 'Infants' Responses to Gender-Inconsistent Events', *Infancy* 3:4 (2002), pp. 531–42; D. Poulin-Dubois, L. A. Serbin, J. A. Eichstedt, M. G. Sen and C. F. Beissel, 'Men Don't Put On Make-Up: Toddlers' Knowledge of the Gender Stereotyping of Household Activities', *Social Development* 11:2 (2002), pp. 166–81. • **23.** '#RedrawTheBalance', EducationEmployers/YouTube, 14 March 2016, https://youtu.be/kJP1zPOfq_0 (accessed 8 November 2018). • **24.** S. B. Most, A. V. Sorber and J. G. Cunningham, 'Auditory Stroop Reveals Implicit Gender Associations in Adults and Children', *Journal of Experimental Social Psychology* 43:2 (2007), pp. 287–94. • **25.** K. Arney, 'Are pink toys turning girls into passive princesses?', *Guardian*, 9 May 2011, https://www.theguardian.com/science/blog/2011/may/09/pink-toys-girls-passive-princesses (accessed 8 November 2018). • **26.** P. Orenstein, *Cinderella Ate My Daughter: Dispatches from the Front Lines of the New Girlie-Girl Culture* (New York, HarperCollins, 2011). • **27.** 'Gender reveal party ideas', Pampers website (USA), https://www.pampers.com/en-us/pregnancy/pregnancy-announcement/article/ultimate-guide-for-planning-a-gender-reveal-party (accessed 8 November 2018). • **28.** C. DeLoach, 'How to host a gender reveal party', *Parents*, https://www.parents.com/pregnancy/my-baby/gender-prediction/how-to-host-a-gender-reveal-party (accessed 8 November 2018). • **29.** K. Johnson, 'Can you spot what's wrong with this new STEM Barbie?' *Babble*, https://www.babble.com/parenting/engineering-barbie-stem-kit-disappoints (accessed 8 November 2018); D. Lenton, 'Women in

Engineering – Toys: Dolls Get Techie', *Engineering and Technology* 12:6 (2017), pp. 60–63. • **30.** J. Henley, 'The power of pink', *Guardian*, 12 December 2009, https://www.theguardian.com/theguardian/2009/dec/12/pink-stinks-the-power-of-pink (accessed 8 November 2018). • **31.** A. C. Hurlbert and Y. Ling, 'Biological Components of Sex Differences in Color Preference', *Current Biology* 17:16 (2007), pp. R623–5. • **32.** R. Khamsi, 'Women may be hardwired to prefer pink', *New Scientist*, 20 August 2007, https://www.newscientist.com/article/dn12512-women-may-be-hardwired-to-prefer-pink (accessed 8 November 2018); F. Macrae, 'Modern girls are born to plump for pink "thanks to berry-gathering female ancestors"', *Mail Online*, 27 April 2011, https://www.dailymail.co.uk/sciencetech/article-1380893/Modern-girls-born-plump-pink-thanks-berry-gathering-female-ancestors.html (accessed 8 November 2018). • **33.** A. Franklin, L. Bevis, Y. Ling and A. Hurlbert, 'Biological Components of Colour Preference in Infancy', *Developmental Science* 13:2 (2010), pp. 346–54. • **34.** I. D. Cherney and J. Dempsey, 'Young Children's Classification, Stereotyping and Play Behaviour for Gender Neutral and Ambiguous Toys', *Educational Psychology* 30:6 (2010), pp. 651–69. • **35.** V. LoBue and J. S. DeLoache, 'Pretty in Pink: The Early Development of Gender-Stereotyped Colour Preferences', *British Journal of Developmental Psychology* 29:3 (2011), pp. 656–67. • **36.** Zosuls et al., 'The Acquisition of Gender Labels in Infancy'. • **37.** J. B. Paoletti, *Pink and Blue: Telling the Boys from the Girls in America* (Bloomington, Indiana University Press, 2012). • **38.** M. Del Giudice, 'The Twentieth Century Reversal of Pink–Blue Gender Coding: A Scientific Urban Legend?', *Archives of Sexual Behavior* 41:6 (2012), pp. 1321–3; M. Del Giudice, 'Pink, Blue, and Gender: An Update', *Archives of Sexual Behavior* 46:6 (2017), pp. 1555–63. • **39.** Henley, 'The power of pink'. • **40.** 'What's wrong with pink and blue?', Let Toys Be Toys, 4 September 2015, http://lettoysbetoys.org.uk/whats-wrong-with-pink-and-blue (accessed 8 November 2018). • **41.** A. M. Sherman and E. L. Zurbriggen, '"Boys Can Be Anything": Effect of Barbie Play on Girls' Career Cognitions', *Sex Roles* 70:5–6 (2014), pp. 195–208. • **42.** V. Jarrett, 'How we can help all our children explore, learn, and dream without limits', White House website, 6 April 2016, https://obamawhitehouse.archives.gov/blog/2016/04/06/how-we-can-help-all-our-children-explore-learn-and-dream-without-limits (accessed 8 November 2018). • **43.** V. Jadva, M. Hines and S. Golombok, 'Infants' Preferences for Toys, Colors, and Shapes: Sex Differences and Similarities', *Archives of Sexual Behavior* 39:6 (2010), pp. 1261–73. • **44.** C. L. Martin, D. N. Ruble and J. Szkrybalo, 'Cognitive Theories of Early Gender Development', *Psychological Bulletin* 128:6 (2002), p. 903. • **45.** L. Waterlow, 'Too much in the pink! How toys have become alarmingly gender stereotyped since the Seventies ... at the cost of little girls' self-esteem', *Mail Online*, 10 June 2013, https://www.dailymail.co.uk/

femail/article-2338976/Too-pink-How-toys-alarmingly-gender-stereotyped-Seventies–cost-little-girls-self-esteem.html (accessed 8 November 2018). • **46.** J. E. O. Blakemore and R. E. Centers, 'Characteristics of Boys' and Girls' Toys', *Sex Roles* 53:9–10 (2005), pp. 619–33. • **47.** B. K. Todd, J. A. Barry and S. A. Thommessen, 'Preferences for "Gender-Typed" Toys in Boys and Girls Aged 9 to 32 Months', *Infant and Child Development* 26:3 (2017), e1986. • **48.** Ibid. • **49.** Ibid. • **50.** C. Fine and E. Rush, '"Why Does All the Girls Have to Buy Pink Stuff?" The Ethics and Science of the Gendered Toy Marketing Debate', *Journal of Business Ethics* 149:4 (2018), pp. 769–84. • **51.** B. K. Todd, R. A. Fischer, S. Di Costa, A. Roestorf, K. Harbour, P. Hardiman and J. A. Barry, 'Sex Differences in Children's Toy Preferences: A Systematic Review, Meta-regression, and Meta-analysis', *Infant and Child Development* 27:2 (2018), pp. 1–29. • **52.** Ibid., pp. 1–2. • **53.** N. K. Freeman, 'Preschoolers' Perceptions of Gender Appropriate Toys and Their Parents' Beliefs about Genderized Behaviors: Miscommunication, Mixed Messages, or Hidden Truths?', *Early Childhood Education Journal* 34:5 (2007), pp. 357–66. • **54.** E. S. Weisgram, M. Fulcher and L. M. Dinella, 'Pink Gives Girls Permission: Exploring the Roles of Explicit Gender Labels and Gender-Typed Colors on Preschool Children's Toy Preferences', *Journal of Applied Developmental Psychology* 35:5 (2014), pp. 401–9. • **55.** E. Sweet, 'Toys are more divided by gender now than they were 50 years ago', *Atlantic*, 9 December 2014, https://www.theatlantic.com/business/archive/2014/12/toys-are-more-divided-by-gender-now-than-they-were-50-years-ago/383556 (accessed 8 November 2018). • **56.** J. Stoeber and H. Yang, 'Physical Appearance Perfectionism Explains Variance in Eating Disorder Symptoms above General Perfectionism', *Personality and Individual Differences* 86 (2015), pp. 303–7. • **57.** J. F. Benenson, R. Tennyson and R. W. Wrangham, 'Male More than Female Infants Imitate Propulsive Motion', *Cognition* 121:2 (2011), pp. 262–7. • **58.** G. M. Alexander, T. Wilcox and R. Woods, 'Sex Differences in Infants' Visual Interest in Toys', *Archives of Sexual Behavior* 38:3 (2009), pp. 427–33. • **59.** 'Jo Swinson: Encourage boys to play with dolls', BBC News, 13 January 2015, https://www.bbc.co.uk/news/uk-politics-30794476 (accessed 8 November 2018). • **60.** G. M. Alexander and M. Hines, 'Sex Differences in Response to Children's Toys in Nonhuman Primates (*Cercopithecus aethiops sabaeus*)', *Evolution and Human Behavior* 23:6 (2002), pp. 467–79. • **61.** Both Cordelia Fine in *Delusions of Gender* and Rebecca Jordan-Young in *Brain Storm* have commented humorously and at length on the monkey studies and their exaggerated role in offering insights into toy preference issues. • **62.** J. M. Hassett, E. R. Siebert and K. Wallen, 'Sex Differences in Rhesus Monkey Toy Preferences Parallel Those of Children', *Hormones and Behavior* 54:3 (2008), pp. 359–64. • **63.** Ibid., p. 363. • **64.** Hines, *Brain Gender*. • **65.** S. A. Berenbaum and M. Hines, 'Early Androgens Are

Related to Childhood Sex-Typed Toy Preferences', *Psychological Science* 3:3 (1992), pp. 203–6. • **66.** M. Hines, V. Pasterski, D. Spencer, S. Neufeld, P. Patalay, P. C. Hindmarsh, I. A. Hughes and C. L. Acerini, 'Prenatal Androgen Exposure Alters Girls' Responses to Information Indicating Gender-Appropriate Behaviour', *Philosophical Transactions of the Royal Society B: Biological Sciences* 371:1688 (2016), 20150125. • **67.** M. C. Linn and A. C. Petersen, 'Emergence and Characterization of Sex Differences in Spatial Ability: A Meta-analysis', *Child Development* 56:6 (1985), pp. 1479–98. • **68.** D. I. Miller and D. F. Halpern, 'The New Science of Cognitive Sex Differences', *Trends in Cognitive Sciences* 18:1 (2014), pp. 37–45. • **69.** Hines et al., 'Prenatal Androgen Exposure Alters Girls' Responses'. • **70.** M. S. Terlecki and N. S. Newcombe, 'How Important Is the Digital Divide? The Relation of Computer and Videogame Usage to Gender Differences in Mental Rotation Ability', *Sex Roles* 53:5–6 (2005), pp. 433–41. • **71.** Shenouda and Danovitch, 'Effects of Gender Stereotypes and Stereotype Threat'.

CHAPTER 10: SEX AND SCIENCE

1. Women in Science website, http://uis.unesco.org/en/topic/women-science; 'Women in the STEM workforce 2016', WISE website, https://www.wisecampaign.org.uk/statistics/women-in-the-stem-workforce-2016 (accessed 8 November 2018). • **2.** A. Tintori and R. Palomba, *Turn On the Light on Science: A Research-Based Guide to Break Down Popular Stereotypes about Science and Scientists* (London, Ubiquity Press, 2017). • **3.** 'Useful statistics: women in STEM', STEM Women website, 5 March 2018, https://www.stemwomen.co.uk/blog/2018/03/useful-statistics-women-in-stem; 'UK physics A-level entries 2010–2016', Institute of Physics website, http://www.iop.org/policy/statistics/overview/page_67109.html • **4.** 'Primary Schools are Critical to Ensuring Success, by Creating Space for Quality Science Teaching', in *Tomorrow's World: Inspiring Primary Scientists* (CBI, 2015), http://www.cbi.org.uk/tomorrows-world/Primary_schools_are_critical_t.html (accessed 8 November 2018). • **5.** 'Our definition of science', Science Council website, https://sciencecouncil.org/about-science/our-definition-of-science (accessed 8 November 2018). • **6.** 'Science does not purvey absolute truth, science is a mechanism. It's a way of trying to improve your knowledge of nature, it's a system for testing your thoughts against the universe and seeing whether they match', *Explore*, http://explore.brainpickings.org/post/49908311909/science-does-not-purvey-absolute-truth-science-is (accessed 8 November 2018). • **7.** 'Essays', Science: Not Just for Scientists, http://notjustforscientists.org/essays (accessed 8 November 2018). • **8.** R. L. Bergland, 'Urania's Inversion: Emily Dickinson, Herman Melville, and the Strange History of Women Scientists in Nineteenth-Century America',

Signs: Journal of Women in Culture and Society 34:1 (2008), pp. 75–99. • **9.** J. Mason, 'The Admission of the First Women to the Royal Society of London', *Notes and Records: The Royal Society Journal of the History of Science* 46:2 (1992), pp. 279–300. • **10.** L. Schiebinger, *The Mind Has No Sex? Women in the Origins of Modern Science* (Cambridge, MA, Harvard University Press, 1991). • **11.** Ibid. • **12.** R. Su, J. Rounds and P. I. Armstrong, 'Men and Things, Women and People: A Meta-analysis of Sex Differences in Interests', *Psychological Bulletin* 135:6 (2009), p. 859. • **13.** J. Billington, S. Baron-Cohen and S. Wheelwright, 'Cognitive Style Predicts Entry into Physical Sciences and Humanities: Questionnaire and Performance Tests of Empathy and Systemizing', *Learning and Individual Differences* 17:3 (2007), pp. 260–68. • **14.** Ibid. • **15.** Baron-Cohen, *The Essential Difference.* • **16.** Ibid. • **17.** S. J. Leslie, A. Cimpian, M. Meyer and E. Freeland, 'Expectations of Brilliance Underlie Gender Distributions across Academic Disciplines', *Science* 347:6219 (2015), pp. 262–5. • **18.** S. J. Leslie, 'Cultures of Brilliance and Academic Gender Gaps', paper delivered at 'Confidence and Competence: Fifth Annual Diversity Conference', Royal Society, 16 November 2017; see 'Annual Diversity Conference 2017 – Confidence and Competence', Royal Society/ YouTube, 16 November 2017, https://www.youtu.be/e0ZHpZ31O1M, at 25:50 (accessed 8 November 2018). • **19.** K. C. Elmore and M. Luna-Lucero, 'Light Bulbs or Seeds? How Metaphors for Ideas Influence Judgments about Genius', *Social Psychological and Personality Science* 8:2 (2017), pp. 200–208. • **20.** Ibid. • **21.** L. Bian, S. J. Leslie, M. C. Murphy and A. Cimpian, 'Messages about Brilliance Undermine Women's Interest in Educational and Professional Opportunities', *Journal of Experimental Social Psychology* 76 (2018), pp. 404–20. • **22.** Quinn and Liben, 'A Sex Difference in Mental Rotation in Young Infants'. • **23.** M. Hines, M. Constantinescu and D. Spencer, 'Early Androgen Exposure and Human Gender Development', *Biology of Sex Differences* 6:1 (2015), p. 3; J. Wai, D. Lubinski and C. P. Benbow, 'Spatial Ability for STEM Domains: Aligning Over 50 Years of Cumulative Psychological Knowledge Solidifies Its Importance', *Journal of Educational Psychology* 101:4 (2009), p. 817. • **24.** S. C. Levine, A. Foley, S. Lourenco, S. Ehrlich and K. Ratliff, 'Sex Differences in Spatial Cognition: Advancing the Conversation', *Wiley Interdisciplinary Reviews: Cognitive Science* 7:2(2016), pp. 127–55. • **25.** L. Bian, S. J. Leslie and A. Cimpian, 'Gender Stereotypes about Intellectual Ability Emerge Early and Influence Children's Interests', *Science* 355:6323 (2017), pp. 389–91. • **26.** M. C. Steffens, P. Jelenec and P. Noack, 'On the Leaky Math Pipeline: Comparing Implicit Math–Gender Stereotypes and Math Withdrawal in Female and Male Children and Adolescents', *Journal of Educational Psychology* 102:4 (2010), p. 947. • **27.** Ibid. • **28.** E. A. Gunderson, G. Ramirez, S. C. Levine and S. L. Beilock, 'The Role of Parents and Teachers in the Development of Gender-Related

Math Attitudes', *Sex Roles* 66:3–4 (2012), pp. 153–66. • **29.** Freeman, 'Preschoolers' Perceptions of Gender Appropriate Toys'. • **30.** V. Lavy and E. Sand, 'On the Origins of Gender Human Capital Gaps: Short and Long Term Consequences of Teachers' Stereotypical Biases', Working Paper 20909, National Bureau of Economic Research (2015). • **31.** S. Cheryan, V. C. Plaut, P. G. Davies and C. M. Steele, 'Ambient Belonging: How Stereotypical Cues Impact Gender Participation in Computer Science', *Journal of Personality and Social Psychology* 97:6 (2009), p. 1045. • **32.** Ibid. • **33.** G. Stoet and D. C. Geary, 'The Gender-Equality Paradox in Science, Technology, Engineering, and Mathematics Education', *Psychological Science* 29:4 (2018), pp. 581–93. • **34.** S. Ross, 'Scientist: The Story of a Word', *Annals of Science* 18:2 (1962), pp. 65–85. • **35.** M. Mead and R. Metraux, 'Image of the Scientist among High-School Students', *Science* 126:3270 (1957), pp. 384–90. • **36.** Ibid. • **37.** D. W. Chambers, 'Stereotypic Images of the Scientist: The Draw-a-Scientist Test', *Science Education* 67:2 (1983), pp. 255–65. • **38.** K. D. Finson, 'Drawing a Scientist: What We Do and Do Not Know after Fifty Years of Drawings', *School Science and Mathematics* 102:7 (2002), pp. 335–45. • **39.** Ibid. • **40.** P. Bernard and K. Dudek, 'Revisiting Students' Perceptions of Research Scientists: Outcomes of an Indirect Draw-a-Scientist Test (InDAST)', *Journal of Baltic Science Education* 16:4 (2017). • **41.** M. Knight and C. Cunningham, 'Draw an Engineer Test (DAET): Development of a Tool to Investigate Students' Ideas about Engineers and Engineering', paper given at American Society for Engineering Education Annual Conference and Exposition, Salt Lake City, June 2004, https://peer.asee.org/12831 (accessed 8 November 2018). • **42.** C. Moseley, B. Desjean-Perrotta and J. Utley, 'The Draw-an-Environment Test Rubric (DAET-R): Exploring Pre-service Teachers' Mental Models of the Environment', *Environmental Education Research* 16:2 (2010), pp. 189–208. • **43.** C. D. Martin, 'Draw a Computer Scientist', *ACM SIGCSE Bulletin* 36:4 (2004), pp. 11–12. • **44.** L. R. Ramsey, 'Agentic Traits Are Associated with Success in Science More than Communal Traits', *Personality and Individual Differences* 106 (2017), pp. 6–9. • **45.** L. L. Carli, L. Alawa, Y. Lee, B. Zhao and E. Kim, 'Stereotypes about Gender and Science: Women ≠ Scientists', *Psychology of Women Quarterly*, 40:2 (2016), pp. 244–60. • **46.** A. H. Eagly, 'Few Women at the Top: How Role Incongruity Produces Prejudice and the Glass Ceiling', in D. van Knippenberg and M. A. Hogg (eds), *Leadership and Power: Identity Processes in Groups and Organizations* (London, Sage, 2003), pp. 79–93. • **47.** A. H. Eagly and S. J. Karau, 'Role Congruity Theory of Prejudice toward Female Leaders', *Psychological Review* 109:3 (2002), p. 573. • **48.** Carli et al., 'Stereotypes about Gender and Science'. • **49.** C. Wenneras and A. Wold, 'Nepotism and Sexism in Peer-

Review', in M. Wyer (ed.), *Women, Science, and Technology: A Reader in Feminist Science Studies* (New York, Routledge, 2001), pp. 46–52. • **50.** F. Trix and C. Psenka, 'Exploring the Color of Glass: Letters of Recommendation for Female and Male Medical Faculty', *Discourse and Society* 14:2 (2003), pp. 191–220. • **51.** S. Modgil, R. Gill, V. L. Sharma, S. Velassery and A. Anand, 'Nobel Nominations in Science: Constraints of the Fairer Sex', *Annals of Neurosciences* 25:2 (2018), pp. 63–78. • **52.** C. A. Moss-Racusin, J. F. Dovidio, V. L. Brescoll, M. J. Graham and J. Handelsman, 'Science Faculty's Subtle Gender Biases Favor Male Students', *Proceedings of the National Academy of Sciences* 109:41 (2012), pp. 16474–9. • **53.** E. Reuben, P. Sapienza and L. Zingales, 'How Stereotypes Impair Women's Careers in Science', *Proceedings of the National Academy of Sciences* 111:12 (2014), pp. 4403–8.

CHAPTER 11: SCIENCE AND THE BRAIN

1. H. Ellis, *Man and Woman: A Study of Human Secondary Sexual Characters* (London, Walter Scott; New York, Scribner's, 1894). • **2.** N. M. Else-Quest, J. S. Hyde and M. C. Linn, 'Cross-national Patterns of Gender Differences in Mathematics: A Meta-analysis', *Psychological Bulletin* 136:1 (2010), p. 103. • **3.** 'Has an uncomfortable truth been suppressed?', Gowers's Weblog, 9 September 2018, https://gowers.wordpress.com/2018/09/09/has-an-uncomfortable-truth-been-suppressed (accessed 8 November 2018). • **4.** Ibid. • **5.** L. H. Summers, 'Remarks at NBER Conference on Diversifying the Science & Engineering Workforce', Office of the President, Harvard University, 14 January 2005, https://www.harvard.edu/president/speeches/summers_2005/nber.php (accessed 8 November 2018). • **6.** 'The Science of Gender and Science: Pinker vs. Spelke: A Debate', *Edge*, https://www.edge.org/event/the-science-of-gender-and-science-pinker-vs-spelke-a-debate (accessed 8 November 2018). • **7.** Y. Xie and K. Shaumann, *Women in Science: Career Processes and Outcomes* (Cambridge, MA, Harvard University Press, 2003). • **8.** Ibid. • **9.** D. F. Halpern, C. P. Benbow, D. C. Geary, R. C. Gur, J. S. Hyde and M. A. Gernsbacher, 'The Science of Sex Differences in Science and Mathematics', *Psychological Science in the Public Interest* 8:1 (2007), pp. 1–51. • **10.** J. Damore, 'Google's Ideological Echo Chamber', July 2017, available at https://www.documentcloud.org/documents/3914586-Googles-Ideological-Echo-Chamber.html (accessed 8 November 2018). • **11.** D. P. Schmitt, A. Realo, M. Voracek and J. Allik, 'Why Can't a Man Be More Like a Woman? Sex Differences in Big Five Personality Traits across 55 Cultures', *Journal of Personality and Social Psychology* 94:1 (2008), p. 168. • **12.** M. Molteni and A. Rogers, 'The actual science of James Damore's Google memo', *Wired*, 15 August 2017, https://www.wired.com/story/

the-pernicious-science-of-james-damores-google-memo (accessed 8 November 2018); H. Devlin and A. Hern, 'Why are there so few women in tech? The truth behind the Google memo', *Guardian*, 8 August 2017, https://www.theguardian.com/lifeandstyle/2017/aug/08/why-are-there-so-few-women-in-tech-the-truth-behind-the-google-memo (accessed 8 November 2018); S. Stevens, 'The Google memo: what does the research say about gender differences?', Heterodox Academy, 10 August 2017, https://heterodoxacademy.org/the-google-memo-what-does-the-research-say-about-gender-differences (accessed 8 November 2018). • **13.** 'The Google memo: four scientists respond', *Quillette*, 7 August 2017, http://quillette.com/2017/08/07/google-memo-four-scientists-respond (accessed 8 November 2018). • **14.** Ibid. • **15.** Ibid. • **16.** G. Rippon, 'What neuroscience can tell us about the Google diversity memo', *Conversation*, 14 August 2017, https://theconversation.com/what-neuroscience-can-tell-us-about-the-google-diversity-memo-82455 (accessed 8 November 2018). • **17.** Devlin and Hern, 'Why are there so few women in tech?' • **18.** R. C. Barnett and C. Rivers, 'We've studied gender and STEM for 25 years. The science doesn't support the Google memo', *Recode*, 11 August 2017, https://www.recode.net/2017/8/11/16127992/google-engineer-memo-research-science-women-biology-tech-james-damore (accessed 8 November 2018). • **19.** M.-C. Lai, M. V. Lombardo, B. Chakrabarti, C. Ecker, S. A. Sadek, S. J. Wheelwright, D. G. Murphy, J. Suckling, E. T. Bullmore, S. Baron-Cohen and MRC AIMS Consortium, 'Individual Differences in Brain Structure Underpin Empathizing–Systemizing Cognitive Styles in Male Adults', *NeuroImage* 61:4 (2012), pp. 1347–54. • **20.** S. Baron-Cohen, 'Empathizing, Systemizing, and the Extreme Male Brain Theory of Autism', *Progress in Brain Research* 186 (2010), pp. 167–75. • **21.** J. Wai, D. Lubinski and C. P. Benbow, 'Spatial Ability for STEM Domains: Aligning Over 50 Years of Cumulative Psychological Knowledge Solidifies Its Importance', *Journal of Educational Psychology* 101:4 (2009), p. 817. • **22.** Ibid. • **23.** M. Hines, B. A. Fane, V. L. Pasterski, G. A. Mathews, G. S. Conway and C. Brook, 'Spatial Abilities Following Prenatal Androgen Abnormality: Targeting and Mental Rotations Performance in Individuals with Congenital Adrenal Hyperplasia', *Psychoneuroendocrinology* 28:8 (2003), pp. 1010–26. • **24.** I. Silverman, J. Choi and M. Peters, 'The Hunter-Gatherer Theory of Sex Differences in Spatial Abilities: Data from 40 Countries', *Archives of Sexual Behavior* 36:2 (2007), pp. 261–8. • **25.** S. G. Vandenberg and A. R. Kuse, 'Mental Rotations, a Group Test of Three-Dimensional Spatial Visualization', *Perceptual and Motor Skills* 47:2 (1978), pp. 599–604. • **26.** Quinn and Liben, 'A Sex Difference in Mental Rotation in Young Infants'. • **27.** Hines et al., 'Spatial Abilities Following Prenatal Androgen Abnormality'. • **28.** M. Constantinescu, D. S. Moore, S. P. Johnson and M. Hines, 'Early Contributions to Infants' Mental Rotation Abilities', *Developmental Science*

21:4 (2018), e12613. • **29.** T. Koscik, D. O'Leary, D. J. Moser, N. C. Andreasen and P. Nopoulos, 'Sex Differences in Parietal Lobe Morphology: Relationship to Mental Rotation Performance', *Brain and Cognition* 69:3 (2009), pp. 451–9. • **30.** Halpern, et al. 'The Pseudoscience of Single-Sex Schooling'. • **31.** Koscik et al., 'Sex Differences in Parietal Lobe Morphology'. • **32.** K. Kucian, M. Von Aster, T. Loenneker, T. Dietrich, F. W. Mast and E. Martin, 'Brain Activation during Mental Rotation in School Children and Adults', *Journal of Neural Transmission* 114:5 (2007), pp. 675–86. • **33.** K. Jordan, T. Wüstenberg, H. J. Heinze, M. Peters and L. Jäncke, 'Women and Men Exhibit Different Cortical Activation Patterns during Mental Rotation Tasks', *Neuropsychologia* 40:13 (2002), pp. 2397–408. • **34.** N. S. Newcombe, 'Picture This: Increasing Math and Science Learning by Improving Spatial Thinking', *American Educator* 34:2 (2010), p. 29. • **35.** M. Wraga, M. Helt, E. Jacobs and K. Sullivan, 'Neural Basis of Stereotype-Induced Shifts in Women's Mental Rotation Performance', *Social Cognitive and Affective Neuroscience* 2:1 (2007), pp. 12–19. • **36.** I. D. Cherney, 'Mom, Let Me Play More Computer Games: They Improve My Mental Rotation Skills', *Sex Roles* 59:11–12 (2008), pp. 776–86. • **37.** Ibid. • **38.** J. Feng, I. Spence and J. Pratt, 'Playing an Action Video Game Reduces Gender Differences in Spatial Cognition', *Psychological Science* 18:10 (2007), pp. 850–55; M. S. Terlecki and N. S. Newcombe, 'How Important Is the Digital Divide? The Relation of Computer and Videogame Usage to Gender Differences in Mental Rotation Ability', *Sex Roles* 53:5–6 (2005), pp. 433–41. • **39.** R. J. Haier, S. Karama, L. Leyba and R. E. Jung, 'MRI Assessment of Cortical Thickness and Functional Activity Changes in Adolescent Girls Following Three Months of Practice on a Visual-Spatial Task', *BMC Research Notes* 2:1 (2009), p. 174. • **40.** A. Moè and F. Pazzaglia, 'Beyond Genetics in Mental Rotation Test Performance: The Power of Effort Attribution', *Learning and Individual Differences* 20:5 (2010), pp. 464–8. • **41.** E. A. Maloney, S. Waechter, E. F. Risko and J. A. Fugelsang, 'Reducing the Sex Difference in Math Anxiety: The Role of Spatial Processing Ability', *Learning and Individual Differences* 22:3 (2012), pp. 380–84. • **42.** O. Blajenkova, M. Kozhevnikov and M. A. Motes, 'Object-Spatial Imagery: A New Self-Report Imagery Questionnaire', *Applied Cognitive Psychology* 20:2 (2006), pp. 239–63. • **43.** J. A. Mangels, C. Good, R. C. Whiteman, B. Maniscalco and C. S. Dweck, 'Emotion Blocks the Path to Learning under Stereotype Threat', *Social Cognitive and Affective Neuroscience* 7:2 (2011), pp. 230–41. • **44.** A. C. Krendl, J. A. Richeson, W. M. Kelley and T. F. Heatherton, 'The Negative Consequences of Threat: A Functional Magnetic Resonance Imaging Investigation of the Neural Mechanisms Underlying Women's Underperformance in Math', *Psychological Science* 19:2 (2008), pp. 168–75. • **45.** B. Carrillo, E. Gómez-Gil, G. Rametti, C. Junque, Á. Gomez, K. Karadi, S. Segovia and A. Guillamon, 'Cortical Activation during Mental Rotation in Male-to-Female and Female-to-Male Transsexuals under Hormonal Treatment', *Psychoneuroendocrinology* 35:8 (2010), pp. 1213–22. • **46.** S. A. Berenbaum and M. Hines, 'Early Androgens Are Related to Childhood Sex-Typed Toy Preferences', *Psychological Science* 3:3 (1992), pp. 203–6. • **47.** J. R. Shapiro and A. M. Williams, 'The Role of Stereotype Threats in Undermining Girls' and Women's Performance and Interest in STEM Fields', *Sex Roles* 66:3–4 (2012), pp. 175–83. • **48.** M. Hines, V. Pasterski, D. Spencer, S. Neufeld, P. Patalay, P. C. Hindmarsh, I. A. Hughes and C. L. Acerini, 'Prenatal Androgen Exposure Alters Girls' Responses to Information Indicating Gender-Appropriate Behaviour', *Philosophical Transactions of the Royal Society B: Biological Sciences* 371:1688 (2016), 20150125. • **49.** 'Women in Science, Technology, Engineering, and Mathematics (STEM)', Catalyst website, 3 January 2018, https://www.catalyst.org/knowledge/women-science-technology-engineering-and-mathematics-stem (accessed 10 November 2018).

CHAPTER 12: GOOD GIRLS DON'T

1. S. Peters, *The Chimp Paradox: The Mind Management Program to Help You Achieve Success, Confidence, and Happiness* (New York, Tarcher/Penguin, 2013). • **2.** B. P. Doré, N. Zerubavel and K. N. Ochsner, 'Social Cognitive Neuroscience: A Review of Core Systems', in M. Mikulincer and P. R. Shaver (eds-in-chief), *APA Handbook of Personality and Social Psychology* (Washington, American Psychological Association, 2014), vol. l, pp. 693–720. • **3.** J. M. Allman, A. Hakeem, J. M. Erwin, E. Nimchinsky and P. Hof, 'The Anterior Cingulate Cortex: The Evolution of an Interface between Emotion and Cognition', *Annals of the New York Academy of Sciences* 935:1 (2001), pp. 107–17. • **4.** J. M. Allman, N. A. Tetreault, A. Y. Hakeem, K. F. Manaye, K. Semendeferi, J. M. Erwin, S. Park, V. Goubert and P. R. Hof, 'The Von Economo Neurons in Frontoinsular and Anterior Cingulate Cortex in Great Apes and Humans', *Brain Structure and Function* 214:5–6 (2010), pp. 495–517. • **5.** J. D. Cohen, M. Botvinick and C. S. Carter, 'Anterior Cingulate and Prefrontal Cortex: Who's in Control?', *Nature Neuroscience* 3:5 (2000), p. 421. • **6.** G. Bush, P. Luu and M. I. Posner, 'Cognitive and Emotional Influences in Anterior Cingulate Cortex', *Trends in Cognitive Sciences* 4:6 (2000), pp. 215–22. • **7.** Eisenberger et al., 'The Neural Sociometer'. • **8.** Eisenberger and Lieberman, 'Why Rejection Hurts'. • **9.** N. I. Eisenberger, 'Social Pain and the Brain: Controversies, Questions, and Where to Go from Here', *Annual Review of Psychology* 66 (2015), pp. 601–29. • **10.** Lieberman, *Social: Why Our Brains Are Wired to Connect*. • **11.** N. Kolling, M. K. Wittmann, T. E. Behrens, E. D. Boorman, R. B. Mars and M. F. Rushworth, 'Value, Search, Persistence and Model Updating in Anterior Cingulate Cortex', *Nature Neuroscience* 19:10 (2016), p. 1280. • **12.** T. Straube,

S. Schmidt, T. Weiss, H. J. Mentzel and W. H. Miltner, 'Dynamic Activation of the Anterior Cingulate Cortex during Anticipatory Anxiety', *NeuroImage* 44:3 (2009), pp. 975–81; A. Etkin, K. E. Prater, F. Hoeft, V. Menon and A. F. Schatzberg, 'Failure of Anterior Cingulate Activation and Connectivity with the Amygdala during Implicit Regulation of Emotional Processing in Generalized Anxiety Disorder', *American Journal of Psychiatry* 167:5 (2010), pp. 545–54; A. Etkin, T. Egner and R. Kalisch, 'Emotional Processing in Anterior Cingulate and Medial Prefrontal Cortex', *Trends in Cognitive Sciences* 15:2 (2011), pp. 85–93. • **13.** M. R. Leary, 'Responses to Social Exclusion: Social Anxiety, Jealousy, Loneliness, Depression, and Low Self-Esteem', *Journal of Social and Clinical Psychology* 9:2 (1990), pp. 221–9; J. F. Sowislo and U. Orth, 'Does Low Self-Esteem Predict Depression and Anxiety? A Meta-analysis of Longitudinal Studies', *Psychological Bulletin* 139:1 (2013), p. 213; E. A. Courtney, J. Gamboz and J. G. Johnson, 'Problematic Eating Behaviors in Adolescents with Low Self-Esteem and Elevated Depressive Symptoms', *Eating Behaviors* 9:4 (2008), pp. 408–14. • **14.** W. Bleidorn, R. C. Arslan, J. J. Denissen, P. J. Rentfrow, J. E. Gebauer, J. Potter and S. D. Gosling, 'Age and Gender Differences in Self-Esteem – A Cross-cultural Window', *Journal of Personality and Social Psychology* 111:3 (2016), p. 396; S. Guimond, A. Chatard, D. Martinot, R. J. Crisp and S. Redersdorff, 'Social Comparison, Self-Stereotyping, and Gender Differences in Self-Construals', *Journal of Personality and Social Psychology* 90:2 (2006), p. 221. • **15.** 'World Self Esteem Plot', https://selfesteem.shinyapps.io/maps (accessed 10 November 2018). • **16.** Schmitt et al., 'Why Can't a Man Be More Like a Woman?' • **17.** J. S. Hyde, 'Gender Similarities and Differences', *Annual Review of Psychology* 65 (2014), pp. 373–98; E. Zell, Z. Krizan and S. R. Teeter, 'Evaluating Gender Similarities and Differences Using Metasynthesis', *American Psychologist* 70:1 (2015), p. 10. • **18.** Eisenberger and Lieberman, 'Why Rejection Hurts'. • **19.** A. J. Shackman, T. V. Salomons, H. A. Slagter, A. S. Fox, J. J. Winter and R. J. Davidson, 'The Integration of Negative Affect, Pain and Cognitive Control in the Cingulate Cortex', *Nature Reviews Neuroscience* 12:3 (2011), p. 154. • **20.** A. T. Beck, *Depression: Clinical, Experimental, and Theoretical Aspects* (New York, Harper & Row, 1967); A. T. Beck, 'The Evolution of the Cognitive Model of Depression and Its Neurobiological Correlates', *American Journal of Psychiatry* 165 (2008), pp. 969–77; S. G. Disner, C. G. Beevers, E. A. Haigh and A. T. Beck, 'Neural Mechanisms of the Cognitive Model of Depression', *Nature Reviews Neuroscience* 12:8 (2011), p. 467. • **21.** P. Gilbert, *The Compassionate Mind: A New Approach to Life's Challenges* (Oakland, CA, New Harbinger, 2010). • **22.** P. Gilbert and C. Irons, 'Focused Therapies and Compassionate Mind Training for Shame and Self-Attacking', in P. Gilbert (ed.), *Compassion: Conceptualisations, Research and Use in Psychotherapy* (Hove, Routledge, 2005), pp. 263–325; D. C. Zuroff, D. Santor and M. Mongrain, 'Dependency, Self-Criticism, and Maladjustment', in S. J. Blatt, J. S. Auerbach, K. N. Levy and C. E. Schaffer (eds), *Relatedness, Self-Definition and Mental Representation: Essays in Honor of Sidney J. Blatt* (Hove, Routledge, 2005), pp. 75–90. • **23.** P. Gilbert, M. Clarke, S. Hempel, J. N. V. Miles and C. Irons, 'Criticizing and Reassuring Oneself: An Exploration of Forms, Styles and Reasons in Female Students', *British Journal of Clinical Psychology* 43:1 (2004), pp. 31–50. • **24.** W. J. Gehring, B. Goss, M. G. H. Coles, D. E. Meyer and E. Donchin, 'A Neural System for Error Detection and Compensation', *Psychological Science* 4 (1993), pp. 385–90; S. Dehaene, 'The Error-Related Negativity, Self-Monitoring, and Consciousness', *Perspectives on Psychological Science* 13:2 (2018), pp. 161–5. • **25.** O. Longe, F. A. Maratos, P. Gilbert, G. Evans, F. Volker, H. Rockliff and G. Rippon, 'Having a Word with Yourself: Neural Correlates of Self-Criticism and Self-Reassurance', *NeuroImage* 49:2 (2010), pp. 1849–56. • **26.** G. Downey and S. I. Feldman, 'Implications of Rejection Sensitivity for Intimate Relationships', *Journal of Personality and Social Psychology* 70:6 (1996), p. 1327. • **27.** Ibid. • **28.** Ö. Ayduk, A. Gyurak and A. Luerssen, 'Individual Differences in the Rejection–Aggression Link in the Hot Sauce Paradigm: The Case of Rejection Sensitivity', *Journal of Experimental Social Psychology* 44:3 (2008), pp. 775–82. • **29.** D. C. Jack and A. Ali (eds), *Silencing the Self across Cultures: Depression and Gender in the Social World* (Oxford, Oxford University Press, 2010). • **30.** B. London, G. Downey, R. Romero-Canyas, A. Rattan and D. Tyson, 'Gender-Based Rejection Sensitivity and Academic Self-Silencing in Women', *Journal of Personality and Social Psychology* 102:5 (2012), p. 961. • **31.** S. Zhang, T. Schmader and W. M. Hall, 'L'eggo my Ego: Reducing the Gender Gap in Math by Unlinking the Self from Performance', *Self and Identity* 12:4 (2013), pp. 400–412. • **32.** Eisenberger and Lieberman, 'Why Rejection Hurts'. • **33.** E. Kross, T. Egner, K. Ochsner, J. Hirsch and G. Downey, 'Neural Dynamics of Rejection Sensitivity', *Journal of Cognitive Neuroscience* 19:6 (2007), pp. 945–56. • **34.** L. J. Burklund, N. I. Eisenberger and M. D. Lieberman, 'The Face of Rejection: Rejection Sensitivity Moderates Dorsal Anterior Cingulate Activity to Disapproving Facial Expressions', *Social Neuroscience* 2:3–4 (2007), pp. 238–53. • **35.** Kross et al., 'Neural Dynamics of Rejection Sensitivity'. • **36.** K. Dedovic, G. M. Slavich, K. A. Muscatell, M. R. Irwin and N. I. Eisenberger, 'Dorsal Anterior Cingulate Cortex Responses to Repeated Social Evaluative Feedback in Young Women with and without a History of Depression', *Frontiers in Behavioral Neuroscience* 10 (2016), p. 64. • **37.** A. Kupferberg, L. Bicks and G. Hasler, 'Social Functioning in Major Depressive Disorder', *Neuroscience and Biobehavioral Reviews* 69 (2016), pp. 313–32. • **38.** Steele, *Whistling Vivaldi*; S. J. Spencer, C. Logel and P. G. Davies, 'Stereotype Threat', *Annual Review of Psychology* 67 (2016), pp. 415–37. • **39.** J. Aronson, M. J. Lustina, C. Good, K. Keough, C. M. Steele and J. Brown, 'When White Men Can't Do Math: Necessary and Sufficient Factors in Stereotype Threat', *Journal of Experimental Social Psychology* 35:1 (1999), pp. 29–46. • **40.** M. A. Pavlova, S. Weber, E. Simoes and A. N. Sokolov, 'Gender Stereotype Susceptibility', *PLoS One* 9:12 (2014), e114802. • **41.** M. Wraga, M. Helt, E. Jacobs and K. Sullivan, 'Neural Basis of Stereotype-Induced Shifts in Women's Mental Rotation Performance', *Social Cognitive and Affective Neuroscience* 2:1 (2007), pp. 12–19. • **42.** M. M. McClelland, C. E. Cameron, S. B. Wanless and A. Murray, 'Executive Function, Behavioral Self-Regulation, and Social-Emotional Competence: Links to School Readiness', in O. N. Saracho and B. Spodek (eds), *Contemporary Perspectives on Social Learning in Early Childhood Education* (Charlotte, NC, Information Age, 2007), pp. 83–107. • **43.** C. E. C. Ponitz, M. M. McClelland, A. M. Jewkes, C. M. Connor, C. L. Farris and F. J. Morrison, 'Touch Your Toes! Developing a Direct Measure of Behavioral Regulation in Early Childhood', *Early Childhood Research Quarterly* 23:2 (2008), pp. 141–58. • **44.** J. S. Matthews, C. C. Ponitz and F. J. Morrison, 'Early Gender Differences in Self-Regulation and Academic Achievement', *Journal of Educational Psychology* 101:3 (2009), p. 689. • **45.** S. B. Wanless, M. M. McClelland, X. Lan, S. H. Son, C. E. Cameron, F. J. Morrison, F. M. Chen, J. L. Chen, S. Li, K. Lee and M. Sung, 'Gender Differences in Behavioral Regulation in Four Societies: The United States, Taiwan, South Korea, and China', *Early Childhood Research Quarterly* 28:3 (2013), pp. 621–33. • **46.** J. A. Gray, 'Précis of *The Neuropsychology of Anxiety: An Enquiry into the Functions of the Septo-hippocampal System*', *Behavioral and Brain Sciences* 5:3 (1982), pp. 469–84; Y. Li, L. Qiao, J, Sun, D. Wei, W. Li, J. Qiu, Q. Zhang and H. Shi, 'Gender-Specific Neuroanatomical Basis of Behavioral Inhibition/Approach Systems (BIS/BAS) in a Large Sample of Young Adults: a Voxel-Based Morphometric Investigation', *Behavioural Brain Research* 274 (2014), pp. 400–408. • **47.** D. M. Amodio, S. L. Master, C. M. Yee and S. E. Taylor, 'Neurocognitive Components of the Behavioral Inhibition and Activation Systems: Implications for Theories of Self-Regulation', *Psychophysiology* 45:1 (2008), pp. 11–19. • **48.** C. S. Dweck, W. Davidson, S. Nelson and B. Enna, 'Sex Differences in Learned Helplessness: II. The Contingencies of Evaluative Feedback in the Classroom and III. An Experimental Analysis', *Developmental Psychology* 14:3 (1978), p. 268. • **49.** C. S. Dweck, *Mindset: The New Psychology of Success* (New York, Random House, 2006); D. S. Yeager and C. S. Dweck, 'Mindsets That Promote Resilience: When Students Believe that Personal Characteristics Can Be Developed', *Educational Psychologist* 47:4 (2012), pp. 302–14. • **50.** M. L. Kamins and C. S. Dweck, 'Person versus Process Praise and Criticism: Implications for Contingent Self-Worth and Coping', *Developmental Psychology* 35:3 (1999), p. 835. • **51.** J. Henderlong Corpus and M. R. Lepper, 'The Effects of Person versus Performance Praise on Children's Motivation: Gender and Age as Moderating Factors', *Educational Psychology* 27:4 (2007), pp. 487–508. • **52.** Ibid.

CHAPTER 13: INSIDE HER PRETTY LITTLE HEAD –
A TWENTY-FIRST-CENTURY UPDATE

1. E. Racine, O. Bar-Ilan and J. Illes, 'fMRI in the Public Eye', *Nature Reviews Neuroscience* 6:2 (2005), p. 159. • **2.** T. D. Satterthwaite, D. H. Wolf, D. R. Roalf, K. Ruparel, G. Erus, S. Vandekar, E. D. Gennatas, M. A. Elliott, A. Smith, H. Hakonarson and R. Verma, 'Linked Sex Differences in Cognition and Functional Connectivity in Youth', *Cerebral Cortex* 25:9 (2014), pp. 2383–94. • **3.** D. Weber, V. Skirbekk, I. Freund and A. Herlitz, 'The Changing Face of Cognitive Gender Differences in Europe', *Proceedings of the National Academy of Sciences* 111:32 (2014), pp. 11673–8. • **4.** F. Macrae, 'Female brains really ARE different to male minds with women possessing better recall and men excelling at maths', *Mail Online*, 28 July 2014, https://www.dailymail.co.uk/news/article-2709031/Female-brains-really-ARE-different-male-minds-women-possessing-better-recall-men-excelling-maths.html (accessed 10 November 2018). • **5.** 'Brain regulates social behavior differences in males and females', *Neuroscience News*, 31 October 2016, https://neuroscience-news.com/sex-difference-social-behavior-5392 (accessed 10 November 2018). • **6.** K. Hashikawa, Y. Hashikawa, R. Tremblay, J. Zhang, J. E. Feng, A. Sabol, W. T. Piper, H. Lee, B. Rudy and D. Lin, 'Esr1+ Cells in the Ventromedial Hypothalamus Control Female Aggression', *Nature Neuroscience* 20:11 (2017), p. 1580. • **7.** D. Joel, personal communication, 2017. • **8.** 'Science explains why some people are into BDSM and some aren't', *India Times*, 7 October 2017. • **9.** K. Hignett, 'Everything "the female brain" gets wrong about the female brain', *Newsweek*, 10 February 2018, https://www.newsweek.com/science-behind-female-brain-802319 (accessed 10 November 2018). • **10.** Fine, *Delusions of Gender*; Fine, 'Is There Neurosexism'. • **11.** C. M. Leonard, S. Towler, S. Welcome, L. K. Halderman, R. Otto, M. A. Eckert and C. Chiarello, 'Size Matters: Cerebral Volume Influences Sex Differences in Neuroanatomy', *Cerebral Cortex* 18:12 (2008), pp. 2920–31; E. Luders, A. W. Toga and P. M. Thompson, 'Why Size Matters: Differences in Brain Volume Account for Apparent Sex Differences in Callosal Anatomy – The Sexual Dimorphism of the Corpus Callosum', *NeuroImage* 84 (2014), pp. 820–24. • **12.** J. Hänggi, L. Fövenyi, F. Liem, M. Meyer and L. Jäncke, 'The Hypothesis of Neuronal Interconnectivity as a Function of Brain Size – A General Organization Principle of the Human Connectome', *Frontiers in Human Neuroscience* 8 (2014), p. 915. • **13.** D. Marwha, M. Halari and L. Eliot, 'Meta-analysis Reveals a Lack of Sexual Dimorphism in Human

Amygdala Volume', *NeuroImage* 147 (2017), pp. 282–94; A. Tan, W. Ma, A. Vira, D. Marwha and L. Eliot, 'The Human Hippocampus is not Sexually-Dimorphic: Meta-analysis of Structural MRI Volumes', *NeuroImage* 124 (2016), pp. 350–66. • **14.** S. J. Ritchie, S. R. Cox, X. Shen, M. V. Lombardo, L. M. Reus, C. Alloza, M. A. Harris, H. L. Alderson, S. Hunter, E. Neilson and D. C. Liewald, 'Sex Differences in the Adult Human Brain: Evidence from 5216 UK Biobank Participants', *Cerebral Cortex* 28:8 (2018), pp. 2959–75. • **15.** T. Young, 'Why can't a woman be more like a man?', *Quillette*, 24 May 2018, https://quillette.com/2018/05/24/cant-woman-like-man (accessed 10 November 2018). • **16.** J. Pietschnig, L. Penke, J. M. Wicherts, M. Zeiler and M. Voracek, 'Meta-analysis of Associations between Human Brain Volume and Intelligence Differences: How Strong Are They and What Do They Mean?', *Neuroscience and Biobehavioral Reviews* 57 (2015), pp. 411–32. • **17.** D. C. Dean, E. M. Planalp, W. Wooten, C. K. Schmidt, S. R. Kecskemeti, C. Frye, N. L. Schmidt, H. H. Goldsmith, A. L. Alexander and R. J. Davidson, 'Investigation of Brain Structure in the 1-Month Infant', *Brain Structure and Function* 223:4 (2018), pp. 1953–70. • **18.** 'Finding withdrawn after major author correction: "Sex differences in human brain structure are already apparent at one month of age"', *British Psychological Society Research Digest*, 15 March 2018, https://digest.bps.org.uk/2018/01/31/sex-differences-in-brain-structure-are-already-apparent-at-one-month-of-age (accessed 10 November 2018). • **19.** D. C. Dean, E. M. Planalp, W. Wooten, C. K. Schmidt, S. R. Kecskemeti, C. Frye, N. L. Schmidt, H. H. Goldsmith, A. L. Alexander and R. J. Davidson, 'Correction to: Investigation of Brain Structure in the 1-Month Infant', *Brain Structure and Function* 223:6 (2018), pp. 3007–9. • **20.** As seen on Pinterest. • **21.** R. Rosenthal, 'The File Drawer Problem and Tolerance for Null Results', *Psychological Bulletin* 86:3 (1979), p. 638. • **22.** S. P. David, F. Naudet, J. Laude, J. Radua, P. Fusar-Poli, I. Chu, M. L. Stefanick and J. P. Ioannidis, 'Potential Reporting Bias in Neuroimaging Studies of Sex Differences', *Scientific Reports* 8:1 (2018), p. 6082. • **23.** V. Brescoll and M. LaFrance, 'The Correlates and Consequences of Newspaper Reports of Research on Sex Differences', *Psychological Science* 15:8 (2004), pp. 515–20. • **24.** C. Fine, R. Jordan-Young, A. Kaiser and G. Rippon, 'Plasticity, Plasticity, Plasticity ... and the Rigid Problem of Sex', *Trends in Cognitive Sciences* 17:11 (2013), pp. 550–51. • **25.** B. B. Biswal, M. Mennes, X.-N. Zuo, S. Gohel, C. Kelly, S. M. Smith, C. F. Beckmann, J. S. Adelstein, R. L. Buckner, S. Colcombe and A. M. Dogonowski, 'Toward Discovery Science of Human Brain Function', *Proceedings of the National Academy of Sciences* 107:10 (2010), pp. 4734–9. • **26.** Van Anders et al., 'The Steroid/Peptide Theory of Social Bonds'. • **27.** M. N. Muller, F. W. Marlowe, R. Bugumba and P. T. Ellison, 'Testosterone and Paternal Care in East African Foragers and Pastoralists', *Proceedings of the Royal Society B: Biological Sciences*

276:1655 (2009), pp. 347–54. • **28.** S. M. van Anders, R. M. Tolman and B. L. Volling, 'Baby Cries and Nurturance Affect Testosterone in Men', *Hormones and Behavior* 61:1 (2012), pp. 31–6. • **29.** W. James, *The Principles of Psychology*, 2 vols (New York, Henry Holt, 1890). • **30.** E. K. Graham, D. Gerstorf, T. Yoneda, A. Piccinin, T. Booth, C. Beam, A. J. Petkus, J. P. Rutsohn, R. Estabrook, M. Katz and N. Turiano, 'A Coordinated Analysis of Big-Five Trait Change across 16 Longitudinal Samples' (2018), available at https://osf.io/ryjpc/download/?format=pdf (accessed 10 November 2018) • **31.** D. Halpern, *Sex Differences in Cognitive Abilities*, 4th edn (Hove, Psychology Press, 2012). • **32.** Halpern et al. 'The Science of Sex Differences in Science and Mathematics'. • **33.** J. S. Hyde, 'The Gender Similarities Hypothesis', *American Psychologist* 60:6 (2005), p. 581. • **34.** E. Zell, Z. Krizan and S. R. Teeter, 'Evaluating Gender Similarities and Differences Using Metasynthesis', *American Psychologist* 70:1 (2015), pp. 10–20.

CHAPTER 14: MARS, VENUS OR EARTH? HAVE WE BEEN WRONG ABOUT SEX ALL ALONG?

1. A. Montañez, 'Beyond XX and XY', *Scientific American* 317:3 (2017), pp. 50–51. • **2.** D. Joel, 'Genetic-Gonadal-Genitals Sex (3G-Sex) and the Misconception of Brain and Gender, or, Why 3G-Males and 3G-Females Have Intersex Brain and Intersex Gender', *Biology of Sex Differences* 3:1 (2012), p. 27. • **3.** C. P. Houk, I. A. Hughes, S. F. Ahmed and P. A. Lee, 'Summary of Consensus Statement on Intersex Disorders and Their Management', *Pediatrics* 118:2 (2006), pp. 753–7. • **4.** C. Ainsworth, 'Sex Redefined', *Nature* 518:7539 (2015), p. 288. • **5.** Ibid. • **6.** V. Heggie, 'Nature and sex redefined – we have never been binary', *Guardian*, 19 February 2015, https://www.theguardian.com/science/the-h-word/2015/feb/19/nature-sex-redefined-we-have-never-been-binary • **7.** A. Fausto-Sterling, 'The Five Sexes', *Sciences* 33:2 (1993), pp. 20–24. • **8.** A. Fausto-Sterling, 'The Five Sexes, Revisited', *Sciences* 40:4 (2000), pp. 18–23. • **9.** A. P. Arnold and X. Chen, 'What Does the "Four Core Genotypes" Mouse Model Tell Us about Sex Differences in the Brain and Other Tissues?', *Frontiers in Neuroendocrinology* 30:1 (2009), pp. 1–9. • **10.** Montañez, 'Beyond XX and XY'. • **11.** L. Cahill, 'Why Sex Matters for Neuroscience', *Nature Reviews Neuroscience* 7:6 (2006), p. 477. • **12.** A. N. Ruigrok, G. Salimi-Khorshidi, M. C. Lai, S. Baron-Cohen, M. V. Lombardo, R. J. Tait and J. Suckling, 'A Meta-analysis of Sex Differences in Human Brain Structure', *Neuroscience & Biobehavioral Reviews* 39 (2014), pp. 34–50. • **13.** Tan et al., 'The Human Hippocampus is not Sexually-Dimorphic'; D. Marwha, M. Halari and L. Eliot, 'Meta-analysis Reveals a Lack of Sexual Dimorphism in Human Amygdala Volume', *NeuroImage* 147 (2017), pp. 282–94. • **14.** Ingalhalikar et al., 'Sex Differences in the Structural

Connectome of the Human Brain'. • **15.** Hänggi et al., 'The Hypothesis of Neuronal Interconnectivity'. • **16.** D. Joel and M. M. McCarthy, 'Incorporating Sex as a Biological Variable in Neuropsychiatric Research: Where Are We Now and Where Should We Be?', *Neuropsychopharmacology* 42:2 (2017), p. 379. • **17.** D. Joel, Z. Berman, I. Tavor, N. Wexler, O. Gaber, Y. Stein, N. Shefi, J. Pool, S. Urchs, D. S. Margulies and F. Liem, 'Sex beyond the Genitalia: The Human Brain Mosaic', *Proceedings of the National Academy of Sciences* 112:50 (2015), pp. 15468–73. • **18.** M. Del Giudice, R. A. Lippa, D. A. Puts, D. H. Bailey, J. M. Bailey and D. P. Schmitt, 'Joel et al.'s Method Systematically Fails to Detect Large, Consistent Sex Differences', *Proceedings of the National Academy of Sciences* 113:14 (2016), p. E1965. • **19.** D. Joel, A. Persico, J. Hänggi, J. Pool and Z. Berman, 'Reply to Del Giudice et al., Chekroud et al., and Rosenblatt: Do Brains of Females and Males Belong to Two Distinct Populations?', *Proceedings of the National Academy of Sciences* 113:14 (2016), pp. E1969–70. • **20.** L. MacLellan, 'The biggest myth about our brains is that they are "male" or "female"', *Quartz*, 27 August 2017, https://qz.com/1057494/the-biggest-myth-about-our-brains-is-that-theyre-male-or-female (accessed 10 November 2018). • **21.** S. M. van Anders, 'The Challenge from Behavioural Endocrinology', pp. 4–6 in J. S. Hyde, R. S. Bigler, D. Joel, C. C. Tate and S. M. van Anders, 'The Future of Sex and Gender in Psychology: Five Challenges to the Gender Binary', *American Psychologist* (2018), http://dx.doi.org/10.1037/amp0000307. • **22.** S. M. Van Anders, 'Beyond Masculinity: Testosterone, Gender/Sex, and Human Social Behavior in a Comparative Context', *Frontiers in Neuroendocrinology* 34:3 (2013), pp. 198–210. • **23.** J.S. Hyde et al (2018) 'The Future of Sex and Gender in Psychology' • **24.** J. S. Hyde, 'The Gender Similarities Hypothesis', *American Psychologist* 60:6 (2005), p. 581; E. Zell, Z. Krizan and S. R. Teeter, 'Evaluating Gender Similarities and Differences Using Metasynthesis', *American Psychologist* 70:1 (2015), p. 10. • **25.** B. J. Carothers and H. T. Reis, 'Men and Women are from Earth: Examining the Latent Structure of Gender', *Journal of Personality and Social Psychology* 104:2 (2013), p. 385. • **26.** H. T. Reis and B. J. Carothers, 'Black and White or Shades of Gray: Are Gender Differences Categorical or Dimensional?', *Current Directions in Psychological Science* 23:1 (2014), pp. 19–26. • **27.** Joel et al., 'Sex beyond the Genitalia'. • **28.** Martin and Ruble, 'Children's Search for Gender Cues'. • **29.** I. Savic, A. Garcia-Falgueras and D. F. Swaab, 'Sexual Differentiation of the Human Brain in Relation to Gender Identity and Sexual Orientation', *Progress in Brain Research* 186 (2010), pp. 41–62; Joel, 'Genetic-Gonadal-Genitals Sex (3G-Sex) and the Misconception of Brain and Gender'. • **30.** J. J. Endendijk, A. M. Beltz, S. M. McHale, K. Bryk and S. A. Berenbaum, 'Linking Prenatal Androgens to Gender-Related Attitudes, Identity, and Activities: Evidence from Girls with Congenital Adrenal

Hyperplasia', *Archives of Sexual Behavior* 45:7 (2016), pp. 1807–15. • **31.** Colapinto, *As Nature Made Him: The Boy Who Was Raised as a Girl.* • **32.** 'Transgender Equality: House of Commons Backbench Business Debate – Advice for Parliamentarians', Equality and Human Rights Commission, 1 December 2016, available at https://www.equality-humanrights.com/en/file/21151/download?token=Z7I8opi2 (accessed 10 November 2018) • **33.** 'Gender confirmation surgeries rise 20% in first ever report', American Society of Plastic Surgeons website, 22 May 2017, https://www.plasticsurgery.org/news/press-releases/gender-confirmation-surgeries-rise-20-percent-in-first-ever-report (accessed 10 November 2018). • **34.** House of Commons Women and Equalities Committee, *Transgender Equality: First Report of Session 2015–16*, HC 390, 8 December 2015, available at https://publications.parliament.uk/pa/cm201516/cmselect/cmwomeq/390/390.pdf (accessed 10 November 2018). • **35.** C. Turner, 'Number of children being referred to gender identity clinics has quadrupled in five years', *Telegraph*, 8 July 2017, https://www.telegraph.co.uk/news/2017/07/08/number-children-referred-gender-identity-clinics-has-quadrupled (accessed 10 November 2018). • **36.** J. Ensor, 'Bruce Jenner: I was born with body of a man and soul of a woman', *Telegraph*, 25 April 2015, https://www.telegraph.co.uk/news/worldnews/northamerica/usa/11562749/Bruce-Jenner-I-was-born-with-body-of-a-man-and-soul-of-a-woman.html (accessed 10 November 2018). • **37.** C. Odone, 'Do men and women really think alike?', *Telegraph*, 14 September 2010, https://www.telegraph.co.uk/news/science/8001370/Do-men-and-women-really-think-alike.html (accessed 10 November 2018). • **38.** T. Whipple, 'Sexism fears hamper brain research', *The Times*, 29 November 2016, https://www.thetimes.co.uk/edition/news/sexism-fears-hamper-brain-research-rx6w39gbw (accessed 10 November 2018); L. Willgress, 'Researchers' sexism fears are putting women's health at risk, scientist claims', *Telegraph*, 29 November 2016, https://www.telegraph.co.uk/news/2016/11/29/researchers-sexism-fears-putting-womens-health-risk-scientist (accessed 10 November 2018).

CONCLUSION: RAISING DAUNTLESS DAUGHTERS (AND SYMPATHETIC SONS)

1. S.-J. Blakemore, *Inventing Ourselves: The Secret Life of the Teenage Brain* (London, Doubleday, 2018). • **2.** L. H. Somerville, 'The Teenage Brain: Sensitivity to Social Evaluation', *Current Directions in Psychological Science* 22:2 (2013), pp. 121–7. • **3.** S.-J. Blakemore, 'The Social Brain in Adolescence', *Nature Reviews Neuroscience* 9:4 (2008), p. 267. • **4.** B. London, G. Downey, R. Romero-Canyas, A. Rattan and D. Tyson,

'Gender-Based Rejection Sensitivity and Academic Self-Silencing in Women', *Journal of Personality and Social Psychology* 102:5 (2012), p. 961; E. Kross, T. Egner, K. Ochsner, J. Hirsch and G. Downey, 'Neural Dynamics of Rejection Sensitivity', *Journal of Cognitive Neuroscience* 19:6 (2007), pp. 945–56. • **5.** Damore, 'Google's Ideological Echo Chamber'. • **6.** Stoet and Geary, 'The Gender-Equality Paradox in Science, Technology, Engineering, and Mathematics Education'. • **7.** J. Clark Blickenstaff, 'Women and Science Careers: Leaky Pipeline or Gender Filter?', *Gender and Education* 17:4 (2005), pp. 369–86. • **8.** A. Tintori and R. Palomba, *Turn on the Light on Science: A Research-Based Guide to Break Down Popular Stereotypes about Science and Scientists* (London, Ubiquity Press, 2017). • **9.** London et al., 'Gender-Based Rejection Sensitivity'. • **10.** J. A. Mangels, C. Good, R. C. Whiteman, B. Maniscalco and C. S. Dweck, 'Emotion Blocks the Path to Learning under Stereotype Threat', *Social Cognitive and Affective Neuroscience* 7:2 (2011), pp. 230–41. • **11.** E. A. Maloney and S. L. Beilock, 'Math Anxiety: Who Has It, Why It Develops, and How to Guard against It', *Trends in Cognitive Sciences* 16:8 (2012), pp. 404–6. • **12.** K. J. Van Loo and R. J. Rydell, 'On the Experience of Feeling Powerful: Perceived Power Moderates the Effect of Stereotype Threat on Women's Math Performance', *Personality and Social Psychology Bulletin* 39:3 (2013), pp. 387–400. • **13.** T. Harada, D. Bridge and J. Y. Chiao, 'Dynamic Social Power Modulates Neural Basis of Math Calculation', *Frontiers in Human Neuroscience* 6 (2013), p. 350. • **14.** I. M. Latu, M. S. Mast, J. Lammers and D. Bombari, 'Successful Female Leaders Empower Women's Behavior in Leadership Tasks', *Journal of Experimental Social Psychology* 49:3 (2013), pp. 444–8. • **15.** J. G. Stout, N. Dasgupta, M. Hunsinger and M. A. McManus, 'STEMing the Tide: Using Ingroup Experts to Inoculate Women's Self-Concept in Science, Technology, Engineering, and Mathematics (STEM)', *Journal of Personality and Social Psychology* 100:2 (2011), p. 255. • **16.** 'Inspiring girls with People Like Me', WISE website, https://www.wise-campaign.org.uk/what-we-do/expertise/inspiring-girls-with-people-like-me (accessed 10 November 2018). • **17.** C. Ainsworth, 'Sex Redefined', *Nature* 518:7539 (2015), p. 288. • **18.** E. S. Finn, X. Shen, D. Scheinost, M. D. Rosenberg, J. Huang, M. M. Chun, X. Papademetris and R. T. Constable, 'Functional Connectome Fingerprinting: Identifying Individuals Using Patterns of Brain Connectivity', *Nature Neuroscience* 18:11 (2015), p. 1664; E. S. Finn, 'Brain activity is as unique – and identifying – as a fingerprint', *Conversation*, 12 October 2015, https://theconversation.com/brain-activity-is-as-unique-and-identifying-as-a-fingerprint-48723 (accessed 10 November 2018). • **19.** D. Joel and A. Fausto-Sterling, 'Beyond Sex Differences: New Approaches for Thinking about Variation in Brain Structure and Function', *Philosophical Transactions of the Royal Society B:*

Biological Sciences 371:1688 (2016), 20150451; Joel et al., 'Sex beyond the Genitalia'. • **20.** L. Foulkes and S. J. Blakemore, 'Studying Individual Differences in Human Adolescent Brain Development', *Nature Neuroscience* 21:3 (2018), pp. 315–23. • **21.** Q. J. Huys, T. V. Maia and M. J. Frank, 'Computational Psychiatry as a Bridge from Neuroscience to Clinical Applications', *Nature Neuroscience* 19:3 (2016), p. 404; O. Moody, 'Artificial intelligence can see what's in your mind's eye', *The Times*, 3 January 2018, https://www.thetimes.co.uk/article/artificial-intelligence-can-see-whats-in-your-minds-eye-w6k9pjsh6 (accessed 10 November 2018). • **22.** M. M. Mielke, P. Vemuri and W. A. Rocca, 'Clinical Epidemiology of Alzheimer's Disease: Assessing Sex and Gender Differences', *Clinical Epidemiology* 6 (2014), p. 37; S. L. Klein and K. L. Flanagan, 'Sex Differences in Immune Responses', *Nature Reviews Immunology* 16:10 (2016), p. 626. • **23.** L. D. McCullough, G. J. De Vries, V. M. Miller, J. B. Becker, K. Sandberg and M. M. McCarthy, 'NIH Initiative to Balance Sex of Animals in Preclinical Studies: Generative Questions to Guide Policy, Implementation, and Metrics', *Biology of Sex Differences* 5:1 (2014), p. 15. • **24.** D. L. Maney, 'Perils and Pitfalls of Reporting Sex Differences', *Philosophical Transactions of the Royal Society B: Biological Sciences* 371:1688 (2016), 20150119. • **25.** http://lettoysbetoys.org.uk • **26.** R. Nicholson, '*No More Boys and Girls: Can Kids Go Gender Free* review – reasons to start treating children equally', *Guardian*, 17 August 2017, https://www.theguardian.com/tv-and-radio/tvandradioblog/2017/aug/17/no-more-boys-and-girls-can-kids-go-gender-free-review-reasons-to-start-treating-children-equally (accessed 10 November 2018); J. Rees, '*No More Boys and Girls: Can Our Kids Go Gender Free?* should be compulsory viewing in schools – review', *Telegraph*, 23 August 2017, https://www.telegraph.co.uk/tv/2017/08/23/no-boys-girls-can-kids-go-gender-free-should-compulsory-viewing (accessed 10 November 2018). • **27.** S. Quadflieg and C. N. Macrae, 'Stereotypes and Stereotyping: What's the Brain Got to Do with It?', *European Review of Social Psychology* 22:1 (2011), pp. 215–73. • **28.** C. Fine, J. Dupré and D. Joel, 'Sex-Linked Behavior: Evolution, Stability, and Variability', *Trends in Cognitive Sciences* 21:9 (2017), pp. 666–73. • **29.** D. Victor, 'Microsoft created a Twitter bot to learn from users. It quickly became a racist jerk', *New York Times*, 24 March 2016, https://www.nytimes.com/2016/03/25/technology/microsoft-created-a-twitter-bot-to-learn-from-users-it-quickly-became-a-racist-jerk.html (accessed 10 November 2018). • **30.** Hunt, 'Tay, Microsoft's AI chatbot, gets a crash course in racism from Twitter'.

Index